Biological Control of Water Pollution

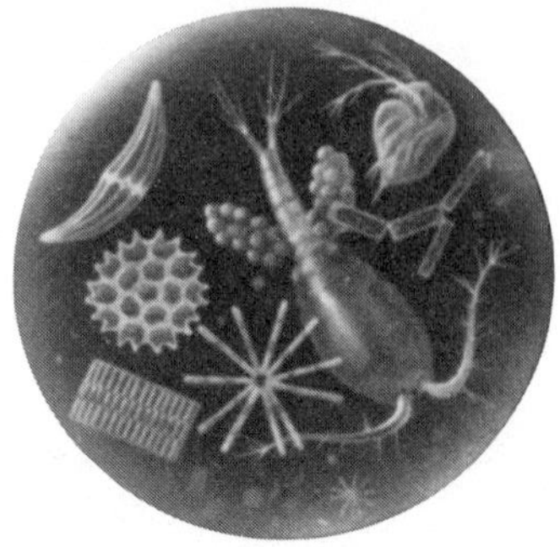

Center for Ecological Studies
in Planning & Design
University of Pennsylvania

Edited by Joachim Tourbier and Robert W. Pierson, Jr.

Biological Control of Water Pollution

Introduction by Edward W. Furia

UNIVERSITY OF PENNSYLVANIA PRESS
1976

Library of Congress Catalog Card Number: 76-8593
ISBN: 0-8122-7709-0
Printed in the United States of America

This publication was made possible by a grant from

THE BOHEN FOUNDATION

The International Conference on Biological Water Quality Improvement Alternatives and this publication were sponsored by the

CENTER FOR ECOLOGICAL RESEARCH IN PLANNING AND DESIGN
DEPARTMENT OF LANDSCAPE ARCHITECTURE AND REGIONAL PLANNING
UNIVERSITY OF PENNSYLVANIA
PHILADELPHIA, PA.

Contents

Introduction

Biological Alternatives in Perspective: More than Academic Curiosities

EDWARD W. FURIA

Attorney and City Planner
Philadelphia

The genesis of the International Conference on Biological Water Quality Improvement Alternatives was characterized more by necessity than by academic curiosity. In November 1974, I asked Joachim Tourbier to help me solve a seemingly impossible water pollution control problem involving a 40 million gallon/day industrial effluent. After discarding conventional techniques such as ion exchange and reverse osmosis because of their $10 to $30 million projected costs, Tourbier began searching the literature for new treatment alternatives which might deal with the extraordinary volume of effluent, but still be economically feasible. He discovered the work of scientists scattered throughout a dozen nations who had been doing unconnected and relatively unpublicized research on the use of higher green plants and animals to purify water. After discovering that Dr. Kaethe Seidel at the Max Planck Institute in Germany had used marsh plants in successful pilot projects that treated municipal and industrial wastewater, we invited her to the United States to advise us on the applicability of her system to the problem. We ultimately designed a five-step system which included a biological component along with conventional techniques. Bulrushes and *Phragmites communis* plants would be used for final polishing of the water; Maccaferri gabions were included for stream bank protection.

Because we personally wanted to gain more knowledge about biological wastewater treatment alternatives and believed that scientists and decision makers in academic, government, and industrial communities would also be very interested in these techniques, we met with Professor Ian McHarg of the University of Pennsylvania and Dr. Ruth Patrick of the Academy of Natural Sciences and proposed that the Center for Ecological Research in Planning and Design sponsor an international conference on the subject, inviting the most prominent experts in the field to present their work at the University of Pennsylvania.

This is the first publication which has attempted to present the considerable body of work which has been done internationally in the field of biological water pollution control; in fact, very few of the scientists who had been involved in biological systems research had ever met or shared their knowledge with each other prior to the conference. This was reason enough to have a conference on the subject and for there to have been such extraordinary international interest in it. But I believe the real reason that several of the world's most important environmental figures accepted an invitation to address the conference—Muskie, Ruckelshaus, McHarg, Patrick, Lambo, Kneese—and the real reason that the national media exhibited such strong interest in it, had nothing to do with the novelty of the idea of purifying water with green plants or the fact that scientists in the field needed more opportunity to talk with each other.[1] The response to the conference can only be understood as an expression of a deep concern both here and abroad that we are failing to deal adequately with the problems of water quality and water supply, and as an expression of the hope that biological systems might provide more effective alternatives or additions to conventional water treatment systems.

Several factors contributed to this concern:

1. The realization that each year 500 new industrial chemicals, some of which are carcinogens, are released into the nation's waterways either directly or through regional sewage treatment plants.
2. Confirming the findings of several other researchers, Drs. Robert Harris and Samuel Epstein made public their studies which revealed that 66 synthetic organic chemicals had been found in the drinking water of New Orleans. Furthermore, they reasoned that chlorination of water for the purpose of purifying it in many cases actually converts harmless chemicals into dangerous ones. They suggested that the approximate 40 percent higher incidence of cancer (than the national average) in New Orleans was partially a result of the presence of these substances.[2]
3. Although conventional water treatment components (e.g., permanganate oxidation, sand filtration, chlorination) can eliminate water-born disease bacteria that cause typhoid, dysentery, and cholera, they do little or nothing to detect or remove the new industrial chemicals; *they were never designed to do so.*
4. The conventional water treatment systems are costly to build and operate and consume precious natural resources including fossil fuels.[3]
5. Because of the high cost of building and operating conventional sewage treatment systems, small communities usually tie into regional treatment plants in an effort to take advantage of economies of scale. Big regional plants, like highways, facilitate rapid residential and industrial development along their interceptors and sewer lines. They can be a major cause of urban sprawl, with its attendant air and water pollution problems.[4] Faced with the prospect of the kind of increased growth associated with regional treatment systems, many communities are searching for treatment alternatives which can be implemented locally at less cost and can be tailored more realistically to local need.
6. Although water treatment problems have become more and more complex, little or no major improvements have been made in conventional water treatment methods for sixty years, despite passage in 1972 of the strongest water pollution control bill in history and an annual expenditure of billions of dollars for new treatment plants.[5]
7. Although the Federal Water Pollution Control Act Amendments of 1972 specifically require that:

> the administrator (of the Environmental Protection Agency) shall encourage waste treatment management which results in the construction of revenue-producing facilities providing for (1) The recycling of potential sewage pollutants through the production of agriculture, silviculture or aquiculture products or any combination thereof.[6]

it is clear that this mandate has largely been ignored by the responsible agency.[7]

In the following chapters, scientists, environmental experts, and city planners from Germany, Holland, the Sudan, Finland, Romania, the United States, and Puerto Rico will discuss how biological systems may be utilized to deal with water pollution problems. Papers are presented suggesting that green plants and biological systems have the capacity to detoxify dangerous synthetic chemicals in industrial wastes, and that vegetation can capture nutrients and be harvested directly as food for man and domestic animals or used for paper manufacture or composting.[8]

More complex systems are presented in Part IV in which food chains are reconstructed from nature, beginning with green plants and phytoplankton and ending with animals of economic value. Biological systems are discussed which rely on forest, marsh, or crop vegetation together with soil microorganisms to improve the quality of aquifer recharge water.[9]

In Chapter 16, Dr. de Jong describes a bulrush and reed sewage treatment system in a Dutch polder which handles the raw sewage of over three thousand tourists every summer. Dr. Kaethe Seidel and Walter Czerwenka explain in Chapter 33 how Rhine River water in Germany is treated for municipal water supplies with a system that relies on artificial aquifer recharge through bulrush lagoons. And in Chapter 13, Dr. Samia Al Azharia Jahn reports how women of the Sudan, in an ancient ritual, purify their drinking water by adding leaves, seeds, or pieces of bark and root from selected plants to clay pots containing water they have collected from the Nile. All of the systems described, of course, use the sun as their principal energy source.

Neither the organizers of the conference nor the scientists and other participants have suggested that biological systems are cure-alls for the world's water quality and water supply problems. There are many questions about these systems and many limitations. Although these systems can be used for highly toxic industrial effluents, there are still situations where biological systems alone would fail because the effluent itself is toxic to plants. In such situations, higher plants would have to be part of systems utilizing conventional techniques. Concern has also been expressed that plants might be less effective in colder northern climates. And although biological systems may be ideal for rural areas and underdeveloped countries where land may be readily available, these systems would probably be unsuitable for many highly urbanized areas where land is scarce.[10] Unlike conventional systems, biological systems must be designed to meet the specific and varying conditions of each industrial or municipal water pollution situation, allowing for differences in climate, soil content, and influent quality. Also, at least at this stage in development of biological water pollution control systems, there is a need for applied research for pilot projects which will determine how effective biological systems can be in a broad range of industrial and municipal treatment situations.

I would also point out that biological systems—no matter how effective—can never be a substitute for designing industrial processes which prevent pollutants from entering streams and rivers in the first place. Similarly, in most situations, industrial waste should be separated from domestic wastes no matter what treatment method is used, and finally, wilderness streams and rivers should continue to be identified and protected from contamination. These, of course, are water quality considerations unrelated to the choice or development of water pollution control systems.

In spite of these cautions, the emergence of biological systems may prove to be one of the most important in the field of water pollution control in many years. Scientists at the conference reported that biological systems were capable of a broad range of water quality improvement activities. Certain plant and associated microorganism activities most promising for municipal application are:

1. Removal of inorganic substances (nitrates, phosphates, sodium, potassium, calcium, etc.) from wastewater.
2. Degradation of highly toxic organic substances like phenol.

3. Neutralization of alkaline and acid waste waters.

4. Improvement of the quality of water polluted by food processing wastes.

5. Aeration of water through photosynthesis.

6. Aeration by plants of water by taking oxygen into their upper stalks and giving it off through their submerged lower stalks.

7. Provision of habitat for other living things (crustaceans, insects, and fish) which themselves purify water.

8. Reduction of the volume of waste waters (by transpiration) by taking water into their stalks and releasing it as a gas to the atmosphere (transpiration).

9. Mechanical filtration of suspended solids through plant root structures.

10. Attenuation of pathogenic organisms.

As is often the case with scientific advances, knowing that a better alternative exists for solving a particular problem is only the beginning. It usually takes years for innovations to be implemented. Several economic, political, and institutional constraints exist which will probably prevent the rapid implementation of biological water pollution control alternatives. In spite of the extraordinary amount of public funds which will be spent on pollution control projects in 1975 and 1976—probably $15 billion—there is little likelihood that the designers of those pollution control projects will seriously consider biological alternatives. Even if the sanitary engineering fraternity, an old and established one, were disposed to try these new techniques, the lead time required for application for federal funds, for EPA review of applications, and for the ultimate commitment of federal monies could still delay the broad applications of these systems for years.[11] There is also a problem of public acceptance: how quickly can Americans accept the idea of human waste for crop fertilizer or marsh nutrient? Furthermore, the fact that biological systems are inexpensive compared to conventional systems means they will probably present fewer profit opportunities for treatment plant designers. This is unfortunate, but realistically this will also delay implementation of the system.

Neither the technological problems of designing or testing biological systems nor the political and institutional restraints on their implementation can detract from the overwhelming sense of opportunity one experiences in reading the chapters that follow. If these systems can be made to work, they may offer partial solutions to several of the perplexing problems confronting modern industrial societies: the need for energy consumption in the face of limited supplies; the waste of nutrients in the face of starvation; destructive land-use patterns caused by regional sewage treatment systems built in the name of economies of scale; and the degradation of surface and ground waters.

Notes

1. See "Bulrushes that Eat Pollutants," *Business Week,* March 10, 1975, p. 80; P. Gwynne, "How Weeds Clean Water," *Newsweek,* April 14, 1975, p. 85; Bayard Webster, "Bulrushes Being Used in Artificial Marshes to Filter Water," *New York Times,* March 9, 1975, p. 1, National Topics section; W. Froelich, "Scientists Take 'Backward Look' on Cleaning Water," *Sunday Bulletin* (Philadelphia), March 9, 1975; D. McKelvie, "Furia to Head Water Improvement Seminar," *Philadelphia Inquirer,* February 26, 1975; K. W. Burkhart, "Dirty Water in the Bulrushes," *New Republic,* June 14, 1975, pp. 16-19.

2. See Harris, Chapter 9, and Epstein, Chapter 10.

3. de Jong reports in Chapter 16 that the capital cost for such a system is one-sixth of the cost of a comparable conventional activated sludge treatment plant and that the annual operating cost would be one-fourth of the cost of operating a conventional plant. In addition, the principle maintenance expense of a marsh filtration system using reeds and bulrushes—harvesting and composting—can be reduced by making use of the harvested reeds.

4. See, for example, E. Furia, "Comments on the Valley Forge Sewage Treatment System Environmental Impact Statement" (U.S. Environmental Protection Agency, Region III, Philadelphia, September 1974).

5. Federal Water Pollution Control Act Amendments of 1972 (Public Law 92-500).

6. Public Law 92-500, Title II Sect. 201 (d).

7. See Gravitz, Chapter 35.

8. See Chapters 21 and 22.

9. See Part V.

10. J. Tourbier and R. Westmacott, "Agricultural Landscapes in a New Age" (Research Proposal to the ASLA Foundation, 1972).

11. See Muskie, Chapter 2.

Part I

A Rationale for the Biological Approach

1

Biological Alternatives to Water Pollution

IAN L. McHARG

University of Pennsylvania

Any consideration of alternatives must begin from the datum of the status quo. It may then be useful to describe the assumptions and processes which now produce what is euphemistically described as drinking water. First let us consider the assumptions. The implicit assumption, the inalienable right governing water management, is that thou shalt defecate into thy neighbor's drinking water and, further, if for any reasons of delicacy, hygiene, or prudence you prefer not to do so, be assured that thy municipality shall do it for thee. So water treatment and sewage treatment are, for all practical purposes, synonymous. We can observe, parenthetically, that this rule does not hold for all primitive societies where often taboos seek to separate excrement from drinking water. Primitive, then, can also mean the capability of drinking water from rivers, lakes, streams, and springs. This is not the way of advanced technological societies. Beginning from a modest commitment to excrement as an essential component of water supply, with advancing technology there followed an enrichment of the attendant nutrients. The development of industry, not least the chemical industry, ensured a wide variety of new additives. Among the earliest of these were the nitrates and phosphates; the former, derived from sewage and fertilizer, is an essential ingredient for blue babies or cyanosis. But a wider population could be served by DDT, dieldrin, aldrin, and the organic phosphates. These are not benign, nor are arsenic, lead, or cyanide; seleniuum, cadmium, or other benefits from modern industry. Until recently these additives to drinking water were uncoordinated, but with the commitment to regionalization of sewage treatment, industry was invited to contribute its toxins to the nation's resources of drinking water. This is advanced technology.

In the competition among major cities for reducing the time factor between the two critical orifices in the human water cycle, the City of Philadelphia assumes a high position, perhaps

Mr. McHarg is Chairman of the Department of Landscape Architecture and Regional Planning at the University of Pennsylvania. His work and teaching urge the rational use of land according to its natural capacity to accomodate man's activities without stress to nature or man. He is partner in the environmental planning firm of Wallace, McHarg, Roberts, and Todd and is author of the book, Design with Nature.

even primary. This success derives from the inspired location of the sewage and water treatment plants on the Delaware River. The sewage receives secondary treatment when it does not rain; it is poured into the Delaware River where the United States Geological Survey monitors the continued absence of dissolved oxygen, year after year, with meticulous instrumentation. As the Delaware is tidal, the sewage moves downstream until slack tide when it returns upstream, passing the sewage treatment plant, and proceeds to the intake basins of the water treatment plant which can only be filled at high tide. Thence follows water treatment in a fully computerized facility. This, we observe, is a modern facility utilizing these high technology innovations called sedimentation, coagulation, filtration and, finally, chlorination. Now we all know that sanitary engineers, valiantly fighting their way into the twentieth century, never accomplished water treatment. This was done by microorganisms. These creatures have been successfully engaged in water treatment for hundreds of millions of years. Their commitment to this endeavor is not related to the objective of cleaning water, but to eating and survival. One would assume that these indispensible creatures and their ways would be of the greatest significance to those engaged in treating water. Not so, for the sanitary engineer does not need to know these essential decomposers. His objective is to kill all organisms. How else can you drink a dilute soup of dead bacteria in a chlorine solution? And so hyper- super- extra- ultra-chlorination ensues at such levels as to guarantee that the consumer at the end of the water line receives only dead bacteria with his toxins. And, not least among these toxins, as we have recently learned, are the carcinogens produced by the chlorine. This is modern technology at its best. Only when drinking water is delivered at 98.4°F will there be another significant advance.

But what of modern industry? It clearly operates on a modified version of the implicit inalienable right. Thou shalt poison thy neighbor's drinking water and, if for any reason of delicacy, hygiene, or prudence, you prefer not to do so, thy regional sewage system shall do it for thee.

Once upon a time I was invited by Fortune Magazine to address the Presidents of fifty of the largest corporations in the United States at the Four Seasons restaurant in New York. "Gentlemen," I said, "I wish to make a bargain with you on behalf of the American people. I know that this is presumption on the part of a Scottish immigrant, but hear me out. I have no doubt that you all bathed and changed your shirt and underclothes before coming here. I am sure you have a clean pocket handkerchief. I would not be surprised if many of you added a dab of underarm deodorant. Moreover, I have observed you at cocktails and dinner, noted your handling of knives, forks, spoons, table napkins, even finger bowls. I am sure that the American people would agree with me that your standards of personal hygiene are impeccable. But as my bargain requires concessions on your part, the American people, too, then must make concessions to you. On their behalf, I offer you a relaxation in these demanding standards of personal hygiene to which you subscribe. You may forego the changes of underwear, finger bowls, underarm deodorants; you may belch or even fart in public. This relaxation may be welcome to you and yet will not endanger the American environment. For this relaxation the American people require an improvement in your standards of corporate hygiene. We require that you now cease and desist from voiding thousands of millions of gallons and tons of your excrement into the American environment. The time has come for American industry to be toilet trained. You are incontinent. This condition may either be a sign of infantilism or senescence. Should it be the latter, we can only await your demise with impatience. If the former, we can help you into adult continence whereby you control the deposition of your excrement to minimize offense to the American people and the American environment, in particular, that you cease and desist from poisoning the nation's drinking water."

Of course, government is not immune to similar criticism. Indeed, the gravest threats to the nation's health and well-being are accomplished at public expense. Atomic testing, atomic reactors—notably accidents to these—contribute radioactivity to the atmosphere, to the hydrosphere, and our drinking water. The government also contributes more modest toxins. An analysis of well logs in the Denver metropolitan region disclosed many wells with excessive levels of arsenic, cyanide, lead, and selenium. When these were mapped, they pointed like an arrow to the source and the culprit, the Rocky Mountain National Arsenal. It would be unfair to omit the United States Army Corps of Engineers, entrusted with the enforcement of the 1899 Refuse Act. They failed to initiate prosecutions under the Act for over three score years and ten. They must be held responsible for the pollution of America's water systems. In other military societies the appropriate remedy for such celestial negligence is ceremonial disembowelment. Industry and government have been coequally negligent in the area of water treatment. Yet, it would be unfair and ungracious if the positive accomplishment of conventional water treatment and distribution were not recognized and applauded. It is precisely this system which has banished the major waterborne diseases, the pestilences of typhoid, gastroenteritis, cholera, dysentery, and others. It would be just as unfair to fail to note that many primitive societies are scourged by such diseases today and are less successful in solving these problems than the western countries. Yet, the successes are long past and the processes for water and sewage treatment, long efficacious, are incapable of dealing with the products of modern industry and chemistry and, finally, the finest tool of the system, chlorination, is now accused of contributing carcinogens to our water supply.

Wherein lies the remedy? Surely it must begin with recognition of the fact that clean, potable water is a by-product of biological systems. The entire subject of water must be seen to be a problem of biology rather than engineering. This would recognize that nature has been in the business of water and sewage treatment since the beginning of life, most effectively, and that solutions to the problems of water lie in understanding aquatic biology and, dominantly, limnology. But, knowing that water systems are affected by terrestrial processes (indeed water quality represents a synthesis of aquatic and terrestrial processes), then the scientific basis for the management of water resources must be ecology. In fact, we can say that the biological alternatives which concern us in this volume are ecological alternatives to nonecological existing methods. The definition of ecology is the study of interaction of organisms and the environment, which includes other organisms. Applying this concept to the provision of potable water would suggest that, as such, water is ultimately a product of aquatic organisms, and as these organisms require specific environmental conditions to survive and prosper, the conditions of terrestrial and aquatic systems be preserved or managed to maintain the survival and success of the essential organisms which produce potable water. We do not have to create or discover the model which would govern such an ecological approach. It is the limnological model. One of its creators and most effective practitioners, Dr. Ruth Patrick (and her colleagues at the Philadelphia Academy of Natural Science), have devoted several chapters in this collection to an exposition of this theoretical concept. It is enough to say here that this model can identify water quality from the number of species and discriminate between levels of quality using certain organisms—their presence, absence or relative abundance—as specific indicators.

Moreover, the factors which determine these conditions of aquatic "health" are generally known and can be manipulated. I subscribe totally to the limnological theory as the essential perception for ecological planning for water quality. How could such a view be institutionalized? I will simply paraphrase parts of a research project which I undertook as an agent of the American Institute of Planners for the Environment Protection Agency entitled, "Towards a Comprehensive Plan for Environmental Quality."

The first matter concerns an administrative structure for the planning, management, and regulation of water resources. Acceptance of ecology-limnology as the scientific basis for this objective implies an administrative structure. The unit of surface water hydrology is the watershed, the unit of subsurface hydrology is the ground water basin, a unit of ecological planning is the physiographic region—an area, homogeneous with respect to geological history and therefore homogeneous with respect to physiography, hydrology, soils, plants, and animals. The United States is subdivided into thirty-four such physiographic regions. The problem of selecting the appropriate unit is not simple. While the major ground water basins are frequently within physiographic regions, the major rivers often transect several regions. A solution to the complex problem of an administrative structure for water planning and management, consonant with ecology-limnology, should first involve acceptance of the watersheds of the major rivers as the primary planning units. It would recognize the Continental Divide and thus divide the country into the two administrative units of east and west. Within each major subdivision the great rivers—for example, the Hudson, Delaware, Susquehanna, Potomac, and James in the east—would provide the major administrative elements. The next subdivision would be the physiographic region within the watershed or, as in the case of the coastal plain, the physiographic region itself. The smallest administrative unit would be the watershed within a physiographic region.

Applied to the Philadelphia area, the major administrative unit would be the Eastern Office. The Delaware River Watershed provides the next unit, subdivided into the physiographic regions of Allegheny Plateau: Valley and Ridge Province; Great Valley; Reading Prong; Triassic Piedmont Lowlands; Piedmont Uplands; and Coastal Plain. Each of these regions is homogeneous and has characteristic problems associated with hydrology, limnology, and land use. These regional problems cumulatively effect water quality in the watershed. The Allegheny Plateau, with The Catskills and Poconos, is primarily used for recreation and second homes. It has excellent water in lakes and streams. The problem here is preservation and enhancement. In the Valley and Ridge Province, transectd by the Delaware and its tributaries and characterized by the location of coal mining, the overwhelming problem is acid mine drainage, and water management must seek to solve this problem. The next province, the Great Valley, is preponderantly limestone with significant ground water resources. It is the location of extensive agriculture. Here the characteristic problems are pollution from fertilizers, herbicides, pesticides, sedimentation, and nitrates. The Reading Prong has little influence upon water quality but the Piedmont, site of extensive urbanization, contributes massive human wastes and concentrates industrial toxins. Finally the Coastal Plains constitutes the most valuable water resource of the watershed reposing in the aquifer system. It presents the problem of managing ground water. This suggests that the characteristic problems of each physiographic region require an appropriate staff of scientists, planners, and administrators. Each region is specific, yet all are concerned with cumulative effects reflected in the entire watershed, its water quality and quantity.

Given an administrative structure consonant with the morphology of hydrologic-limnologic systems, it is then possible to plan and regulate within existing legislation. Sections 201 and 208 of the Federal Water Pollution Control Act Amendments (Public Law 92-500) provide effective vehicles for water planning and management, particularly the latter. Section 201 is devoted to the provision of sewage treatment facilities; section 208 is concerned with nonpoint sources of pollution and with the effects of urbanization and land use. If this latter legislative power required ecological planning, including the limnological model, and were such planning structured within watersheds and physiographic regions, there would be reason for confidence of ultimate success. However the process today is hampered by arbitrary political subdivisions and by an assumption that the objectives of the Act can be satisfied by the orthodoxy of sanitary engineering. Acceptance of ecology as the scientific basis, and limnology as the specific science for water managment, hydrological units as administrative units and ecological planning as the instrument for fulfilling the objectives of the Water Pollution Control Act would provide massive improvement but yet would still not be sufficient. The Environmental Impact Study procedure, while valuable, is ad hoc, adventitious, and negative. That is, the sum of environmental impact analyses for projects in a region do not contribute to an understanding of that region as an interacting biophysical system.

Planning and management of water resources requires that the workings of the biophysical cultural region, in this case the watershed, be understood so that prediction can be made of the consequences of certain contemplated acts. Moreover, in accordance with the Act, these contemplated actions must be assessed for their effect upon water quality and quantity. The primary requirement then is to understand the operation of the system, be it a major watershed or a physiographic region. This suggests that modelling of watersheds and regions must be undertaken to provide the necessary predictive capability. Such inventories and models then permit both positive planning towards established social ends and the avoidance of social costs. The avoidance of negative impacts, towards which the present procedures are oriented, would be complemented by a capability of inducing positive acts, i.e., development of propitious areas and positive steps to ameliorate water quality. It would require only an interpretation of Sections 201 and 208 to undertake such ecological planning studies involving both inventories of natural phenomena and processes, and the creation of preditive models, insofar as science can now provide these.

I would recommend a major improvement to existing planning method, that is, the uniform employment of a planning method having the attributes of being overt, explicit, and replicatable. In other words, the data employed should be available to the public, the weights or values attributed to factors should also be overt and explicit so that the planning process should be replicatable. Such a planning method would then be employed east and west in every watershed and physiographic region.

Yet another improvement to present processes would involve devolution of water planning and management to the smallest units of government, recognizing differences in perception of the environment, values attributed to its several aspects, and alternate forms of remedy. This suggests that municipalities produce ordinances regulating land use and management, based upon the health and welfare provisions of the Constitution, the provisions of the National Environmental Policy Act, and appropriate state legislation specifically related to place and people. This requries the same understanding and capability necessary for planning watersheds and regions. However, if Section 208 induced ecological inventories and models as central to ecological planning, these data could be made available to municipalities. If performance specifications can be written, local ordinances can be formulated specific to the municipality, the natural system it includes, and the needs, desires, expectations, and aversions of its population.

This returns us to the primary concerns of this book: the biological control of water pollution. The insistence upon performance specifications is tolerant to alternatives. There are many simple and effective methods to be considered and employed. These range from preservation of existing resources to a wide range of remedial techniques. What distinguishes those techniques presented in this collection is their ecological-limnological basis. They recognize that organisms are essential for potable water in nature or by human management, that water planning and management must be ecological-limnological, that sanitary engineering cannot achieve the objectives of the nation.

The world of water and sewage treatment is dominated by the subject of waste. Yet this is a misnomer. There are no wastes, only matter in various states. Most of that named waste is better described as nutrients. We need to learn how to complete the cycle, notably the return stroke involving the decomposing microorganisms in water and soil and their symbionts. And it is these microorganisms who are the heroes of this book, the creatures who have been thoughtlessly engaged in water treatment for aeons and who continue to be indispensible. The EPA disclosure that carcinogens are produced as a product of chlorination is only the death knell to the concept of killing for water potability. It must be replaced by a biological alternative that utilizes the essential microorganisms. The future of water treatment lies in ecology-limnology, as this book so clearly affirms. Let us formally abandon our right to defecate in our neighbor's drinking water. Let us embark upon a national toilet training program, and above all, let us make a formal commitment to limnology and ecology as the basis for this endeavor.

2

The Economy, Energy, and Clean Water Legislation

SENATOR EDMUND S. MUSKIE

Democrat (Maine)

In today's political climate, any discussion of environmental issues is risky at best. Not only must those of us who continue to believe in environmental objectives explain our position more carefully but also we must defend those objectives in the context of an unfavorable economic climate.

We must begin by documenting the point that environmental programs did not generate either the econmic crisis or the energy shortage. We must begin by demonstrating that environmental controls do not cost jobs but rather create jobs. And we must begin by restating that basic objective of improving the quality of human life while not detracting from an improved standard of living, especially for those who have not enjoyed the abundance which so often has caused deterioration of our environment.

So let me place environment in context. A recent study for the Environmental Protection Agency came to the following conclusions:

First, "The stimulus of increased expenditures on pollution control equipment in the early years of the decade is expected to raise the rate of economic growth through 1976 above the rate of increase otherwise projected for those years.

Second, "The unemployment rate, in keeping with the pattern of overall economic growth, is projected to be 0.4% lower in 1975 and 0.3% lower in 1976 than it would have been without pollution controls . . . ."

Third, "The impact on prices over the decade show only slight increases which are almost phased out by 1982."

Now what do these figures mean? In essence they indicate that environmental requirements will have a stimulative economic impact—more jobs—more GNP—and slightly higher prices.

SENATOR MUSKIE has served the people of Maine in the U.S. Senate since 1959 and as their Governor from 1955 until his election to the Senate. In the Senate, he is Chairman of the Subcommittee on Environmental Pollution and is chief author of the Federal Water Pollution Control Act Amendments of 1972.

And what will the investment accomplish? You have heard broad statements about ecological and biological integrity. You have seen evidence of improved recreational and aesthetic benefits. You have probably seen statistics on the benefits to public health. I cannot place a dollar value on these benefits. Nor can I tell you precisely what the value is of a clean stream or a clean sky. I am not prepared to place a dollar value on the estimated 15,000 deaths a year which result from dirty air. I am not even certain that our estimates of the value of a productive fishing resource are particularly accurate.

I am satisfied that man lives in a delicate balance with his environment and that our objectives—healthy air and clean water—are essential to maintaining that balance.

I am satisfied that our short term investment on environmental improvement will provide immediate economic benefit and in the longer run, essential ecological protection.

To those who assert that we are going too fast, I must respond that we are going too slow. Five years after Earth Day—four years after enactment of the Clean Air Act—three years after enactment of the Clean Water Act we are still negotiating over how clean is clean. Only last year was the issue of water pollution funding finally resolved. Only now can we get on with the task of cleaning the nation's water.

It is as if we are beginning again. The recent Supreme Court decision, declaring illegal the Administration's failure to allot the full program authorization, and our recessionary economy, with its ever-increasing unemployment, have accomplished what couldn't be accomplished before; they have stirred the sleeping giant, the 1972 Clean Water Act—put to sleep by the Administration's disregard of the law of the land—and headed that law again toward its stated objective: "to restore and maintain the chemical, physical, and biological integrity of the Nation's waters."

The Supreme Court decision, dated February 18, 1975, made the following statement: "As conceived and passed in both Houses, the legislation was intended to provide a firm commitment of substantial sums within a relatively limited period of time in an effort to achieve an early solution to what was deemed an urgent problem." An early solution to an urgent problem indeed!

It is now two and one-half years later. President Nixon, with a stroke of his pen, impounded the funds to do the job *and* delayed the program two and one-half years, thus seriously impairing our ability to meet our national goals. The decision in the end turned on a technicality—whether the Administrator had the authority to withhold certain amounts from allotment—and was decided on the basis of overwhelming legislative history to the contrary. But its impact will be far-reaching, both on the environment and on the economy.

During Senate consideration of the Conference report, I said in a passage that was cited by the Supreme Court decision as evidence of Congressional intent:

> to achieve the deadlines we are talking about in this bill, we are going to need the strongest kind of evidence of the Federal Government's commitment to pick up its share of the load. We cannot back down, with any credibility, from the kind of investment in waste treatment facilities that is called for by this bill. And the conferees are convinced that the level of investment that is authorized is the minimum dose of medicine that will solve the problem we face.

We didn't get the minimum dose, and consequently the deadlines are not being met—at least the 1977 deadlines for communities are not. Municipalities have been able to raise the local share, many states have been willing to participate financially, but the Federal commitment hasn't been there to match it.

Plans have been shelved, workers have been idled, factories have slowed down production of component equipment, and—this is the most insidious result—communities have lost hope, and lost belief, in the program. Projects that were ready to go in October of 1972 are no longer ready. There is no incentive to maintain momentum when there is no follow through. Why go through the motions, they ask?

We passed a law specifically designed to prod the communities into action and to push the states to develop aggressive programs. The prod and the push was to be assured Federal assistance. But when they did their part, when they conducted their infiltration-inflow studies, when they established user charge systems, the Federal Government held back.

How can we cause innovation, how can we keep the pressure on technological improvement, when the Federal Government does not live up to its part of the bargain? We have listened to the argument that the money couldn't be spent, that there were not enough projects ready to go, and we know that is not true. In Maine alone, we could obligate, within one year, the full entitled allotment of water pollution funds and still have to hold off on some projects.

And, we have been told that impoundment would have little impact on the program to clean up industry. The contrary is true. Industries which had planned to hook into municipal systems have been stymied. They don't know which way to go—whether to build their own system or whether to wait for the municipal system. Other industries have argued that it is unfair to force them to make a financial commitment when the Federal Government is not prepared to meet its own obligations. But now the Supreme Court has ruled and the Administrator of EPA has agreed to make the full $9 billion available for obligation. We are, indeed, beginning again.

It is a time for decisive action; if the money is going to do any good, it must be quickly delivered. It is also a time to remind ourselves of our long-term objectives.

The contributors to this volume remind us that it is not time to take the pressure off the technological community to find better and less expensive ways to do the job. We are, first and foremost, interested in reducing and then eliminating pollution from our water, restoring and maintaining the quality of our lakes and streams. The more pollution we eliminate with each dollar invested, the better off we are.

But the water pollution program, conceived, and designed to meet one major national concern—cleaning up the water—has taken on a new major role as well: stimulating a stagnant economy. The economy needs the money now; we must move those projects that are ready. But the environment ultimately needs long-term solutions as well—solutions which may not be available in time to help stimulate our present sagging economy, but which, nonetheless, must be developed, refined, and perfected.

When we were working on the Clean Water Act some said that the country couldn't afford the cost of water pollution control. Now it appears that the country can't afford *not* to do it, for economic as well as environmental reasons.

These are two vital purposes—to restore our economy and to restore our environment—and they come together in the clean water program. As Chairman of the Senate Budget Committee and Chairman of the Subcommittee on Environmental Pollution, I have unique vantage point and an awesome responsibility. My role illustrates the scope of the legislation, and how its components reach out to touch many segments in our society. The extent to which the immediate and future needs of each effort can be merged into an active and vigorous program will provide a blueprint for solving major problems of the seventies. The seventies will show the health of the economy as dependent upon, not exclusive of, many major social forces, including education and health as well as environment.

As a United States Senator, my major objective is to get America back to work again; we desperately need to pump money into the sagging areas of the economy. We have a program ready to go, a program already committed to a major national purpose. By allotting and quickly obligating available water pollution funds, we can mitigate some of the construction industry's 15–20 percent unemployment rate.

The program will have important direct beneficial effects—the obligation of the $9 billion in impounded funds will create 360,000 direct new jobs, 180,000 in actual construction. But its major impact may be in its indirect effects. Factories will have to gear up and take on workers to provide equipment. The equipment will have to be transported and installed. Municipalities will have to hire consultants and staff for implementation. The 360,000 new workers will have money to spend on homes, appliances, and other consumer goods, further stimulating the economy. Steel and concrete will have to be made. The ripple effect is manifold and essential.

Of course, there will be some delays. All of the states will not be ready to spend all of the money; that's expected. But the appearance of movement and the promise of funding will have the effect of a stirring giant. Everyone will know the giant is going to move. Training programs will be revived. I would suspect that the full release of the money, if announced, will even move the market up a point or two.

We need a specific national commitment to accelerate the rate at which we approve projects, obligate these available funds, get construction under way, get people back to work and reduce the pollution load on our streams. EPA Administrator Russell Train and I have discussed the need for this kind of commitment. As a result, I am convinced that EPA can and will increase the level of obligation of water pollution funds. And I am convinced that Administrator Train shares this goal.

In order to accelerate the obligation of funds some projects which are not ready to go will have to be replaced on state priority lists with projects on which construction can begin before the end of 1976. Enforcement of permits for municipal discharges, heretofore ignored because communities lacked the Federal funds to move ahead, must now be enforced if communities delay submitting applications or are slow in getting approved projects under construction. Communities should establish penalty provisions in contracts to assure prompt performance by engineers and contractors. Reviews and analyses of projects and project applications which have been sequential must become concurrent. And projects which are routine must move so that projects which are controversial can be subjected to adequate public review.

Projects which will significantly improve water quality, such as those which eliminate raw waste discharges, must receive priority in order that projects designated primarily to serve new growth can complete environmental impact analysis.

As Chairman of the Public Works Subcommittee on Environmental Pollution, I want to keep pollutants from reaching the water—or the air, for that matter. We developed the 1972 Clean Water Act to challenge technology to do better, to develop new and better treatment systems for all sources.

We should be on guard, however, against repeating past mistakes as we move quickly to gain the economic benefits of the construction grant program. I do not want to see the technological community, both public and private, interpret the rush to move the construction grant program as an opportunity to reinforce their old patterns, to rely on their old solutions. For example, we adopted secondary treatment as a minimum concept. But we need, as a nation, to move beyond that minimum. The current waste treatment methodology has been around, without major improvement, for sixty years. The methods described in this volume hold great promise, and we need more people looking into more alternatives. We have spurred technology before, and technology has responded. Technology must respond again. It must go beyond existing methods.

The Act provides for this. The Administrator of the Environmental Protection Agency is directed to "encourage waste treatment management which results in . . . the recycling of potential sewage pollutants through the production of agriculture, silviculture, or aquiculture products . . ." Systems which cooperate with nature are cheaper in the short run, and in the long run use less of our valuable resources because they are renewable. There will, however, be a tendency, as the full amount of money is allocated, to rely on easy solutions—traditional engineering and construction practices. This would be contrary to the Clean Water Act's purposes.

Through the coming years as the program moves into full implementation, the Subcommittee on Environmental Pollution will continue to push for the development of that kind of innovative technology. Last year we discussed the question in a hearing in Hawaii, and I said:

> secondary treatment is not necessarily the last word in methods to deal more effectively with water purity. Indeed . . . I am personally a little disturbed that in all the years since 1916 we have developed nothing better than secondary treatment as a way of dealing with these problems . . . I don't think we ought to be inhibited by the questions of expenditure of monetary and other resources. If we can find a better method, we ought to go do it, hopefully at less cost and resource demands.

As a Senator representing the State of Maine, I am especially interested in systems with a biological or natural base because, in the long run, they tend to employ more people from the local area and they are less capital and energy intensive. My state has hundreds of small communities which tend to rely on treatment systems that were really developed for big cities. These communities are employing capital-intensive, energy-intensive systems, designed for areas that have a minimum of available land. In other words, they must put up something they don't have—money and machinery—and not something they do have—land. These communities have retained their natural support system which the cities have abandoned, but they are often not using it. We need to provide them with options tailored to their specific locations and needs.

Our economic problem is immediate, to be sure, but in addressing it we should not provide only short-term solutions. Treatment plants that employ people only for construction are not the optimum solution. When the construction is completed, the jobs are done. The long-term employment implications must be factored in.

As Chairman of the new Senate Budget Committee, I want to keep our expenditures in reasonable proportion to our income. I am especially interested in those treatment systems which produce financial, as well as environmental benefits, benefits which recover valuable resources. Further, I want to see systems which can expand as a community expands, at community, not Federal expense.

This round of Federal assistance should not have to be repeated. The Federal Government should not and cannot commit itself towards an endless capital expense. The land application system developed in the community of Muskegon, Michigan, recently sold a corn crop for $400,000 despite an early killing frost, and that community predicts that for the calendar year 1975 they will have paid their full operating expenses with the receipts of their corn sale. Muskegon is capable of handling growth without new construction. Their system is adaptable. In fact, two industries have sited in that area because of the adaptability of their system, and those industries are going to pay their full share of the costs of operation. Muskegon also uses land and air as cleansing agents—both renewable resources. There is no resource depletion.

In the greatest food-producing country the world has ever seen, we continue to consider nutrients as pollutants. Phosphorous, nitrogen, and potassium have been the targets of water pollution clean-up efforts for years, and then viewed as a sludge disposal problem, while the cost of fertilizer has escalated to unacceptable levels. The average day's worth of municipal waste is 40 billion gallons per day. With an average of 40 parts per million of nitrogen and 10 parts per million of potassium and phosphates, it is laden with nutrients.

Each year in the United States we discharge over $200 million worth of valuable nutrients into our waterways, and call them pollutants; 486 million pounds of organic nitrogen worth $121 million; 121 pounds of available phosphorous worth $68 million; and the same amount of potassium worth $12 million. And these are deliberately conservative figures. The value of these three major chemical fertilizer components does not tell the whole story because of the micronutrients: zinc, copper, and lead, which are frequently added at the farmer's expense to agricultural land. Healthy soil is composed of myriad organic compounds which need to be replenished; these are present in a municipal waste stream. Finally, in many areas of the country we cannot overlook the value of water, which is imported to agricultural lands at great cost.

We need to return these materials to nature, for nature's use, for economic as well as environmental reasons. We need to continue to push the program toward self-sufficiency and permanence with long-term solutions based upon sound ecological concepts like the kind that are discussed in this volume and required by the Clean Water Act.

As I have said, we are beginning again, and we are provided with new opportunities. We have great problems to meet and overcome, and we have a law to do it, enacted by an overwhelming majority of the people's representatives and embodying fundamentally sound ecological principles. We need all of the most expert help we can get: from scientists and engineers; from lawyers and laborers; from politicians and administrators; all of us need to work together again to restore our economy and our environment and create an order which embodies the ethic of the naturalist, Aldo Leopold:

> A thing is right when it tends to preserve the integrity, stability and beauty of the biotic community. It is wrong when it tends otherwise.

3

The Safe Drinking Water Act of 1974: A Call for Action

CONGRESSMAN PAUL ROGERS

Democrat (Florida)

The goal of safe water is so admirable that in retrospect I wonder how we went so long without addressing ourselves to the problem. Water is the most basic resource. I think—if the reader will pardon the turn of a phrase—progress was our most important problem.

Industry and municipalities increased their use of water as a depository of wastes and we quickly reached the finite line where Nature could no longer bear the burden, could no longer mend our mistakes. Ironically, in addition to the ill which this caused our environment and wildlife it was our need for water for personal consumption which eventually brought our attention to our own deeds. At the same time there also developed a logical sequence of thought that beyond fish and wildlife, and beyond aesthetics, polluted water was being used for human consumption. Evidence began to mount and concern became widespread. The Committee on Public Health and Environment of which I am a member, called hearings to discuss safe drinking water legislation which I had introduced.

That was more than four years ago. It is hard to understand why it took four years to get this legislation passed into law. There were unbelievable pressures brought to bear from the Administration each time we discussed this legislation. Almost annually we called up the Administration in the personnage of the Environmental Protection Agency (EPA). And each year we received less than enthusiastic testimony. I am sure you realize that during those unusual years, the EPA, like all other departments and agencies, was given marching orders from the White House and from the Office of Management and Budget (OMB).

In fact, the first time we had the pleasure of having William Ruckleshaus testify on the Safe

MR. ROGERS represents the 11th congressional district in the State of Florida to the U.S. House of Representatives. Having served his district since 1957, Congressman Rogers is chief author of the Safe Drinking Water Act of 1974 and is currently Chairman of the Public Health and Environment Subcommittee in the House.

Drinking Water Act, he tacitly agreed that one of the main reasons he could not give an endorsement to the legislation was because OMB had not taken its position on the bill. In retrospect, his position of not being able to support the bill was the closest the Administration came to supporting it. After that, it gave basically a negative response. This continued as we moved the bill through the Subcommittee, the full Committee and the Rules Committee. But we did manage to get it passed, and with reluctance, the President signed it into law on December 16, 1974.

During these years we had as one of our greatest obstacles the fact that most people, including members of the Committee and the Congress, sincerely questioned the need for the legislation. Small outbreaks were often reported beyond their locale, and in almost every case the problem did indeed appear localized. An untrained manager of the local water works, most probably serving on a part-time basis, forgot to add this or that and the city's users developed minor gastrointestinal problems. The outbreak in California, where some 16,000 people were stricken, was the most eye-catching.

But during the same period, there were several studies done which helped greatly in developing an appreciation of the problem. The one conducted by the U.S. Public Health Service, the General Accounting Office, and the Water Supply Division of the Environmental Protection Agency seemed to fit into each new set of hearings by the Committee. Each report received attention in the popular press until, along with environmental groups, there formed a coalition of public interest groups who focused in on the problem of the quality of drinking water.

The legislation did run into great difficulty when it arrived at the Rules Committee, the final obstacle before reaching the House floor for a vote. We were help up and put off for more than a month. We were rapidly running out of time, and a pocket veto loomed large.

Then came the Harris study on the drinking water of New Orleans, and the corresponding report from Cincinnati. As unfortunate as it was, the situation in New Orleans finally brought national attention to the need for legislation. The Clean Water Act became law in late 1974. Celebrate we did. But now, so recently after the President signed the Safe Drinking Water Act into law, I again have concern over the fate of the law. The law called for the establishment of an advisory council. A public interest coalition of health and environmentalists submitted six recommendations for the council. None was appointed and the list which was accepted included only two who testified on the bill and one who was actually against the bill. We were at least fortunate in that one of the members invited to serve withdrew and Dr. Sidney Wolfe was appointed. I still do not feel that the provision of the law pertaining to the makeup of the council has been strictly followed, but now that the council is at work, I think it will do a good job.

In a statement presented at the signing ceremony, President Ford said that he had two concerns about this legislation. First the Federal involvement in an area which he considered a state and local domaine, and secondly, the cost. After reviewing the President's budget proposal, I can only say that Mr. Ford has moved to ease his concern over costs. The law authorized $156 million over a three year period. As written, we hoped to get the program underway by helping the States establish their programs and then get the federal government out.

But the President's budget is considerably below the authorization level. For fiscal year 1975, the bill calls for $23.5 million compared to the proposed budget figure of $7.779 million. For fiscal year 1975–76, the budget reflects a $33 million difference or roughly a 45% cut from the authorization levels. The President's budget called for approximately one hundred additional personnel, plus another eighty-one in fiscal year '76. But there were no funds or position created for enforcement, and just as important, there were no funds for demonstration projects.

I consider the demonstration projects one of the most important parts of this legislation.

Without these projects, we will be helping to promote water systems which we know are not doing the job when we should be developing new and innovative methods which will handle the problems of today. Unless we do promote new technology, we will be using stone-age tools to handle problems caused by space-age technology.

It goes without saying that we are developing chemicals and substances far beyond our capability to even determine their safety, much less engineer systems which can dispose of them. Our water systems now are not too different from those designed one hundred years ago to stop cholera, typhoid, and dysentery.

Yet even in a slow year we develop more than three hundred new substances many of which enter ground sources used for drinking water. This is simple shortsightedness. We know these substances are in the water, or at least our instrumentation permits us to identify more and more of them, and yet the budget would have us overlook this problem. The people in Duluth, Minnesota, will tell you the importance of demonstration projects.

A parallel concern of mine is the overall lack of knowledge we have on carcinogens. We find them in New Orleans. We know the story of Duluth and Cincinnati—of the entire section served by the Mississippi River, but what are we doing? In the best traditions of the bureaucratic manual we see the EPA conducting a survey. I can only imagine that if I were a resident of New Orleans and concerned about the elements found in the drinking water, I would not be at all appeased by finding out that other cities in this nation suffer from carcinogens in their drinking water.

In December, 1974, I wrote to Russell Train, Administrator of EPA and suggested that a survey of the type he initiated was a good thing. We do need to know more about the general condition of the drinking water in terms of such ingredients as heavy metals, pesticides, and chemicals. But I also thought that there was a very good chance that these elements were introduced somewhere along the Mississippi or Ohio Rivers and that the EPA should therefore be trying to halt this practice while the survey was being conducted by invoking provisions of the Clean Water Act. Also, in December I offered a suggestion that the National Cancer Institute, the EPA and the Food and Drug Administration (FDA), intensify their joint efforts through the National Toxological Center in Arkansas to determine the toxicity of the carcinogens found in the New Orleans drinking water.

The FDA has jurisdiction over the Arkansas facility, which is relatively new. And I have it on good authority that the staff there is heavily burdened because of the recently increased demand for carcinogen testing. In looking at the budget for this facility, I find that there are approximately 220 employees and they are operating on a budget of about $11 million. The 1976 budget calls for no increase in positions and a pay increase which is considerably less than the cost of living. I inquired as to whether this staff could do the studies needed for the target cities which the EPA will be surveying. There are only two ways this would have been possible—by increasing the budget and staff or by junking the work they are now doing and starting work on the data which is supplied from the target cities.

I do not think this reflects a proper priority. As I understand it, there is about a two-year period needed to develop conclusive evidence on carcinogens. This would lead me to believe that we are, at best, three years from even understanding the potential threat of the carcinogenic elements we know exist in the Mississippi River today.

I am fully aware that men of science are cautious when speaking of carcinogens and tolerance levels. And I understand that we cannot, through law, condense the time frame which is required for scientific evaluation. However, I think we are unnecessarily delaying the time in which we can either tell the American public that their water is safe, or move to correct the situation should the fears shared by many be realized as a result of the EPA survey. I think the

American public deserves such an assurance.

Finally, I would like to touch on one other point. The law included provisions that the government would assure the public of an adequate quantity of drinking water—quantity, not to be confused with quality. In many cities of this nation we see demand and supply teeter perilously close. In Washington, D.C., there are reports that the city's peak demand exceeds the Potomac River's flow on low flow days.

More alarming to me is the situation developing in our Western States as the rush for energy has apparently blinded many to the potential of emptying those states' ground and underground water supply. According to a report from Helena Hunnington Smith, one company is planning to drill no less than fifty-eight wells which would provide more than a quarter of a million acre-feet of underground water to cool its plants. This and other plans for developing the Western states' coal supplies represent, I feel, a substantial threat to this area's water supply.

For this reason, I have written to Elmer Staats, the Comptroller General of the United States, and asked that the General Accounting Office begin a study on the problem of potable water supplies for the present and future needs of these Western States. I have also asked Mr. Staats to make recommendations as to whether it appears that additional legislation is needed to control this and similar situations, or if the matter can best be handled by regulation.

Certainly all of us have a continued concern about the quality of this nation's water systems. As we began work, it was evident that we were trying to correct the mistakes of the past and prevent future deterioration of our water resources. But I contend that the potential danger of depleting our water resources in the name of energy is even more frightening than the misuse of water under the banner of progress during the past one hundred years. At least we have identified the problem before it has begun in the case of the Western States energy proposals. But we must move quickly.

Because of the interest and activity of scientists, planners and engineers, and the public at large, we have accomplished much in the past two years with the passage of the Clean Water Act and the Safe Drinking Water Act. Still, there is much to do and the time is short.

4

The Protection and Improvement of the World's Drinking Water Quality

DR. THOMAS A. LAMBO

World Health Organization
Geneva, Switzerland

Why are we concerned about drinking water quality?

Man's earliest known prayers were incantations for rain, for purifying the stream, for potency, and for good health. His first effortless travel was down bouyant streams, his sense of belonging to a community was bound up with sharing a source of water, and his early settlements were dictated by the distribution and presence of water. He was quick to resent members of other communities who would use his source of supply. After the passion of love, water rights have caused more domestic and political troubles to the human species than anything else.

Awareness of the need for taking precautions with water is equally very old. A Sanscrit text written about 2000 B.C. says "It is good to keep water in copper vessels, to expose it to sunlight and filter it through charcoal." Another text, from about the same period, advises that "impure water should be purified by boiling over a fire, or being heated in the sun, or by dipping a heated iron into it and then allowing to cool, or that it may be purified by filtration through sand and coarse gravel".

There are various ways by which water affects the health of man but we are here concerned with how substances in it affect his health when water is drunk directly, or in food, or by using it in personal hygiene.

Today, much is known about drinking water quality and health hazards, although a lot remains to be known. The first amongst these

DR. LAMBO is the Deputy Director General of the World Health Organization in Geneva, Switzerland. A native of Nigeria, Dr. Lambo received the African Research Award in 1970 and has published extensively in neurology and ethnography. His previous work for WHO encompassed mental health manpower development and noncommunicable disease. Dr. Lambo has served on the United Nations' Advisory Commission on application of science and technology to development.

are the biological hazards. The principal biological agents transmitted through water are: pathogenic bacteria, viruses and parasites, the contamination of the water occurring through pollution of the water source itself or during its conveyance from source to consumer by excreta of man and animals, sewage and sewage effluents, and washings from soil. Pathogenic bacteria transmitted directly by water or indirectly through water to food constitute one of the principal sources of morbidity and mortality in developing countries. Cholera, bacillary dysentery, typhoid fever, paratyphoid fever, gastro-enteritis, and infantile diarrhoea are some of the more common bacterial diseases, capable of being transmitted through water or food prepared with such water. These diseases could be eradicated or controlled through high-quality water supply.

In 1970 a series of outbreaks of "El Tor" cholera occurred in areas not normally affected, such as the Eastern Mediterranean region and the USSR, as well as in a number of African countries. In 1971 it spread to nine more African countries causing high mortality, and small outbreaks or sporadic cases of cholera occurred in six European countries. Person-to-person transmission of cholera does occur but by far the most important mode of dissemination is through the environment, especially water. Typhoid and paratyphoid fevers are still widely disseminated throughout the world. In Europe the explosive outbreak of typhoid fever in Zermatt in 1963 was a great warning. Outbreaks of salmonellosis, though usually food-borne, may occasionally be spread by water.

Viruses most commonly present in water and sewage are the enteroviruses, adenoviruses and reovirus, and the virus of infectious hepatitis. Although the virus of infectious hepatitis has not yet been isolated and identified, there is ample epidemiological evidence that outbreaks of this infection, which has a global distribution, are caused by infected waters. It can also be spread by shellfish contaminated with sewage effluent.

Of the many parasites that may be ingested *Entamoeba histolytica* is the causal agent of both intestinal amoebiasis and extraintestinal forms of the disease such as amoebic liver abscess. It is widespread throughout the warm countries of the world. Good filtration is effective and essential, especially since amoebic cysts are resistant to chlorine in the doses normally applied in water treatment.

As we are dealing today with the quality of drinking water, I shall not digress in outlining health hazards from biological agents transmitted through water contact other than ingestion, such as the health hazards from bathing in polluted waters, schistosomiasis, and the diseases transmitted by water-associated insect vectors such as malaria, onchocerciasis, yellow fever, and filariasis. Nevertheless, they merit mention.

The other groups of substances in water that constitute a health hazard arise from chemical and radioactive pollution. If present above a certain level, some chemical pollutants (e.g., arsenic, nitrate, lead) will constitute a direct toxic hazard when consumed. We have a few examples where chemical water pollution has direct and immediate impact on man's health, either from direct consumption of water or from the use of aquatic food products from polluted water.

Typical examples are methyl mercury poisonings, the first reported major incident of mercury poisoning arising from the discharge of industrial wastes occurring in the Minamata Bay area in Japan where the first case of an unidentified disease of the central nervous system was noticed in April 1956. By February 1971 the total number of cases was 121, including 22 cases of congenital Minamata disease. It was ascertained that the disease was caused by methyl mercury, a waste product of the acetaldehyde production, in which inorganic mercury catalyst is used. The methyl mercury was present in the water itself at an undetectable low level proportion, but was taken up by fish and shellfish and became biologically concentrated. Minamata disease was again reported in 1964-65 in the area of the Agano river, Niigata Prefecture,

Japan; this time 49 persons were affected of whom six died.

Another water pollutant incriminated in an epidemic, known as "Itai-Itai" disease, is cadmium. The disease was observed first in Toyama City in Japan and was related to the exposure to cadmium released from a nearby mining complex. Here, however, the relationship between the exposure to cadmium, which occurred apparently through rice which was grown in fields irrigated by polluted water, is not so clear-cut as in the case of methyl mercury. Many other factors such as nutritional status, vitamin D deficiency, and specific social and cultural conditions in the area were undisputed environmental factors which might have been important in the etiology of the disease as was cadmium.

Concentrations of arsenic in surface water bodies are usually low, but high concentrations have been reported in some drinking water supplies in Latin America and the Western Pacific and have been associated with endemic arsenic poisoning and the so-called "black foot" disease.

More recently, in a number of developed countries, attention has been drawn to a variety of organic chemical compounds discharged with industrial wastes. They include small quantities of various pesticides such as DDT and aldrin; and industrial chemicals such as diphenyl-ether, chloroethers, polychlorinated biphenyls, and many others. Some of these substances at higher levels have been shown to be carcinogenic for animals. Another class of compounds which are invariably found in surface waters are polycyclic aromatic hydrocarbons, some of which are again known to be carcinogenic under high level occupational exposure or in animal experiments.

The chemical industry, such as rubber works and dye works may release to sewer systems a variety of organic amino compounds; other sources of potential hazards are the chemical and pharmaceutical industry, textile dyeing plants, and plastics works. All these substances are not well removed by conventional biological treatment and sand filtration, although up to 99% or more are removed by activated carbon filtration, if flow rates are low. In considering the health effects of water pollutants, it should be remembered that many water pollutants also appear in air and food, which are often more important sources of intake than water, and the assessment of pollutant levels of water should always be made in relation to the actual intake of drinking water and to the body burden resulting from other sources in a given locality.

For many trace elements and organic compounds, information is insufficient for a scientific assessment of their health significance. Thus, although there is circumstantial evidence of an inverse statistical association between the hardness of drinking water and the death rate from cardiovascular diseases, there is still no definite information as to which water components may be protective or harmful, whether these are major constituents or trace elements, or whether it is their presence or absence that is responsible for the effects. We need more sensitive scientific tools and more refined clinical and epidemiological observations.

Other substances such as fluorides are beneficial and may be essential to health if present in small concentrations, though toxic in larger concentrations. Certain other substances or characteristics affect the acceptability of water for drinking purposes. International and national criteria and standards have been established to provide a basis for the control of human exposure to many of these substances through consumption of polluted water. Consumption is, however, only one possible pathway to exposure.

Man can be exposed to water pollutants through other types of direct contact—e.g., in recreation or the use of water for personal hygiene. The possible health implications of these nondrinking uses of water (including agriculture and industrial uses) are less well understood and no international criteria guidelines exist for the control of such exposure. Chemical pollutants can also disturb the aquatic ecosystems or accumulate in aquatic organisms used in human food. The various chemical and biochemical transformations that pollutants may undergo in the aquatic environment also deserve attention. Degradation or transformation products may appear that are more toxic than the original pollutant. Little is known of these physical, chemical, and biological processes, yet an understanding of these processes is essential to assess the health implications.

In the industrialized countries, the problem of water quality is compounded by increasing demands for water, industrial pollution, the limits of existing water resources and energy constraints.

It has been estimated that the total demand for water may double every twenty years, and in some countries it is estimated that doubling may take place every ten years. The frightening and exponential growth of population and industry not only increases demand for the water quantity, but also puts high requirements on water quality. There is a conflicting requirement here because increased water demand and use degrades the overall water quality, particularly of surface water resources, as a consequence of the growth of the discharge of wastewater into the same water bodies (streams and lakes) from which we abstract water for use. Since the ground water resources are limited, there is a tendency to use more surface water as a source of community water supply and for other uses. New types of wastes are continually produced, many of which are known to adversely affect the natural and man-made systems of water purification; some of these new pollutants cannot be efficiently removed by conventional treatment processes.

In some rivers in the United States, at periods of low flow, between 3.5 and 18.5% of the water is estimated to have passed through domestic waste systems. If the volume of industrial effluents is also taken into account, it would be expected that some 20–40% of the river water at low flow in some areas may be reused water. The lower reaches of the Rhine, serving as a water source for six million people, may often contain 40% of sewage effluent, a figure that may rise to nearly 100% in periods of extremely low flow. In times of drought the water supply source for Agra, India, consists almost entirely of partially treated sewage from New Delhi, 190 km away. These phenomena occur in many other countries.

Another major use of water is in the production of energy where huge quantities of cooling water are required and discharged into the surface waters at increased temperature, causing what is called thermal pollution which may have serious ecological consequences. In fact, with respect to nuclear power production, thermal pollution is of more concern under normal operational conditions than the pollution by radioactive wastes.

In the developing countries the rapid and staggering growth of population, the constant influx of population into urban areas, usually unplanned and with poor infrastructure, and the shortage of virtually everything that is needed for the provision of safe and adequate community water supply to the population, namely, capital, skilled manpower, appropriate technology, and managerial skills, goes to aggravate the problem that is already grave. The problem as regards water quality is, in the main, one of biological pollution.

A global survey of community water supply conditions in developing countries carried out by the World Health Organization (WHO), shows that as from 1970 only 50% of the urban populations of these countries had access to piped water supplies in their houses or courtyards; another 19% had access only from street standposts and the remaining 31% of the urban population had no access at all to public piped municipal water supply. Amongst even those who had access to piped water supply, nearly half had only intermittent supplies. What this implies in terms of health hazards hardly needs elaboration. Allow me to emphasize once again that good quality water supply would lead to the disappearance or at least control of most of the major killers in these countries. In the rural areas where well over 70—80% of the population of the developing countries still live, access to relatively uncontaminated water was available only to approximately 14% of that population. Any sensible strategy to develop and innovate the rural communities must start with the provision of good water supply. In many developing countries the surveillance of drinking water quality is conspicuously absent.

What are the strategies which may lead us to a better solution of our present predicament? It is clear that whatever action is taken should command public support. Communications with the public are therefore vital. As problems arise it is necessary to take the public into confidence, explaining actions being taken, including studies to fill gaps in knowledge, where such a gap is clearly identified. A scientific approach is needed that will exclude both unnecessary scares or a false sense of security. It also imposes on scientists and public servants the willingness to listen to the public. Confidence of the public in the Water Authority is crucial. Public motivation is sometimes not health-related. A health measure may sometimes be difficult to "sell" purely from an explanation of the health considerations but might be accepted if other impacts of the measure in terms of convenience, economics, etc., which have a greater motivation potential, are explained. In other cases, a problem which is technologically difficult to solve can often be ameliorated through changes in patterns of water use, economy in use, distribution of peakloads, storage, etc. All such behavioural changes imply full public cooperation which will only come when the confidence that I talked about exists.

The strategies that we adopt must be appropriate to the situation. Here, the concepts of cost-benefit and the balancing of risks, as well as public acceptability, come into play. I have already talked about the importance of public support. The strategy must also be economically sound—it must have a high cost-benefit ratio. This implies also taking calculated risks. There is, however, a higher risk associated with not doing anything. We must apply the knowledge already available. Do we abandon, for instance, disinfection by chlorination because some haloforms that may be potentially hazardous are formed in minute quantity when chlorination of water is practised? Present indications are that these substances at the levels at which they are formed do not constitute a health hazard.

If chlorination were stopped, the resulting hazard of exposure to water-borne diseases would be far greater. Here in the United States studies have already been instituted to find out the levels and extent of the prevalence of these compounds, where they are formed, whether they can be prevented and, if not, whether they can be easily removed by conventional or other processes—while at the same time continuing the practice of chlorination. This seems to me a sound approach.

Both industrial wastes treatment and the purification of water polluted by industrial wastes often require advanced technology. Such wastes commonly contain substances that are difficult or very costly to remove. In some cases the only solution is to change industrial processes or restrict their operation by legislative or similar actions. The extent of wastewater discharges from industrial processes may be reduced by such techniques as recirculation and the use of countercurrent washing procedures. Considerable advances are being made in the recovery of byproducts. Occasionally such recovery may be economically justifiable; even if this is not the case this may be less expensive than treating wastewaters containing some chemical pollutants.

A particular problem is posed by various organic compounds where the only solution at the present state of technology is the use of activated carbon filtration. Other techniques are now in the stage of development including the use of oxidizing agents such as ozone and hydrogen peroxide. The removal of the inorganic materials can be achieved by ion exchange, reverse osmosis, and electrodialysis and other physical chemical processes. These are feasible but very expensive and may not be applicable at all to the conditions in developing countries. The tendency in such countries—India, for example—has been to try to adopt simple processes such as oxidation ponds to treat effectively industrial wastes. There is need for newer and simpler technology and there is room for a retesting of old procedures and technologies in other countries.

The energy constraint that I mentioned earlier has also focussed attention on the need for increased use of biological treatment processes. One such is slow sand filtration. A myth has grown up that this process is old-fashioned and therefore inefficient, that new techniques have rendered it obsolete, and because it is simpler than many more recent innovations it is necessarily inferior to them. None of these objections to biological filtration is warranted. In many circumstances it is still the most appropriate choice when treatment methods are being selected, and the designer who automatically turns to other methods is often acting in ignorance of the continuing potentialities of the process. No other single process can effect such an improvement in the physical, chemical, and bacteriological quality of normal surface waters as that accomplished by biological (that is, slow sand) filtration. Yet what is paradoxical is that this water treatment process—the oldest of them all—is one of the least understood; less scientific research has been carried out into its theoretical and practical applications than into the more recent but less effective methods. It is for this reason that WHO thought it fit to publish a guide on slow sand filtration.[1]

I have mentioned slow sand filtration merely as an example of biological treatment processes. The use of oxidation ponds for the treatment of sewage is another biological process that is sound from ecological, energy conservation and public health points of view. Its high efficiency in conditions of low industrial pollution and high ambient temperatures makes it particularly suitable for developing countries.

Strategy does not rest with choice of right treatment processes alone. Concepts of conservation, reclamation, and reuse must be brought fully into play. Conservation consists in making the best use of existing water resources and preserving them from deterioration. Reclamation is the act of restoring used water as nearly as possible to its previous quality. By judicious reclamation it is possible to reserve first-class water sources for first-class uses, the demand for second-class water being met from reclaimed water. As much as seventeen years ago, the United Nations Economic and Social Council stated that "no higher quality water unless there is a surplus of it, should be used for a purpose that can tolerate a lower grade." There is also an increasing need to consider the optimum distribution of purification between the wastewater treatment plant, the river itself (self purification), and the treatment plant that produces potable water.

The World Health Organization is a specialized agency of the United Nations. Its membership is open to all countries. As of today, 141 States are full members, with three associate members. The objective of WHO is the attainment by all peoples of the highest possible level of health; health itself being defined as not merely absence of disease or infirmity, but a postive state of physical, mental, and social well-being. WHO is charged with a number of functions under its constitution to achieve this objective. It acts as the directing and coordinating authority on international health work; maintains liaison with other agencies; assists governments in strengthening their health services; proposes conventions, agreements and regulations with respect to international health matters; and promotes research. Much of the Organization's work lies in direct assistance to member states. In addition, the Organization disseminates information by publishing documents, guides, manuals, and standards.

Some of WHO's publications include: The monograph, "Operation and Control of Water Treatment Processes" (1964); the publication entitled "Health Hazards of the Human Environment" (1972) was produced in conjunction with the Human Environment Conference that was held in Stockholm that year; the publication, "Slow Sand Filtration" (1974); the Technical Report Series, No. 517—a report of a WHO Meeting of Experts that dealt with *Reuse of Effluents: Methods of Wastewater Treatment and Health Safeguards* (1973); another publication in preparation is a "Guide on the Surveillance of Drinking Water Quality." One of the issues of the World Health Statistics Report gives statistical information on the status of water supply and wastewater disposal in approximately ninety developing countries in the world as of the year 1970.

WHO's Environmental Criteria Programme is a good illustration of information distribution. Under this Programme, the dose-effect and dose-response relationships of environmental pollutants are sought to be established by an evaluation and assessment of available toxicological, epidemiological, and clinical data. The Programme will eventually cover some seventy chemical and physical pollutants and is being implemented in very close collaboration with member states and with the support of the United Nations Environment Programme. Its major feature is to look at the exposure of man in an integrated way—taking into account that chemicals may reach his body and various organs by air, water, food, at his workplace, and at home—and to evaluate the most important contribution to the overall exposure. The products of this Programme will be the so-called Criteria Documents which will, it is hoped, help member states and their health authorities in establishing more rational standards for water quality, air quality, and standards for work conditions.

At the same time, WHO has intensified its efforts to assist member states in establishing and developing environmental monitoring systems which they require for their programmes to protect human health from adverse environmental influences and, at the same time, to facilitate the international exchange of information on levels and trends of environmental pollution and other hazards, the resulting human exposure, and the associated effects on man's health; and the use of this information for the assessment and improvement of environmental conditions of relevance to public health. A large portion of this programme is devoted to the monitoring of water, both used for drinking and other applications with public health implications.

In conclusion, please allow me to emphasize the need to look at total development of man and his environment. We should seek to understand wholes as indivisible entities in a perpetually dynamic state. To my mind, man is the central concern of all our enquiries, be it in population, environment and water quality, health and disease, industrialization, and urbanization—and true development becomes meaningful only if man, who is both instrument and beneficiary, is also its justification and its end.

In all this, we must never take our gaze off man and remain insensitive to other challenges of our time as alienation, the sick conscience, the impoverished soul in the affluent society, and the uncanny feeling that there is a doom for a civilization which tends to discourage any expression of human creativity that does not support cumulative technological advances. I believe the theory of development, of which the quality of life is a measure, offers a valuable approach to the understanding of situations in which inequalities—for instance of property and power—are maintained in equilibrium.

Notes

1. L. Huisman and W. E. Wood, *Slow Sand Filtration* (Geneva: World Health Organization, 1974)

5

An Environmental Overview

WILLIAM D. RUCKELSHAUS

Former Administrator of the U.S. Environmental Protection Agency

It is terribly important to the world that steps are taken to preserve our planet in the face of an expanding population and expanding demand coupled with diminishing resources. People, not only in this country but indeed in the world, must be open to new ways of pursuing things. We cannot under any set of circumstances be prisoners of the past.

If we think of the way in which we treat sewage in this country it will provide an example of the purposes of this volume. Here, of course, we use water as a transporting agent in the treatment of sewage as is common in many developed countries of the world. One of the main reasons we have to build such large and expensive sewage treatment plants is because of the volume of water that is used in transporting sewage. We take the sewage from households and commercial establishments, mix them with water, and send this mixture through pipes into a very large installation whose sole function is to separate what we have mixed in the transporting agent in our individual homes and commercial establishments.

I suppose we ought to be asking ourselves now as we should have been asking ourselves for many years, "Why did we do it this way?" Why do we use the volume of water that we now use to transport sewage if in fact the volume itself is one of the reasons for the tremendous expenditure in the separation? The average person in this country flushes the toilet ten times a day. Many do not realize that. Each time that happens, five gallons of water are used; hence, fifty gallons are utilized in this manner each day by each individual. To help alleviate this consumption, experiments are now being tried in certain places in America. One is a toilet that uses one quart of water to perform the same function that five gallons now perform in most toilets—certainly an improved use of water. It is necessary to ask ourselves if we need to use all this water if we are going to continue to use water itself as a transporting agent. Indeed, is there some way in which we can continue to use water as a transporting agent but not mix sewage with the

MR. RUCKELSHAUS was the first administrator of the U.S. Environmental Protection Agency from 1969 to 1973. In this post he served as the chief enforcer of Federal laws regulating pollution. In 1973 he served as acting Director of the Federal Bureau of Investigation and then as Deputy U.S. Attorney General. Presently, Mr. Ruckelshaus is engaged in the private practice of law in Washington, D.C.

water in homes or commercial establishments? There are people looking into the possibility of not mixing the sewage, instead using some kind of receptacle to transport it to the central sewage treatment plant. It seems clear that this is an approach in which we ought to be spending a good deal more of our time and effort as a nation.

As a result of a Supreme Court decision, in fiscal year 1976 we will be spending $9 billion for the construction of sewage treatment plants, interceptor sewers and, in some instances, lateral sewers. I can remember that when I first joined the EPA in 1969 we spend as a nation $200 million. Now we are spending forty-five times as much. It is the largest public works program in the country, larger even than the highway program. Twenty million dollars has been appropriated to develop alternative ways of treating sewage. I think that is a woefully meager sum when we consider the amount of money that we are now spending to construct sewage plants to treat sewage precisely the way we have been treating it for many decades. It's hard to say whose fault this is. It could be a failure on the part of our society to take into account the amount of research necessary to answer some of the obvious questions that are going to come up as we start spending amounts like the $9 billion. It seems that these same kinds of questions need to be asked across the board.

In my estimation, one of the problems of transferring technology is the environmental movement itself. We reflect on the origins of what has come to be called environmentalism in this country and find in part it was a reaction against a technology which people viewed as having run amuck. It was a vague but nevertheless real sort of longing among many Americans for a past that was simpler and less cluttered. I've always felt that the defeat of the Supersonic Transport (SST) was an example of a reaction against technology by people in this country. Personally, I thought the SST deserved for other reasons to be defeated, namely it was economically unsound. Much of the reason for the defeat of the requested appropriation for the construction of two prototypes SSTs was related to this public reaction against technology. There was some irony in the defeat of the SST in that there was a program in the SST prototype development effort aimed at assessing the impact on the ozone layer of some of the pollutants that would have been emitted by the SST. When the SST itself was defeated, so also was the $28 million that had been set aside to get the answers to some of these questions.

In my opinion what we need as a society is not less technology but more, although clearly we need to control it and not have it control us. And I think there is some very legitimate criticism that can be made of technological developments in this country. We have given little if any thought to the immediate effect of that technology, and much less to the secondary and tertiary effects which we now know are bound to follow any technological developments. If we have not arrived at a sufficient degree of sophistication as shown in the SST situation, we should not stop the development of technology when we find potential problems. Rather we should develop the ability to find the answers to those problems in the process of the development of that technology. If a prospect turns out to be negative with respect to its ultimate use, then we should have the strength of governmental will to stop the development. In fact, we must distinguish between technology and man's past inability to measure the secondary and tertiary effects of technology. For instance, I think the automobile is not innately bad. Rather, it was our failure to fit it into a comprehensive transportation system aimed at improving the quality of our lives. Just because the automobile contributes to smog in Los Angeles, to noise in New York, to congestion in Washington, I don't believe we should abolish it. I think instead we should control it. We should control its use much more carefully than we have in the past in an effort to put it into a rational transportation plan that seeks to enhance the quality of everyone's life. It we need more technology we must develop it. Why isn't it more wisely and widely demanded? If a better sewage system exists why don't we use it?

In the first place I believe that when we talk of technology transfer we have got to ask ourselves what we mean by technology. We need to open our minds, stretch them if you will, to the definition of technology. Ideally it would mean three things. The first instance is the traditional definition of hardware in the case of the treatment of wastes. The hardware is utilized at the end of the pipe in order to care for the material that the plant or the industries themselves are discharging. That could be true whether we are handling or attempting to deal with pollutants, or problems of water supply, or sewage and water treatment of many kinds. There are, secondly, process changes. Changes that can be made within a given process that would lead to the need for less hardware at the end of the pipes. If we can direct the attention of people to the processes by which goods are made, then many changes can be effected that will minimize the need to treat wastes and manufacture fertilizer. Lastly, and most importantly, technology should be the transfer of knowledge concerning devlopment, whether industrial, resource, or human development, and how it can take place consistent with the protection of the environment in which it occured.

The United Nation's environmental program has established a separate division for fostering around the world what Maurice Strong calls ecodevelopment. Ecodevelopment is the process of insuring that the resultant activity which takes place is consistent with the protection of the ecosystem in which it is undertaken. I have recently completed a study for the German Marshall Fund on the potential for the establishment of an international trust fund on the environment. Two of the functions that were identified as being the most important for such an institution to perform deal with the transfer of technology, particularly the transfer of knowledge about how development can take place consistent with environmental integrity. Under this definition of technology we can see what it is that needs to be transferred. Knowledge—whether it's knowledge about hardware or processes—must be harmonized with the environment. I suppose we then have to ask ourselves the question: if we have this knowledge, why have we not used it more wisely and more widely? Let me warn you at the outset that we must not fall for any panacea as far as technology is concerned. There is no magic bullet in any of these areas for solving all of our problems.

I can remember a process that was, at least in part, developed at Pennsylvania State University involving on-land treatment of various kinds of effluents including sewage. Many of the converts to land treatment felt that this was the answer to all our sewage treatment problems. There is a demonstration program in Muskegan, Michigan, to treat the sewage of that city by spraying it onto sandy soil and then growing crops on that soil. Yet, in Melbourne, Australia, spray irrigation with sewage has not had the same success as in Michigan. Having traveled there in December, 1974, I discovered that land disposal in Melbourne is not in reality as clean as it seems to be in the eyes of the convert. And there were indeed those converts who believed in the technique of on-land disposal of sewage with such a religious zeal that in one 1972 House version of the Clean Water Bill, a provision was included that would have biased the entire country toward treating sewage in this manner by providing extra funds to communities selecting that technique. My own belief is that this method should be considered one of the alternatives when any sewage treatment process is being selected for a given municipality or area. It is unwise to attempt to direct a national law in favor or any of a number of technological approaches without more information than we have for this specific example. To adopt one process which appears to solve all the problems is a very unwise policy.

On the shelves of many federal agencies there is a great deal of knowledge that exists and is never used. Why not? Obviously I do not have all the answers but I can suggest a few. In the first place, there is human nature. This is translated into the tendency on the part of human beings to do things today the way they did things yesterday on the assumption that the old way of doing things is better. What has grown to be almost a cliché now is the charge that many sanitary engineers want to use only old plans to build the sewage treatments plants needed next year. There seems to be some validity in the charge despite the many engineering firms in this country which are probably willing to adopt new and innovative techniques. If we look at where we are in the sewage treatment program that I just described, we see it has grown from a relatively small federal grant program into the largest public works program in the country. We are not developing new and innovative techniques in the use of this tremendous public expenditure. It is a sad commentary on the amount of thought that has been given to this whole area. If the old ways are not good enough, clearly we must explore alternatives.

The second reason why we are failing to use existing knowledge is the gap between the scientist and the policy maker. There were discussions of some of the problems that were perceived as existing in EPA between the knowledge the administrator possessed as a decision maker and the knowledge he did not possess. Having been a decision maker in EPA, I want to say that we never made a decision that wasn't founded on the soundest scientific knowledge we could get. The trouble was that we did not receive very much scientific knowledge. Much of the money that is spent by the government in research has resulted in some significant advances but has never resulted in their use throughout society. I think we need to develop better mechanisms in our governmental structure and in our society as a whole. We could take some of this money and use it to make sure that the results of research are applied. We cannot leave it up to the scientists themselves to insure it is properly used. It takes a different kind of person, a different kind of mechanism, and a different sort of research operation to focus attention on the real problems of transferring new knowledge as it comes into existence. There are numerous examples of effective techniques being employed locally but for some reason not used anywhere else. I will never forget visiting a chicken farm in France where the manure was used to generate all of the energy needed to operate the farm. It was a remarkable closed system that had been created. To the best of my knowledge there are some two hundred farms using this system in France but nowhere else in the world.

The transfer of technology presents this country with a marvelous opportunity, one that I believe exists today: the opportunity to transfer technology to the development in the Western Mountain states of the U.S. I have spent a good deal of time in those states in the last several years. They contain 3% of the population and roughly 50% of all the resources in the country. What is predicted for those mountain states is an explosion of development. People there are extremely concerned about the focus this development will take. The Interior Department recently predicted that by 1985 there will be in those seven states fifty new cities with a population of 50,000 or more. This is bound to concern people, whether they moved there recently to get away from precisely what has been following them or whether they have lived there for many generations and cherish a way of life which they see seriously threatened. Development will undoubtedly take its place in those cities regardless of its form. If we are capable as a people of controlling that development so as not to destroy the amenities and values of the people that live in that region, we will have done something very special for the rest of the world.

Unlike much of the underdeveloped world, there exists in the western and mountain states of this country a political and technological infrastructure that can be used for intelligent, well-planned development. We can avoid making the same mistakes we made in the past. If we can demonstrate proper development in this country, it should be possible for us to transfer it to the less developed countries of the world.

I spent two weeks in Stockholm in June of 1972 at the first World Environmental Conference. I remember being struck by the negative way which the developing countries of the world view the whole environmental movement. I understood intellectually before going to Stockholm that they felt this way; after about a week there I knew viscerally how strongly the peoples of the developing world believe that the environment is something to be addressed only after they reach the level of material affluence that has been achieved in the developed countries. It is therefore crucial for the future of the world that we set a good example of technology transfer in the western U.S. To a certain extent we are lucky today because there is more receptivity among the people of America to change and innovate than has traditionally been the case. It is my own belief, particularly in traveling in the country during the last election, that there is a greater understanding of the complexities of modern America among the people of this country than there is among the politicians. This is not perhaps all that unusual, but on the other hand one must believe it is true or the politicians would not be saying the things they are saying.

It is true that the technology is here to remove almost all water pollutants. It is also true that the technology exists to find substitutes for the more persistant and damaging pollutants. The only question remaining is what is society willing to pay for clean water?

As a society, we must move away from judging pollution measures strictly in terms of whether they are tough or weak. We must ask ourselves the truly relevant question: Are they wise? Take the example of the Clean Air Act of 1970, an act which I believe has resulted in a genuine start toward cleaning up our air. I can remember that in January of 1973 I made a trip to Los Angeles. I had been ordered by the District Court in Los Angeles to enforce the transportation controls laid out in Section 110 of the act. This section simply said that if by mid-1975 a reduction in the levels of pollutants necessary to protect the public health, is not achieved, then transportation controls must be imposed in order to achieve those standards. The Court in Los Angeles ordered me as Administrator of EPA to enforce the transportation control section of the Clean Air Act which, according to our analysis for Los Angeles, would have meant removing 93% of the automobiles from the road. Since the Court had ordered me to do so or be held in contempt it boiled down to an issue of their mobility versus my freedom. I opted for my freedom and announced in Los Angeles that, as of June 1975, 93% of the automobiles must be off the road. I might add that I announced this in the airport and immediately flew out. No matter how good the intention of the application of the Clean Air Act in Los Angeles, it was obvious that a distortion had occurred. While there is no question that if only 7% of the present automobiles were on the road the air would be cleaner, there would be other social problems that would have to be addressed. Realism dictates that an adjustment will have to be made to the rate at which we achieve our goals of clean air and water but we should not lower or abandon those goals.

I also think that we are able to handle in the short term the problems involved in the analysis of how to control pollution. The confidence that is gained from that exercise can be used to address a second generation of problems. An example of this is the question of how to handle chemicals that are carcinogenic. We need to develop a mechanism that gives us a very careful analysis of the chemicals that are now in widespread use. Some are carcinogenic at high levels, some at lower levels, but in testing animals for cancer we can find no threshold below which there seems to be no effect. There are varying estimates as to how many of these chemicals are now in widespread use in the environment—in industrial process or chemical processes—but by anybody's guess there are many. We need, I believe, as a society a Toxic Substances Law that will give us the protocol for the pretesting of new or potentially dangerous chemicals in order to avoid the kind of societal wrench that occurs when we discover their awful qualities after they have been in wide use. We need to develop a beneficial analysis that provides an administrator of an agency like the Environmental Protection Agency with adequate flexibility so that when dealing with such a chemical he is not faced with the choice of either banning it or permitting its indiscriminate use. In many cases neither choice is a wise one.

Part II

Limnological Overview

6

Overview of Aquatic Ecosystems

RUTH PATRICK

Francis Boyer Chair of Limnology
Academy of Natural Sciences of Philadelphia

Natural aquatic communities have certain basic similarities, although they may differ greatly in degree of complexity as to species carrying out a given function and to some degree in the efficiency of energy transfer.

Typically there are four or five stages of energy transfer—that is, there are detritivores, which are organisms that consume nonliving organic matter; primary producers, which are organisms that fix carbon by the use of the sun's energy in the presence of chlorophyll; herbivores, which are organisms that eat plants; carnivores, which are organisms that feed on other organisms; and omnivores, which are organisms that feed upon a great variety of foods. The more we learn about their diets the more we realize that most species have preferences for a given food type, but when pushed may eat other types of foods.

Typically, about 10% of the energy fixed in any one level of the food web is transferred to the next level. Some scientists estimate that the amount of nutrients transferred from the primary producers (or plants) to herbivores is a little greater than that transferred through subsequent levels of the food web. There are usually many species performing each of these functions, representing several different major groups of organisms and many different families, genera, and species.

The formation of these communities according to degree of complexity is dependent upon the species pool available to invade the area; the diversity of the substrate, such as rocks, sand, rubble, and mud; the size of the area to be invaded (i.e., is the stream very small or the pond a small pool); and upon the physical and chemical conditions of the water. This includes such things as temperature, rate of flow or current pattern, pH, and many different kinds of chemicals that influence the types of species and the numbers of organisms that may grow in a given area. Not only is the concentration of these various physical and chemical aspects of water important, but also the variability and predictability of such changes.

Under natural conditions the functioning of these communities is a continuum over time, although the important species performing a given function may change greatly with the season or other environmental shifts. These communities typically have considerable flexibility because of the variation of the requirements for the species composing them. Some species have rather specific habitat requirements and short turnover times. This allows a rapid replacement of species to take place as the ecological conditions of the water change. Consequently, similar species with different optimums perform the function formerly performed by another species. Other species in a community have a wide range of tolerance and are able to adjust to changing seasonal conditions. These are specifically referred to as eurytrophic species.

The numbers and kinds of species composing natural communities may vary depending upon whether one is considering rivers, lakes, ponds, pools, springs, etc. However, in similar ecological conditions we have found that the numbers of species that are important in the performance of a function remain similar over time. Although the functions performed by an ecosystem are discrete, it is quite difficult to associate a given species with a given function. The reason for this is that a given species may have different feeding strategies and preferences at different stages of its life cycle. Also, its feeding strategies may be affected by the availability of the prey which in turn are affected by varying density-independent factors or predator pressure.

The degree of development of the food web's various aspects determines the assimilative capacity of the stream. For example, whether a given amount of nutrients results in fish and organisms that have a high preference value to man, or whether they develop into nuisance growths, is largely dependent upon the transfer of nutrients and the predator pressure on the organisms in which these nutrients reside. In other words, if the standing crop of species is approximately equal to the numbers of individuals necessary to satisfy predator pressure, the vagaries in the environment, and the reproduction of the species, large standing crops will not develop. However, if for some reason more individuals are produced than these various factors demand, large standing crops result which then have to be recycled before they can be consumed.

7

Nutrient Cycles in Natural Systems: Microbial Involvement

THOMAS L. BOTT

Stroud Water Research Center
Academy of Natural Sciences of Philadelphia

Nutrient cycles can be thought of as the conversion of nutrients from an inorganic form to an organic form and their return again to the inorganic state. The cycling of materials also involves changes from the particulate state to the dissolved, returning again to the particulate. In addition, nutrient cycling may involve exchange between subsystems of the biosphere. Nutrients move from terrestrial environments to aquatic habitats and from the atmosphere to aquatic environments, and there are subsequent returns. There may also be exchanges within a subsystem, such as between the sediment and overlying water in a lake.

Although bacteria and fungi comprise only a small part of the total biomass in any environment, they are the essential organisms in nutrient cycling. Without microbial decomposition activity, nutrients would be permanently locked in plant and animal matter. The bacteria and fungi possess great metabolic versatility and utilize substrates, as nutrient and energy sources, that no other organisms can. Also, the bacteria in particular possess a high surface area: volume ratio and have extremely rapid metabolic rates. A single genus or closely related genera may be solely responsible for certain nutrient transformations; for example, the nitrification sequence. Note, however, that although they perform the biochemical conversions, the bacteria and fungi do not act alone in the mineralization of materials. There are a great many interactions with other microbes (such as the protozoa) and other forms (insect larvae or zooplankton, for example) whereby rates of activity are accelerated or depressed. Higher forms, however, are dependent on the activity of the bacteria and fungi for continued existence.

It should be noted that the pool of a given class of materials is continually changing in size in any environment. The presence of a given intermediate is important, but flux into and from a pool is even more important if we are to understand the functioning of ecosystems. Variation in size of given dissolved pools may occur in a fraction of a day.[1] Thus, it is important to determine the turnover rate, turnover time, and half-life of a given material.

Carbon Cycle

A conceptualization of the carbon cycle is shown in Fig. 7-1. The various subsystems (atmosphere, terrestrial habitat, aquatic habitat, and sediment) are differentiated and a separation between aerobic and anaerobic environments is shown. In nature this separation is not fixed. In many habitats it may be found at the sediment-water interface where there may be an oxidized surface microzone. In other situations (stratified or eutrophic lakes), the demarcation may occur in the hypolimnion and parts of the water column may be anaerobic.

Carbon inputs to the aquatic environment are many and varied. Carbon dioxide may diffuse into and from the aquatic habitat depending upon the partial pressure of the gas, pH, water temperature, and degree of turbulence. It has been estimated that one lake may have attained as much as 30 percent of its total carbon input in this manner.[2] In each body of water an equilibrium is established between free carbon dioxide, bicarbonate, and carbonate ions. Inorganic carbon may also enter as the carbonate ion in leachate from the watershed. Photosynthetic activity affects this equilibrium by removing free carbon dioxide (and when the supply is exhausted, that stored as bicarbonate), thereby raising the pH and bringing about the precipitation of carbonate. Some algae also use bicarbonate ion directly.

Inputs of dissolved organic carbon from the watershed may also be important. In urban areas particularly, runoff water or effluents from sewage treatment facilities or industrial operations may contain significant quantities of dissolved organic carbon. In rural watersheds, inputs from feedlots, composting areas, or fertilized fields may carry significant quantities of dissolved organic carbon into the aquatic environment. Other inputs to this pool of material may be derived within the aquatic environment. Under certain conditions, algae have been shown to excrete from 3–30% of their photosynthate as dissolved organic carbon.[3] Brown, red, and green algae excrete from 23–40% of their gross primary productivity per year as dissolved organic carbon and may lose another 30% to this pool on their decomposition.[4] Similar values have also been reported for the littoral algae *Fucus* and *Laminaria*.[5] Glycolic acid, glycerol, proline, various sugars and phenolic substances have been identified as excretory products.

Some dissolved organic carbon compounds may chelate metals and thereby affect their availability. Nonpolar organic molecules, some of which could be toxic, such as pesticides or dialkyl phthalates, may also chelate.[6] The dissolved organic carbon pool serves as a nutrient and energy source primarily for heterotrophic bacteria and fungi, although growth factors and nutrients for heterotrophic growth are also supplied for algae. It is generally accepted that the bulk of the dissolved organic carbon pool is comprised of refractory polymeric compounds (humic substances). This is because the readily metabolized substrates are used rapidly by bacteria, thereby keeping them in low concentrations.

Several workers have studied the utilization of specific dissolved organic compounds, primarily sugars or amino acids, by naturally occurring heterotrophic microbial populations.[7] From these investigations it is clear that the bacteria keep the concentrations of many materials at extremely low levels by possessing low enzyme saturation constants. Low temperatures increase the turnover time for the substrates studied, but other environmental factors are certainly also involved. Some of the carbon from the dissolved organic matter is assimilated by the bacteria and enters the food web in this manner. In deep reservoirs receiving large inputs of organic carbon, it was reported that the production of heterotrophic bacteria could be greater than that of phytoplankton.[8] In Rybinsk reservoir, bacterial production was 117% that of phyto-

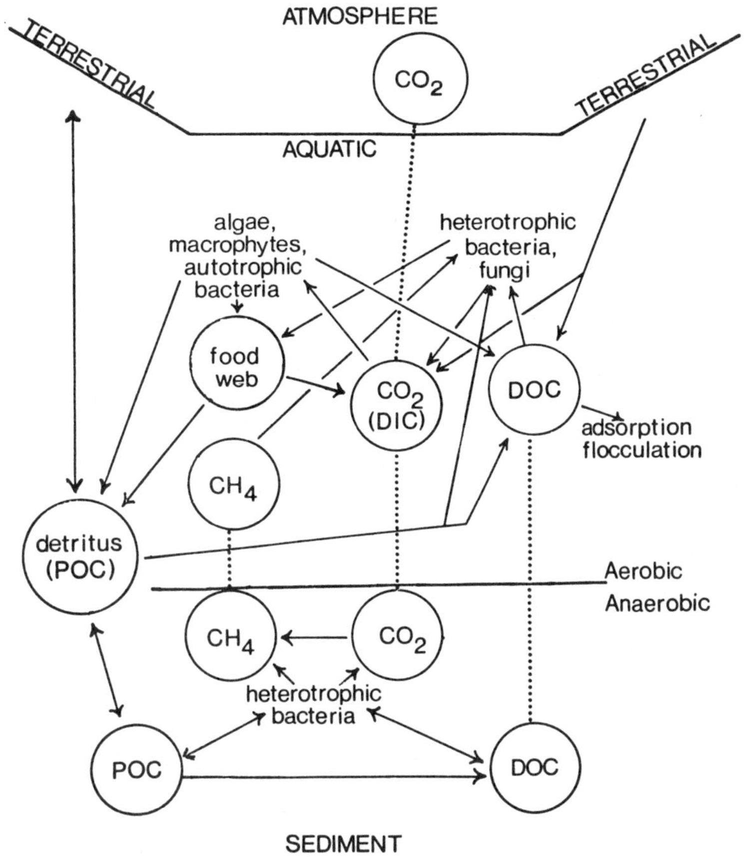

FIG. 7-1. The carbon cycle in an aquatic environment. *D*, dissolved; *P*, particulate; *O*, organic; *I*, inorganic; *C*, carbon; CO_2, carbon dioxide; CH_4, methane.

plankton over one year and 267% during a second year. Depending on the substrate, from 8 to 60% may be respired.[9] Carbon dioxide from microbial activity and that respired by other organisms in the food web is an important source of dissolved inorganic carbon.

Losses from the pool of dissolved organic carbon may also occur through incorporation into calcium carbonate crystals in hard waters and in marine habitats.[10] A 1971 report showed that high concentrations of ammonia-N, protein, and hexose sugars could be tightly bound to river bottom sediments, demonstrating that adsorption to clays and other substances can contribute to carbon losses.[11] In addition, it was demonstrated that from 3 to 30% of the dissolved organic matter in leaf leachate may be removed abiotically by flocculation as metallo-organic complexes.[12]

The pool of dissolved inorganic carbon is converted to biomass through the photosynthetic activity of algae, higher plants, and also the chemoautotrophic and photosynthetic bacteria. C. R. Goldman has reported that 1 mg C/m²/hr may be fixed by photoplankton in an Antarctic lake and that 300 mg C/m²/hr may be fixed during the growing season in a eutrophic shallow lake with periodic blooms of blue-green algae.[13] The annual phytoplankton production rates for oligotrophic lakes has been approximated as 7–25 g C/m²/yr; for naturally eutrophic lakes as 75–250 g C/m²/yr;

and culturally enriched lakes as 350–700 g $C/m^2/yr$.[14] The primary productivity of the benthic algae in the headwater reaches of the White Clay Creek, Chester Co., Pa., has been intensively studied by the staff at the Stroud Center over the past several years and values ranged from 155 to 1550 mg $C/m^2/day$[15] (Bott, unpublished data). These values are similar to those reported for the Upper Raritan River[16] and for the Neuse River system in North Carolina[17] when appropriate conversions are made. Factors affecting primary productivity are species composition, flow regimes, nutrient supply, temperature, and light.

Photosynthetic and chemoautotrophic bacteria also use carbon dioxide as a sole carbon source, but their use is small on a global basis compared to utilization by algae and macrophytes. However, in certain environments their activity may be important. Two groups of photosynthetic sulfur bacteria—the purple (*Thiopedia* and *Chromatium*) and green (*Chlorobium* and *Pelodictyon*)—derive energy from light and use hydrogen sulfide, thiosulfate, hydrogen, and organic substances (some strains) as electron donors to reduce carbon dioxide. These organisms develop under anaerobic conditions where hydrogen sulfide (H_2S) is plentiful and light penetration is sufficient for photosynthesis. Several workers have demonstrated the importance of these photosynthetic bacteria in two environments, Lake Belovod[18] and Fayetteville Green Lake, New York.[19] The activity of these bacteria was important to the productivity of the systems studied and zooplankton were shown to effectively utilize this nutrient resource. The nonsulfur purple bacteria *(Rhodospirillum* or *Rhodopseudomonas)* also use light energy, but do not use reduced sulfur compounds as electron donors. The chemoautotrophic bacteria obtain the energy to reduce carbon dioxide from the oxidation of a variety of substances, hydrogen sulfide (*Thiobacillus* sp.), nitrite (*Nitrobacter* sp.), or hydrogen (*Hydrogenomonas* sp.). In addition, estimates show that heterotrophic bacteria derive approximately 6% of their carbon requirements from inorganic carbon.[20]

Particulate organic carbon inputs from the watershed may be considerable. Along small streams that pass through wooded areas, leaf litter inputs may be particularly important to the energetics of the system. Leaf litter inputs to the headwaters of the White Clay Creek were estimated to be 11 metric tons, whereas algal biomass produced in the system was approximated as 8 metric tons. Note that the exchange with the terrestrial environment may be in both directions. Studies at the Stroud Water Research Center have shown that during storm events a considerable amount of particulate matter is moved in the watershed, but over comparatively short distances.[21]

The particulate organic carbon pool also includes material generated within the system such as feces or other dead plant and animal matter. The chemical composition of the particulate organic carbon pool may include cellulose, chitin, lignin, and other polyaromatic material as well as hemicellulose, pectin, lipids, and proteins. The fungi, particularly the aquatic hyphomycetes and phycomycetes, are well adapted to degrading particulate organic materials because their mode of growth permits penetration into particulate materials. The aquatic hyphomycetes are commonly found in small well-aerated streams. It should be pointed out that conversion from particulate organic carbon to dissolved organic carbon or incorporation of nutrient from detritus into bacterial or fungal biomass is not necessarily a one-step transformation. More often the products from the metabolism of one organism will be acted on by another and the breakdown will be accomplished by a series of steps.

If particulate organic matter becomes sedimented into an anaerobic environment, fermentation processes and anaerobic respiration become operative. Organic acids, ketones, alcohols, and carbon dioxide or methane are some end products that will occur in contrast to the carbon dioxide and water that results from complete mineralization in aerobic environments and the rate of anaerobic conversion of a given substrate is slower. As the pH becomes lower from the accumulation of acid end products, decomposition slows down and such is the origin of petroleum, coal and peat reserves. Methane may be produced either by the reduction of carbon dioxide by the organism *Methanobacterium omelianskii* or by heterotrophic methane-producing bacteria. The high methane content in Lake Kiva, a deep African rift lake, has recently been ascribed primarily to bacterial methane production by *M. omelianskii* rather than to anaerobic decomposition of organic matter or to volcanic action. If the methane diffuses into the overlying water, it may be oxidized to carbon dioxide by the methane-oxidizing organism *Methylomonas*.[22]

At the Stroud Center we have been particularly interested in leaf litter decomposition. To study this, leaves were labeled with $^{14}CO_2$ photosynthetically, after which they were either cut or allowed to fall naturally at the end of the growing season. Experiments were conducted in plexiglass microcosms containing approximately 2.5 liters water and into which were placed known weights of leaves with known radioactive content. Loss of particulate organic carbon to the pool of dissolved organic carbon through leaching and the release of carbon dioxide from dissolved and particulate organic carbon material was followed. Four species of leaves were used: beech, maple, oak, and tulip poplar. The leaching pattern from each species differed slightly and the time for the process to be nearly complete ranged from less than 6 and up to 50 hours which was probably a reflection of the chemical composition of each. Microorganisms present in the channels rapidly used the radioactive carbon in the leachate. Microbial utilization of the leachate depended upon temperature, population size, nutrient levels, etc., but in general, within 3 to 4 days the readily utilizable dissolved organic carbon compounds were incorporated into bacterial biomass. Labeled carbon dioxide was evolved as the microorganisms utilized the dissolved organic materials in the leachate. After this initial activity, a highly recalcitrant portion of leachate persisted through the remainder of each experiment.[23] A half-life of approximately 2 days was reported for the total dissolved organic carbon in leaf leachate from hickory and maple leaves in an experimental hard water system when the water temperature was 10°C.[24] Similar kinetics were observed for dissolved organic nitrogen utilization. A recalcitrant portion of each pool was observed.

Nitrogen cycle

A conceptualization of the nitrogen cycle is shown in Fig. 7-2. Nitrogen is the most abundant gas in the atmosphere, comprising 76% by weight. A certain amount of nitrogen will be fixed (combined with other elements) in the atmosphere through photochemical reactions, but the bulk of nitrogen fixation is mediated by symbiotic and free-living microorganisms. In the aquatic environment certain free-living blue-green algae (e.g., *Aphanizomenon, Anabaena,* or *Nostoc*), photosynthetic bacteria (*Chlorobium* sp.), or the heterotrophic aerobic bacterium, *Azotobacter,* are important nitrogen-fixing organisms. Nitrogen fixation may also occur in anoxic habitats by the activity of the anaerobic bacterium *Clostridium pasteurianum.* In some studies nitrogen fixation appears to be inhibited by the presence of combined nitrogen compounds such as urea, ammonia, or nitrate, but in other studies this has not been the case.

The nitrogen-fixing enzyme complex will also reduce acetylene to ethylene and this has been used to measure rates of nitrogen fixation in a number of habitats. In a study of nitrogen fixation by sediment and water populations in the western basins of Lake Erie, no fixation was noted in the water in June and July.[25] Peaks of activity were observed in August and October. In August, when a visible bloom of algae was present and the water temperature was 25–26°C, 25–97 ng acetylene/mg/protein/30 min were reduced by natural phytoplankton populations while in November (water temperature 1°C) 4 ng/mg protein/30 min were reduced. Activity was observed even in the presence of bound ammonia and nitrate. A slow, but constant, activity was associated with the sediment; 30 ng/mg protein/day were reduced at all times measures were made.

Nitrogen fixation activity has also been found in 8 of 37 samples taken from Lakes Superior, Huron, and Michigan.[26] Activity ranged from 0.003–0.088 nm acetylene reduced/l/hr with specific activities ranging from 3.6–25 nm/mg N/hr. In Green Bay of Lake Michigan, nitrogen fixation increased toward the southern end of the bay. However, at the mouth of the Fox River which enters at the extreme southern end, activity was depressed even though algal biomass was high. This was attributed to the inputs of fixed nitrogen in the effluents from the city of Green Bay. Further out in the bay, nutrients supported high algal biomass, but fixed nitrogen compounds were low and it was here that nitrogen fixation activity peaked. Values ranged from 0.29–12.6 nm/l/hr and specific activities ranged from 7.6–360 nm/mg N/hr. In Lake Erie, nitrogen fixation ranged from 0.033–35 nm/l/hr with specific activities from 5–130 nm/mg N/hr. Algal biomass was low at the mouth of the Detroit River, presumably due to the presence of toxic substances in the water, but biomass increased with distance from the mouth of the river. Bound nitrogen again may have limited nitrogen fixation activity near the mouth of the Detroit River; nitrogen fixation activity peaked some distance into the lake. Activity was, nevertheless, greatest in the western basin of the lake. The authors estimated that nitrogen fixation was approximately 2% of the nitrogen input to the lake originating in Detroit River water.

Losses of nitrogen from anaerobic habitats occur through the process of denitrification (the ultimate conversion of nitrate and nitrite to nitric oxide (NO), nitrous oxide (N_2O), and molecular nitrogen). *Pseudomonas denitrificans, Thiobacillus denitrificans,* and *Spirillum* sp. are some organisms capable of denitrification. This process is generally inhibited by free oxygen, and an organic energy source is required. There is some recent data demonstrating nitrite reduction to N_2 (denitritification) in the presence of oxygen in nitrogenous wastewater, and the process may be similarly operative in nature.[27] Using ^{15}N as a tracer, it has been estimated that 11% of the total annual nitrogen input to Lake Mendota was lost by denitrification.[28]

Ammonia may diffuse into the aquatic environment from the atmosphere. For instance, it has has been shown that in rural areas ammonia volatilized from cattle feedlots was absorbed into water as far as 2 km downwind from the feedlot.[29] Absorption rates twenty times those near control sites (no cattle feedlots) were reported. More often, however, ammonia inputs (as ammonium ion) will be associated with sewage effluents. Ammonia will also be generated in the system by mineralization of organic nitrogen compounds and by nitrate reduction which may be assimilatory (when the reduced nitrogen is used biosynthetically and incorporated into protoplasm) or dissimilatory (when the nitrate serves as the terminal electron acceptor in the process of anaerobic respiration). Nitrate inputs in ground water are important because nitrate is not retained by the soil to any great extent. However, nitrate-N inputs from precipitation were found to be greater than the contribution from fertilized fields in a study of surface runoff.[30] Similarly, there has been a noted increase in the nitrate content of rainwater in the northeastern United States.[31] Nitrate (following reduction to ammonia) and ammonia are both nitrogen sources for bacteria and fungi, but utilization by algae and macrophytes is more important because most bacteria and fungi utilize various organic nitrogen sources as well.

Particulate organic nitrogen received as detritus from the watershed or generated within the system is present as protein, nucleic acids, or other amino compounds. Dissolved organic nitrogen, present as amino acids and other readily utilized substrates, was rapidly used and assimilated by bacteria.[32] However, there are recalcitrant dissolved organic nitrogen com-

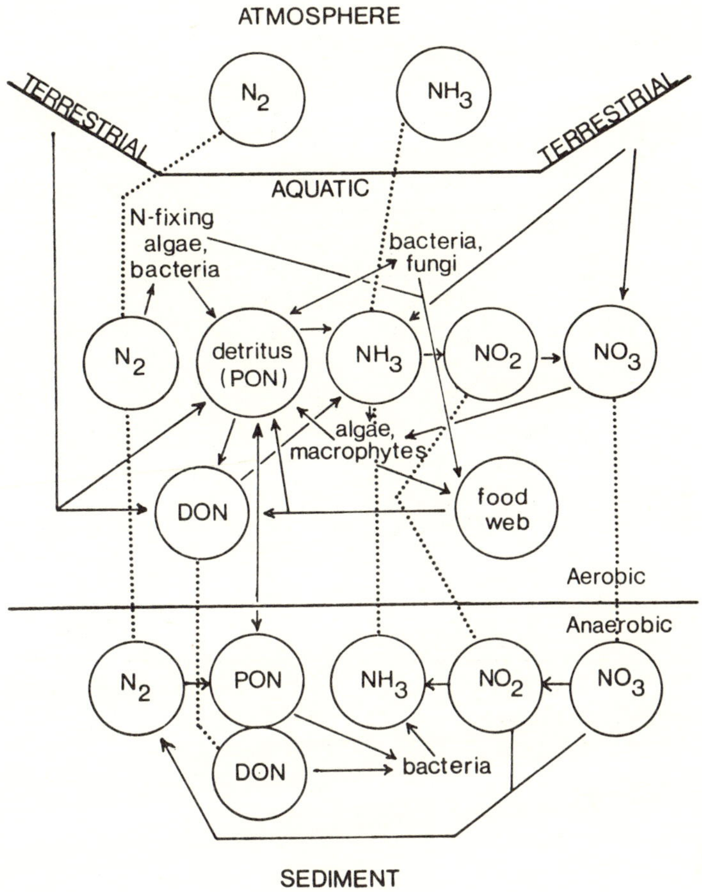

FIG. 7-2. The nitrogen cycle in an aquatic environment. *NH_3*, ammonia; *N_2*, elemental nitrogen; *NO_3*, nitrate; *NO_2*, nitrite; *other symbols*, as in *Fig. 7-1*.

pounds as well.[33] The bacteria and fungi are the important agents in the decomposition of particulate organic nitrogen leading ultimately to ammonia formation.

In the nitrification sequence the strict autotrophic bacteria, *Nitrosomonas* and *Nitrobacter,* derive energy from the oxidation of ammonia to nitrite, and nitrite to nitrate respectively, in aquatic habitats. Other genera of nitrifying bacteria have been identified. These organisms are operative when the oxygen tension is high and the pH has approximately neutral or slightly basic values. There is also evidence that heterotrophic organisms of certain genera are capable of nitrification.

Sulfur Cycle

Sulfur as the sulfate anion reaches aquatic environments either in precipitation or in runoff and waste water from the watershed (Fig. 7-3). Increased sulfur dioxide in the atmosphere has led to an increased sulfate content in rainwater where sulfate may form 60% of the anionic content.[34] The acidity of rainwater in the northeastern United States has increased, with the pH of individual storms ranging from 2.1–5.0 (average 4.0). The total sulfur content of the rain has actually decreased, but the sulfur is now present in a different form. Formerly, the sulfur was probably as unionized particulate

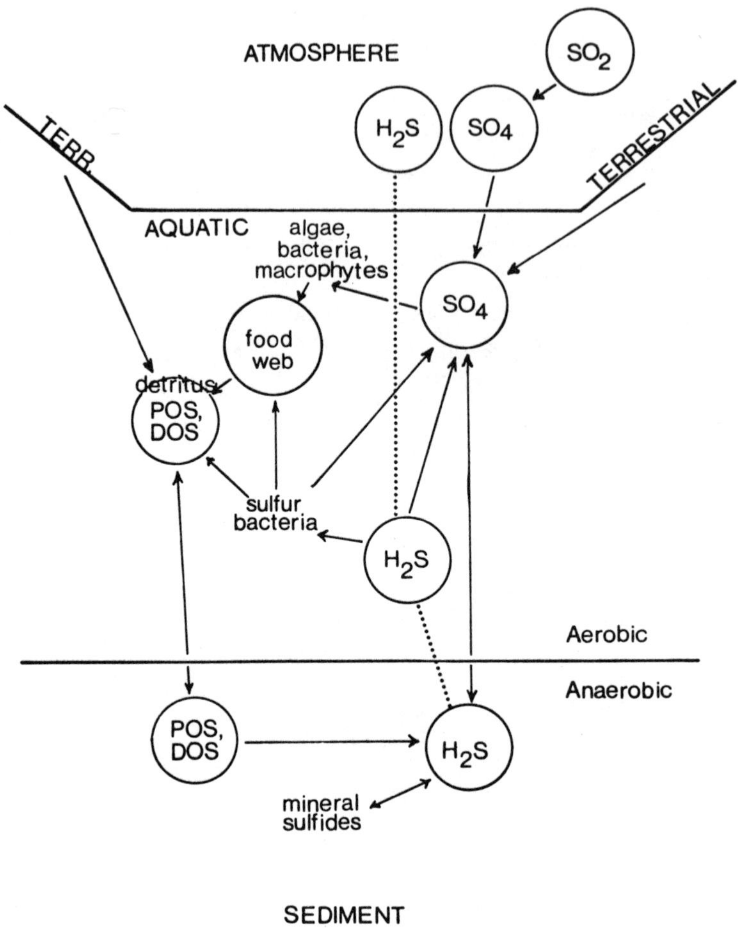

FIG. 7-3. The sulfur cycle in an aquatic environment. *S*, sulfur; *H_2S*, hydrogen sulfide; *SO_2*, sulfur dioxide; *SO_4*, sulfate ion; *other symbols*, as in *Fig. 7-1*.

sulfate or neutralized ionic species. The authors suggest that acid rain is related to the increased use of natural gas and tall smokestacks with particle removers that discharge sulfur as sulfur dioxide at greater heights into the atmosphere.

Sulfate is assimilated by plants, algae, and bacteria and is introduced into the food web as organic material in this way. The organic sulfur in plant and animal remains is broken down to hydrogen sulfide gas (H_2S) in the process of putrefaction which occurs in anaerobic environments. Hydrogen sulfide is also the product of sulfate reduction, which is mediated by the strict anaerobic bacteria of the genera *Desulfovibrio* and *Desulfotomaculum* in the presence of organic compounds which serve as energy sources. Conditions suitable for sulfate reduction are found in many environments; lake and stream sediments, marshes, bogs, swamps, flooded soils, or decomposing sludge. The sulfide produced will react with metal ions, and an accumulation of metal sulfides results.

Hydrogen sulfide gas may also diffuse to aerobic environments and undergo spontaneous oxidation or it may be used in the metabolism of a variety of sulfur oxidizing bacteria. For example, in 1972 hydrogen sulfide produced by bacteria in Great Salt Lake and evolved to the atmosphere was reported second in amount only to that produced by industrial activity in the

Salt Lake City area, the most important of which was copper smelting.[35] It was calculated that ten thousand tons of bacteriogenic hydrogen sulfide were produced annually. Although this was only 10% of that generated by man's activity on an annual basis, the bacterially produced gas may be more important than that from industry during the warm months of the year.

Photosynthetic sulfur bacteria use hydrogen sulfide as a one-electron donor for the reduction of carbon dioxide, releasing sulfur which may accumulate intracellularly (as in *Chromatium*) or extracellularly (as in *Chlorobium*). As discussed above, their activity may be of local importance in strict anaerobic environments enriched with hydrogen sulfide where there is sufficient light (a shallow lake or pond, for example) and sulfur may be introduced into the food web in this way.

Other nonphotosynthetic bacteria also oxidize hydrogen sulfide. Granules of sulfur may collect in the cells of *Beggiatoa* (a heterotroph) or *Thiothrix* (considered to be an autotroph) as a result of this transformation, although other, more oxidized, products may also be produced such as thiosulfate and sulfate. The genus *Thiobacillus* is also important in sulfur oxidation. These forms are usually considered obligate autotrophs although heterotrophic growth has been reported. *T. thioparus* generally prefers a pH around neutrality, but some species are capable of growth under extremely acidic conditions generated from the accumulation of sulfuric acid (H_2SO_4). For example, *T. thiooxidans* has been shown to live in environments with pH less than 1, such as those associated with acid mine drainage. *T. ferroxidans* can also oxidize ferrous ions to ferric ions under acidic conditions. Ferric hydroxide forms and collects as a rust-colored precipitate in areas receiving acid mine drainage. At this point, the sulfur cycle again interacts with the cycling of other minerals.

Phosphorus Cycle

The phosphorus cycle (Fig. 7-4) in a sense is only a partial cycle on a global basis because there is no exchange with the atmosphere and the sequence is generally unidirectional owing to close links with erosional and sedimentation processes. Inorganic and organic phosphates may adsorb to particulates. There is, however, a certain amount of return to the terrestrial environment in the form of guano from marine birds.

Phosphorus enters the aquatic environment primarily in waste water from domestic sewage treatment facilities and in surface runoff from fertilized fields in rural watersheds. When fertilizer is applied to frozen ground, significant amounts (sometimes more than half) of single application may be lost to streams draining the watershed.[36]

In the aquatic environment, dissolved inorganic phosphate is utilized extremely rapidly by algae, bacteria, and macrophytes, and the phosphorus is usually assimilated by the organisms rather than being transformed to another state. In 1958 it was demonstrated in microcosm experiments that most of the ^{32}P labeled orthophosphate offered was assimilated into bacterial and phytoplankton biomass within 5 minutes.[37]

The salt marsh plant *Spartina alterniflora* has been shown to serve as a phosphorus pump.[38] Phosphorus derived from deep within the sediments (1m) is released into the overlying water as dissolved organic phosophorus compounds. This organic phosphorus can be acted on by heterotrophic microbes and either released as an inorganic phosphate or assimilated into tissue.

Various abiotic transformations, or reactions, are also important in the phosphorus cycle. Ferric phosphate is precipitated when the dissolved oxygen tension is high. However, when anaerobic conditions develop and hydrogen sulfide is produced, ferrous sulfide forms and the phosphorus is released. Concentrations of phosphate may accumulate in the hypolimnion of a lake and then be mixed through the water column at the time of spring or fall overturn. Phosphate may also be contributed from dissolution of rock owing to the production of organic, nitric, or sulfuric acids by bacteria.

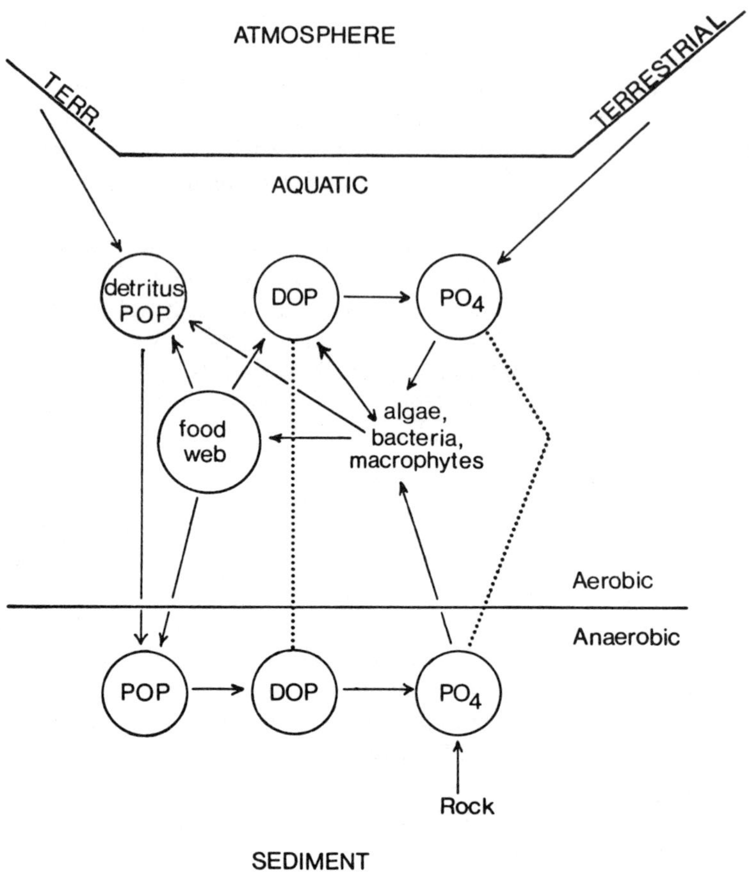

FIG. 7-4. The phosphorus cycle in an aquatic environment. *P,* phosphorus; *PO_4*, phosphate ion; *other symbols*, as in *Fig. 7-1.*

Summary

Thus, we see that in any natural system, nutrient elements may take many pathways through the aquatic environment. We have selected four cycles for discussion, but microbial involvement in biogeochemistry is considerably more extensive. It is important to note at this point that although the cycles have been discussed individually, they do not exist isolated one from the other. For example, organisms utilizing detritus are obtaining carbon, nitrogen, phosphorus, and sulfur and other elements simultaneously. Interactions may also occur because the activity of one organism provides a suitable environment for a second organism; i.e., one report shows that methane oxidation by the organism *Methylomonas* removed the last traces of dissolved oxygen from the water, thus permitting development of the sulfate-reducing strict anaerobe, *Desulfovibrio.*[39] The metabolism of *Thiobacillus denitrificans* involves three cycles, the nitrogen, the sulfur, and the carbon. This organism growing in anaerobic environments reduces nitrate to nitrogen gas, the nitrate serving as a terminal electron acceptor for the oxidation of hydrogen sulfide to sulfate. At the same time, this strict autotroph is assimilating carbon dioxide.

Man may influence the nutrient cycles in many ways. We have already mentioned how man's activity has affected the sulfur cycle, through increased production of sulfur dioxide, leading to acid rain. Man may also overwhelm the capacity of the system to function efficiently. The rate of microbial activity in nature is affected by many factors: temperature, pH, geomorphology of the habitat, numbers of organisms, nutrient and energy supplies, and the presence of stimulatory and/or antagonistic substances. Metabolizable substrates may be introduced in excess of the capacity of the system to assimilate them. It has been noted that nuisance growths of the sewage fungus (which really is an association of various bacteria, fungi, protozoans, and algae along with selected invertebrates) extends further downstream in the winter, from a point of introduction of high organic wastes, than in the summer because competition from unicellular bacteria is reduced in winter.

Man may also introduce into the environment synthetic compounds such as polychlorinated biphenyls (PCB's) or chlorinated hydrocarbon pesticides that are unable to be effectively metabolized at high rates by naturally occurring organisms. Some recalcitrant hydrocarbons have been shown to be co-metabolized, i.e., the compound is converted to another form, but not used as a nutrient or energy source, and the degree of alteration is, therefore, not great. It has been pointed out that 32 lb/acre of DDT and its residues DDD and DDE have accumulated in various Long Island marshes after twenty years of spraying for mosquito control.[40] Phthalate esters and PCB's have been shown to affect certain phytoplankton and zooplankton species at part-per-billion levels.

Man may also affect the cycles we have discussed by disturbing natural inputs. Thus, clear-cutting vegetation to the banks of streams will reduce inputs of leaf litter detritus. We have noted the importance of the salt marsh plant, *Spartina alterniflora,* in the cycling of phosphorus. In addition, the detritus produced from the decay of marsh vegetation is used in the food web, both in the marsh and in the open estuary. The detritus supports large bacterial, fungal, and protozoan standing crops and associated invertebrates, and is a protein-rich nutrient source for filter-feeding and burrowing organisms. Thus, when marshes are destroyed for the construction of highways, housing, industrial facilities, or shopping centers, or when they are managed so that the export is affected, pathways of nutrient cycling are also altered.

Notes

1. G. W. Saunders, "Carbon Flow in the Aquatic System," in *The Structure and Function of Fresh-water Microbial Communities,* J. Cairns, ed. (Blacksburg, Va.: American Micros. Soc. V.P.I., 1969) 31-34; B. A. Manny and R. G. Wetzel, "Diurnal Changes in Dissolved Organic and Inorganic Carbon and Nitrogen in a Hardwater Stream," *Fresh wat. Biol.* 3 (1973): 31-43.

2. D. W. Schindler and E. J. Fee, "Diurnal Variation of Dissolved Organic Carbon and Its Use in Estimating Primary Production and CO_2 Invasion in lake 227," *J. Fish. Res. Bd. Can.* 30 (1973): 1501-10.

3. G. Chapman and A. C. Rae, "Excretion of Photosynthate by a Benthic Diatom," *Mar. Biol.* 3 (1969): 341-51. G. E. Fogg, C. Nalewajko, and W. D. Watt, "Extracellular Products of Phytoplankton Photosynthesis," *Proc. Roy. Soc. B.* 162 (1965): 517-34.

4. K. M. Khailov and Z. P. Burlakova, "Release of Dissolved Organic Matter by Marine Seaweeds and Distribution of Their Total Organic Production to Inshore Communities," *Limnol. & Oceanogr.* 14 (1969): 521-27.

5. J. M. Sieburth and A. Jensen, Production and Transformation of Extracellular Organic Matter from Littoral Marine Algae: A Resume," in *Organic Matter in Natural Waters,* D. W. Hood, ed. (College, Alaska: University of Alaska, 1970) pp. 203-23.

6. G. Ogner and M. Schnitzer, "Humic Substances: Fulvic Acid-dialkyl Phthalate Complexes and Their Role in Pollution," *Science* 170 (1970): 317-18.

7. H. L. Allen, "Chemo-organotrophic Utilization of Dissolved Organic Compounds by Planktic Algae and Bacteria in a Pond," *Int. Rev. ges. Hydrobiol.* 54 (1969): 1-33. C. W. Hobbie and C. C. Crawford, and K. L. Webb, "Amino Acid Flux in an Estuary," *Science* 159 (1968): 1463-64. B. M. Thompson and R. D. Hamilton, "Heterotrophic utilization of Sucrose in an Artificially Enriched Lake," *J. Fish. Res. Bd. Can.* 30 (1973): 1547-52. R. T. Wright and J. E. Hobbie, "Use of Glucose and Acetate by Bacteria and Algae in Aquatic Ecosystems," *Ecology* 47(1966): 447-64.

8. S. I. Kuznetsov, "Recent Studies on the Role of Microorganisms in the Cycling of Substances in Lakes," *Limnol. & Oceanogr.* 13 (1968): 211-24.

9. J. E. Hobbie and C. C. Crawford, "Respiration Corrections for Bacterial Uptake of Dissolved Organic Compounds in Natural Waters," *Limnol. & Oceanogr.* 14 (1969): 528-32.

10. A Otzuki and R. G. Wetzel, "Interaction of Yellow Organic Acids with Calcium Carbonate in Freshwater," *Limnol. & Oceanogr.* 18 (1973): 490-93. K. E. Chave, "Carbonate-organic Interactions in Sea Water," in *Organic Matter in Natural Waters,* pp. 373-85.
11. C. W. Hendricks, "Enteric Bacterial Metabolism of Stream Sediment Eluates," *Can. J. Microbiol.* 17 (1971): 551-56.
12. D. L. Lush and H. B. N. Hynes, "The Formation of Particles in Freshwater Leachate of Dead Leaves," *Limnol. & Oceanogr.* 18 (1973): 968-77.
13. C. R. Goldman, "Aquatic Primary Production," *Am. Zool.* 8 (1968): 31-42.
14. W. Rhode, "Crystallization of Eutrophication Concepts in Northern Europe," in *Eutrophication: Causes, Consequences, and Correctives* (Washington, D.C.: National Academy of Sciences, 1969) 50-64.
15. T. L. Bott, unpublished data.
16. D. A. Flemmer, "Primary Productivity of the North Branch of the Raritan River, New Jersey," *Hydrobiol.* 35 (1970): 273-96.
17. C. M. Hoskin, "Studies of Oxygen Metabolism of Streams of North Carolina," *Publ. Inst. Mar. Sci.* 6 (1959): 186-92.
18. Y. I. Sorokin, "On the Trophic Role of Chemosynthesis and Bacterial Biosynthesis in Water Bodies," in *Primary Productivity in Aquatic Environments,* C. R. Goldman, ed. (Berkeley: Univ. Cal. Press, 1966), pp. 187-205.
19. D. A. Culver and G. J. Brunskill, "Fayetteville Green Lake. V. Studies of Primary Production and Zooplankton in a Meromictic Marl Lake," *Limnol. & Oceanog.* 14 (1969): 862-73.
20. Y. I. Sorokin, "The Heterotrophic Assimilation of Carbonic Acid by Microorganisms," *J. Gen. Biol.* 22 (1961): 265-72.
21. R. Vannote, unpublished data.
22. W. G. Dueser, E. T. Degens and G. R. Harvey, "Methane in Lake Kivu: New Data Bearing on Its Origin," *Science* 181 (1973): 51-3.
23. T. L. Bott and J. Preslan, unpublished data.
24. R. G. Wetzel and B. A. Manny, "Decomposition of Dissolved Organic Carbon and Nitrogen Compounds from Leaves in an Experimental Hard Water Stream," *Limnol. & Oceanogr.* 17 (1972): 927-31.
25. D. L. Howard, et al., "Biological Nitrogen Fixation in Lake Erie," *Science* 169 (1970): 61-62.
26. J. H. Mague and R. H. Burris, "Biological Nitrogen Fixation in the Great Lakes," *Biosci.* 23 (1973): 236-39.
27. J. P. Voets, H. Vanstaen, and W. Verstraete, "Removal of Nitrogen from Highly Nitrogenous Wastewaters," *J. Water Poll. Control Fed.* 47 (1975): 394-98.
28. P. L. Brezonik and G. F. Lee, "Denitrification as a Nitrogen Sink in Lake Mendota," *Wis. Env. Sci. Tech.* 2 (1968): 120-25.
29. G. L. Hutchinson and F. G. Viets, Jr., "Nitrogen Enrichment of Surface Water by Absorption of Ammonia Volatilized from Cattle Feedlots," *Science* 166: (1969) 514-15.
30. G. E. Schuman and R. E. Burwell, "Precipitation Nitrogen Contribution Relative to Surface Runoff Discharge," *J. Env. Qual.* 3 (1974): 366-68.
31. G. E. Likens and F. H. Bormann, "Acid Rain: A Serious Regional Environmental Problem," *Science* 184 (1974): 1176-79.
32. Hobbie, Crawford and Webb, "Amino Acid Flux."
33. Wetzel and Manny, "Decomposition of Dissolved Organic Carbon."
34. Likens and Bormann, "Acid Rain."
35. D. C. Grey and M. L. Jensen, "Bacteriogenic Sulfur in Air Pollution," *Science* 177 (1972): 1099-1100.
36. R. Vannote, personal communication.
37. F. R. Hayes and J. E. Phillips, "Lake Water and Sediments. IV. Radiophosphorus Equilibrium with Mud, Plants and Bacteria under Oxidized and Reduced Conditions," *Limnol. & Oceanogr.* 3 (1958): 459-75.
38. R. H. Reimold, "The Movement of Phosphorus Through the Salt Marsh Cord Grass, *Spartina alterniflora Loisel," Limnol. & Oceanogr.* 17 (1972): 606-11.
39. T. E. Cappenberg, "Ecological Observations on Heterotrophic, Methane Oxidizing and Sulphate Reducing Bacteria in a Pond," *Hydrobiol.* 40 (1972): 471-85.
40. G. M. Woodwell, C. F. Wurster, Jr., and P. A. Isaacson, "DDT Residues in an East Coast Estuary: A Case of Biological Concentration of a Persistent Insecticide," *Science* 156 (1967): 821-24.
41. J. L. Mosser, N. S. Fisher, and C. F. Wurster, Jr., "Polychlorinated Biphenyls and DDT Alter Species Composition in Mixed Cultures of Algae," *Science* 176 (1972): 533-35, have shown that the growth of algae in mixed culture was affected by PCB and DDT additions even though no such effect was noted on pure cultures of the same organisms.

8

The Role of Aquatic Plants In Aquatic Ecosystems[1]

RUTH PATRICK

Francis Boyer Chair of Limnology
Academy of Natural Sciences of Philadelphia

Introduction

Aquatic plants are the primary producers and function in a similar manner as terrestrial plants. They may be algae, mosses, ferns, or flowering plants, and they perform various functions. For example, they may act as shelter for fish and invertebrates; they may be substrates on which such organisms live; they are a food source; and they also enrich the aquatic ecosystem by fixing carbon thus increasing those foods necessary for energy expenditure. They also produce free oxygen, a requirement for all aerobic organisms. This fixation of carbon and the production of oxygen is done by the process of photosynthesis.

In the growth of aquatic plants various nutrients such as nitrogen, phosphorus, carbon, and many trace elements—all found in organic and inorganic compounds—are removed from the water. The research of many workers has shown that various aquatic plants require varying concentrations of these chemicals for optimum growth.[2] Furthermore, they can accumulate more of these nutrients than they need for growth. It has been observed that this surplus storage of luxury consumption occurred in *Asterionella formosa* when the concentration of phosphorus was less than 1 μg/l in the external media.[3] We found in Tinicum Marsh that the uptake of nutrients by aquatic plants was greater in a marsh where the nutrient content of the water was high due to the presence of sewage treatment plants than in a natural marsh.[4] In the eutrophic lakes (Mendota Lake in Gerloff's study) the phosphorus uptake for several aquatic plants such as *Vallisneria americana* and *Heteranthera dubia* is also much higher than in the less nutrient-rich lakes.[5]

This luxury consumption allows plants to maintain optimum or near optimum growth for a period of time after the chemical, i.e., phosphorus, becomes limiting in the external medium. It also results in plants having a more significant effect on improving water quality in organically enriched water.

The algae particularly, and some of the higher

plants such as *Spartina,* are known to give off into the water many chemicals which they take up as simple forms of nitrogen and phosphorus and secrete into the water as complex organic molecules. These complex organic molecules then become food for bacteria and larval stages of many organisms, and so the nutrients are cycled. However, there is one fundamental difference in the affects of algae and other aquatic plants on bodies of water when compared to bacteria. Bacteria by their action on easily decomposed organic matter produce simpler forms of chemicals which then are immediately utilized to promote growth of organisms, particularly algae. As a result, algal blooms quickly develop after bacterial action. In contrast, the detritus produced by algae and other aquatic plants is more slowly decomposed and usually does not produce such a severe impact on the aquatic ecosystem.

Distribution of Aquatic Plants in Streams

When we examine the regions of rivers and compare the growth of various kinds of aquatic plants, we see some interesting changes. In headwater streams where the bed of the stream is largely rocks, sand, and rubble and very little detritus or sediments accumulate, one typically finds the growth of algae extending over the surfaces of rocks, along the edge of the stream where the current is less swift, and on the surfaces of twigs and other matter that might enter the stream. In such areas the water is shallow, the flow is relatively swift, and the gradient of the stream is high. Very few rooted aquatic plants are found under such conditions, and diatoms are the dominant algae. This is the condition often existing in first-and second-order streams and sometimes in third-order streams. However, as the order of the stream increases there is usually increased accumulation of silt, particularly in the pools. Under such conditions if the nutrient level in the stream is fairly high one may find rooted aquatics such as *Potamogeton.* Along the edges of the stream one often finds patches of floating *Lemna* and other types of plants which trail into the stream such as watercress and *Myotis* and various species of the Polygonemaceae. On the rocks in second- and third-order streams one often finds mosses of various types. In soft water, *Fontinalis* sp. is often very common, and in those with a little greater hardness one often finds species of the Hypnaceae. These species of moss cover the rocks and act as fairly good habitats for some kinds of aquatic insects. However, one rarely finds many species of invertebrates feeding upon them, and their value as a food source as compared with algae is much less.

In the southern part of the United States, in similar habitats, one sometimes finds the flowering plant *Podostemon.* As is the case with the mosses, typically very few invertebrates are associated with these forms.

As one progresses down river, particularly in fourth-order and larger streams if they are shallow, rooted aquatics are often found where silt has accumulated along the edges of the stream or in pools. Also, along silted edges one often finds emergent aquatic plants whose roots are in the deposited silt. If the stream is deeper one may find floating mats of aquatics such as *Myriophyllum, Anacharis,* and *Utricularia.* In such deep streams, rarely does the photosynthetic zone reach to the bed of the river, and therefore rooted aquatics do not occur. If, however, in these larger streams the current is slowed down and silt is deposited in the photosynthetic zone, one often finds rooted aquatics on the stream bed. This occurs in oxbows and sloughs of large rivers.

The most common freshwater emergent plants in the upper estuary in the Middle Atlantic States are cattails, wild rice, water lilies, etc. Of course, a great many other species may occur, but these are the most common.

As one proceeds to the lower estuary to the brackish water marshlands, *Spartina* (often referred to as cord grass) is the most common genus in undisturbed areas. In disturbed areas,

Phragmites often becomes quite common. It has been shown that *Spartina alterniflora* typically prefers fresh water, but in that habitat it is often out-competed by other freshwater plants. The fact that it has mechanisms enabling it to tolerate salt water allows it to grow very successfully in brackish waters where freshwater species cannot live, and since there are fewer species that can live in salt water there is less competition. Thus we find *Spartina* to be the dominant genus along a great deal of the Eastern part of the United States. Other plants such as mangroves, saw-grass, etc. assume dominance in the marshlands of the south.

As stated above, aquatic plants may serve one or more functions. Shelter is mainly performed by aquatic plants other than algae. Rooted aquatics, particularly the pond lilies and plants of this type, often are used by fish and large invertebrates as a shelter as predators cannot reach them so easily. Floating aquatic plants may also perform this function. Salmon, for instance, are attracted by chemicals produced by certain species of aquatic plants.[6] Furthermore, the species of plants associated with the early life stages of the salmon imprint upon them a recognition of a given smell and that this odor cannot be drowned out by more dominant odors.[7]

These submerged or emergent aquatic plants may act as a substrate for the growth of other species. This is particularly true for smaller microscopic invertebrates; also, many algae grow attached to the stems of rooted or floating aquatic higher plants. It is well known that many species of animals will eat the tips of these submerged plants in order to get the small epiphytes of diatoms and microscopic invertebrates that are attached to them. Floating aquatic plants such as *Anacharis, Utricularia,* or *Myriophyllum* often furnish habitats for various microscopic invertebrates. For example, in the Savannah River we have found that floating beds of *Anacharis* and *Utricularia* are excellent habitats for caddisflies, damselflies, blackfly larvae, and many different kinds of chironomids.

One of the main functions of aquatic plants is their value as food. Algae have been found to vary greatly in their value as a food source. Diatoms and many other unicellular green algae such as *Chlamydomonas* and *Chlorella* are much more valuable food sources to various invertebrates than some of the filamentous green algae such as *Cladophora.* Among the diatoms we also find a difference as to the value of these species for food. For example, *Melosira varians,* which is a filamentous diatom, seems to be much less desirable as a food source than many species of *Navicula* and *Nitzschia.* It is believed that the shape of these diatoms and the ease with which they can be ingested many be one of the main reasons why this is true.

The higher plants such as many of the rooted aquatics seem to have varying values as a food source. For example, very few species seem to eat *Podostemon.* Perhaps some of the aquatic snails eat this plant, but in our experience we have found relatively little predator pressure on them.

The emergent aquatics are food sources for both terrestrial and aquatic organisms. The roots and lower parts of the stems are often eaten by various kinds of invertebrates in the aquatic world; whereas the flowers, seeds, and leaves are eaten by birds and mammals. Thus we see that these rooted aquatics, although obtaining their nutrients from the water and sediments, disburse their nutrients widely. Hence there is an outflow of nutrients from the system.

The fungi and the bacteria in swamps create a large amount of vitamin B_{12}, which is necessary for the growth of many algae and aquatic invertebrates, particularly in their young stages.[8] This points up the importance of maintaining these wetlands not only for the fact that they absorb many of the nutrients which enter them but also that they produce nutrients essential for the aquatic ecosystem.

In freshwater wetlands which one often finds in the Southeastern parts of the United States, a great quantity of nitrogen, phosphorus, and carbonates as well as other elements are absorbed by the trees and shrubs and thus taken out of circulation for a considerable period of time. The elimination of this wetland-open water system may greatly affect the water quality by preventing the absorption of these nutrients by these wetland plants.

It is well known that many organisms in the aquatic ecosystem change their nutrient requirements at different stages of their life cycle. This has been pointed out for many aquatic insects that are detrital feeders during certain instars and diatom feeders during others.[9] Likewise croakers will eat mainly zooplankton when they are less than 25 mm long; microbenthos—which is largely algae and smaller invertebrates—when they are 25–50 mm long; and shrimp, macroinvertebrates, and detritus when they are greater than 50 mm long.

Just as in the algae, certain emergent aquatic plants are preferred food. For example, those species which are most desirable for waterfowl are fyke rushes (*Eleocharis* sp.), bullrushes (*Scirpus* of various species), wild rice (*Zizania aquatica*), banana lily (*Nymphaea mexicana*), etc.; those that are undesirable species are sawgrass, cut grasses, pickerel weed, grasswert, alligator weed, bladderwort (*Utricularia*), parrot feather (*Myriophyllum*), etc. It is noticeable in the Delaware marshes that certain of the emergent vegetation is used more than others for housebuilding for muskrats. The numbers of huts in *Spartina* are far greater than those found in *Phragmites* marshes. This may be correlated with other factors, but at least superficially these two types of plants and/or the habitats in which they are found seem to have a very different desirability for the building of muskrat huts.

The fixation of carbon brought about by photosynthesis is the reason algae and other aquatic plants are often referred to as primary producers. The fact that they fix carbon makes them a very desirable food source. All of the carbon fixed by submerged plants is retained within the aquatic system whereas that which is produced by emergent plants is also transferred to the terrestrial ecosystem. In a similar way, all of the oxygen produced by photosynthesis by algae is retained within the aquatic system whereas emergent plants discharge oxygen to the terrestrial atmosphere. It has been found that algae are much more efficient oxygenators of water than are more complex plants, the reason being that with such unicellular organisms the oxygen produced is immediately transferred to the water.

The estuaries are probably the most productive areas of the world—that is, from the standpoint of carbon fixed.[10] The rooted aquatics such as *Spartina* fix more carbon, having a net production of 6,580 Kcal./m^2/yr[11] than the algae on the bare mud surfaces which have a gross production of 1,620 Kcal./m^2/yr.[12] However, the net production in algae as compared with gross production is far higher than in the *Spartina*. One of the reasons why the estuaries are so productive is the flooding at high tide which brings in a great many nutrients to the marshland system. The water is relatively shallow, and the photosynthetic zone extends to the substrate. Thus we find that in such estuaries 200,000 g/m^2/yr of fixed carbon may be produced. It is estimated that 5% of this is consumed by grazers and about 95% of this forms detritus. Of these amounts, about 45% is exported and about 55% stays within the marshlands. These are the reasons why these marshland areas are such a desirable food source for many kinds of invertebrates and fish. They become spawning grounds and nursery grounds, and are essential to the commercial fisheries for crabs, shrimp, oysters, and fish.

The Growth of Algae and Rooted Aquatics

The growth of algae and other types of aquatic plants is dependent upon a great number of chemical elements such as nitrogen, phosphorus, carbon, calcium, magnesium, sodium, potassium, sulfur, chlorides, and many trace elements such as iron, manganese, cobalt, molybdenum, zinc, and vanadium.

The fact that both the concentrations required by different species and the amounts for optimum growth vary insures a continuum of plants throughout the year. Also, since growth is dependent upon the interaction of many factors—not only nutrient chemicals but also temperature and light—the correlation of growth with any one factor is extremely difficult. It is rather the matrix of factors which determines optimum conditions. Any one of these factors may be limiting and, if limiting, when added may stimulate growth. This is the reason why it is so very difficult to determine what are the conditions that increase algal growth and correlate them to any one or two density independent factors.

The seasonal succession found in algal species, either in the plankton or the benthos, is due to the varying of these density independent factors. It is well known that *Asterionella formosa* and *Fragilaria crotonensis* grow best in the early spring months when nutrients are higher. As the day length progresses the water becomes warmer and the relative concentration of nutrients changes. These species are often replaced by various species of *Synedra* in the plankton. In the benthos we often find such species as *Gomphonema olivacea, Diatoma vulgaris,* and *Meridion circulare* growing best in the cold winter months. As the day length lengthens and the temperatures become warmer, various species of *Synedra, Melosira varians, Navicula,* and *Nitzschia* increase greatly.

The growth of these algae is not only dependent upon the intake of various nutrients but also on the interaction of species with each other. For example, the growth of *Chlorella vulgaris* is greatly reduced if *Anacystis marina,* a blue-green, is present.[13] It is also somewhat reduced if the green flagellate *Chlamydomonas reinhardii* is present. *Chlamydomonas reinhardii* is strongly affected and cannot grow in the presence of *Anacystis marina* and is also greatly reduced in growth in the presence of *Chlorella vulgaris. Scenedesmus quadricauda* is greatly influenced, and its growth is retarded by the presence of *Anacystis marina, Chlorella vulgaris,* and *Chlamydomonas reinhardii.* Thus we see that the presence or abundance of species is not only determined by the concentration of nutrients which are left after the growth of other species but also by those species that are present at the time it has the potential of growth.

Our recent work (in press) at the Academy of Natural Sciences has brought out the importance of trace metals in influencing what groups of algae may be prevalent.[14] Although nitrogen, phosphorus, carbon, calcium, magnesium, etc. are essential for the production of organisms, the kinds of taxa may be greatly influenced by single trace elements or a mix of such trace elements. For example, if manganese is present at 40 μg/1 – 400 μg/l diatoms will usually dominate. However, if the manganese is reduced to 10–15 μg/l the diatom flora will not be able to compete against the blue-green algae, and a shift in species will occur from diatoms to blue-green algae.

Whereas manganese seems to stimulate diatom growth because of the need of manganese for fat metabolism, in the case of chromium a toxic effect seems to be present. Blue-green algae seem to be much more tolerant of chromium than diatoms. As a result, small amounts of chromium (a few parts per billion) have little influence upon the diatom community. However, at concentrations as high as 0.4mg/l a diatom-dominated community will change over to a blue-green one completely. Varying concentrations between these amounts will variously influence the amounts of blue-greens present.

Vanadium in very small amounts, such as 3–9 μg/l seems to stimulate diatom growth and also diversity. However, at 20 μg/l the diversity is reduced and at 4 mg/l blue-green algae are able to out-compete the diatoms, creating a switch from a diatom-dominated to a blue-green-dominated community.

It should be noted that with these heavy metals, particularly chromium and nickel, *Stigeoclonium lubricum* increases when the diatom growth starts to diminish so that we find that at 95–97 μg/l of chromium *Stigeoclonium lubricum* became exceedingly common and the blue-greens increased. At 0.4 mg/l *Stigeoclonium* remained dominant but the blue-greens became much more abundant. It is well documented in the literature that *Stigeoclonium lubricum* is tolerant and grows well in the presence of higher concentrations of trace metals than do some algae such as diatoms.

Very small amounts of nickel (8 μg/l) seem to influence the structure of the algal community. A diatom-dominated community at this concentration of nickel will start to shift to blue-greens. Likewise, we see an increase of *Stigeoclonium lubricum.* At 0.47 mg/l the diatoms were in very poor condition and the blue-greens, *Microcystis vaginatus* and *Schizothrix calcicola,* were very common, and *Stigeoclonium lubricum* was common. At 1 mg/l these two blue-greens were extremely common as was *Stigeoclonium lubricum.*

The change of the structure of the algal community from a diatom-dominated to a blue-green-dominated one seems to be affected by the amount of accumulation within the species. That is, the amount of micrograms per gram within the biomass more closely correlates with the diminishing success of diatoms and the increase in success of greens and blue-greens than does the concentration in the external media. It would appear that certain concentrations accumulated within the cells have adverse effects upon the physiological processes and thus decrease the ability of the given algae or group of algae to compete. It should also be pointed out that whether or not the cells are dividing actively also seems to influence the uptake and hence the accumulation within the cells. When the algal cells are dividing rapidly, such as in the spring of the year, biological dilution prevents the accumulation per cell from being as great in a given time interval. As a result, different concentrations in the external media may have different effects.

The accumulation of heavy metals by aquatic plants may be many thousand times the amount in the surrounding body of water.[15] This accumulation up to a given amount does not seem to interfere with the normal functioning of many species. As a result, these aquatic plants will remove from the water significant amounts of heavy metals and accumulate them within their tissues. They are usually released when the plant decays. Of course, they may be transferred through the ecosystem by other organisms preying upon them. Often they are removed from the river ecosystem by terrestrial animals or transferred to the ocean by migrating fish.

Conclusions

From this discussion it is evident that algae and other aquatic plants play a major role in the functioning of aquatic ecosystems. They may act as shelters or substrates for many different types of invertebrates and fish. They reoxygenate the water by the process of photosynthesis and they form an important food base for the whole ecosystem. In the process of performing their function within the aquatic ecosystem they assimilate during the growing season large amounts of nutrients and also they may assimilate considerable amounts of trace metals which may be concentrated within the cells. It is also known that they excrete various soluble organic compounds which return elements such as nitrogen and phosphorus to the aquatic ecosystem. These then become food for various kinds of organisms. The amounts of nitrogen and phosphorus returned in this way is usually small compared to the amount removed.

As a result, the immediate effect on water quality is the removal of considerable amounts of nutrients and, in some cases, trace metals which are not known to be nutrients. These nutrients and trace metals are then locked up within the protoplasm of these plants. In the case of emergent vegetation, these nutrients are dispersed to the terrestrial ecosystem as well as being returned to the aquatic ecosystem in the form of detritus and directly as food. The decay of the detritus and the release of nutrients from them by fungal and bacterial activity is usually a slower process than the breakdown of simple chemical compounds often occurring in waters. As a result the impact is not as sudden.

It should be further pointed out that a great deal of detritus is utilized as food and passes through the food web before it is decomposed. It is the assimilative capacity of marshland plants and sediments and of algae and aquatic plants in the open system that is of particular value in improving water quality.

Notes

1. This chapter is based on the author's studies of streams located mainly in the eastern half of the United States.

2. S. P. Chu, "The Influence of the Mineral Composition of the Medium on the Growth of Planktonic Algae. Part I: Methods and Culture Media," *Jour. Ecology* 30 (1942): 284-325. S. P. Chu, "The Influence of the Mineral Composition of the Medium on the Growth of Planktonic Algae. Part II: The Influence of Inorganic Nitrogen and Phosphate Phosphorus," *Jour. Ecology* 31 (1943): 109-48. W. Rhode, "Environmental Requirements of Freshwater Plankton Algae," *Symb. Bot. Upsal.* 10 (1948): 1-149. G. C. Gerloff, G. P. Fitzgerald, and F. Skoog, "The Isolation, Purification, and Nutrient Solution Requirements of Blue-Green Algae," In *Charles F. Kettering Foundation Symposium on the Culturing of Algae* (Dayton, Ohio: 1950) 27-44. J. Skok and W. J. McIlrath, "Distribution of Boron in the Plant Cell in Relation to Boron Deficiency and Boron Availability," *Plant Physiol.* 32 (1957): suppl. xxiii.

3. Skok and McIlrath, "Distribution of Boron."

4. R. R. Grant, Jr. and R. Patrick, "Tinicum Marsh as a Water Purifier," In *Two Studies of the Tinicum Marsh* (Washington, D.C.: The Conservation Foundation, 1970), pp. 105-23.

5. G. C. Gerloff, "Evaluating Nutrient Supplies for the Growth of Aquatic Plants in Natural Waters," In *Eutrophication: Causes, Consequences, Correctives* (Washington, D.C.: National Academy of Sciences, 1969), pp 537-55.

6. A. D. Hasler, *Underwater Guideposts* (Madison: University of Wisconsin Press, 1966).

7. *Ibid.*

8. P. R. Burkholder and L. M. Burkholder, "Vitamin B_{12} in Suspended Solids and Marsh Muds Collected Along the Coast of Georgia," *Limnol. & Oceanogr.* 1 (1956): 202-8. T. J. Starr, "Relative Amounts of vitamin B_{12} in Detritus from Oceanic and Estuarine Environments near Sapelo Island, Georgia," *Ecology* 37 (1956): 658-64.

9. R. Vannote, personal communication.

10. E. P. Odum, "The Role of Tidal Marshes in Estuary Production," *The Conservationist* (N.Y.) 15 (1961): 12-15, 35.

11. J. M. Teal, "Energy Flow in the Salt Marsh Ecosystem of Georgia," *Ecology* 43 (1962): 614-24.

12. L. R. Pomeroy, "Algal Productivity in Salt Marshes of Georgia," *Limnol. & Oceanogr.* 4 (1959): 386-97.

13. V. W. Proctor, "Studies of Algal Antibiosis Using *Haematococcus* and *Chlamydomonas*," *Limnol. & Oceanogr.* 2 (1957): 125-39.

14. R. Patrick, T. Bott, and R. Larson, *The Role of Trace Elements in Management of Nuisance Growths,* (Washington D.C.: U.S. EPA, in press 1975).

15. J. R. Watts and R. S. Harvey, "Uptake and Retention of ^{137}Cs by a Blue-green Alga in Continuous Flow and Batch Culture Systems," *Limnol. & Oceanogr.* 8 (1963): 45-9.

Part III

Drinking Water Problems

9

Carcinogenic Organic Chemicals in Drinking Water

ROBERT H. HARRIS

Associate Director, Toxic Chemicals Program
Environmental Defense Fund, Washington, D.C.

The number of cancer deaths is on the rise in the United States, and the rate of increase appears more rapid than either the rate of increase of population or the rate of increase in the total death rate.[1] More than 25% (53 million) of the 200 million people now living in the United States will develop some form of cancer, and approximately 20% of the American population die of cancer.[2] In 1973, there were over 600,000 new cases of cancer, and approximately 350,000 cancer deaths. Cancer is a leading cause of premature death. (Table 9-1).

The economic impact of cancer is massive. The direct and indirect costs of cancer, including loss of earnings during illness and during the balance of normal life expectancy, have been estimated at a total of $15 billion for 1971.[3]

There is now growing recognition that the majority of human cancers are caused by chemical carcinogens in the environment and hence that they are potentially preventable. Estimates by the World Health Organization, the National Cancer Institute, and numerous cancer specialists suggest that between 60% and 90% of human cancers are environmental in origin.[4] The basis for these estimates largely derives from the epidemiological studies, in large community populations over extended periods, which have revealed wide geographic variations in the incidence of cancer of various organs.[5] Although less certainty exists as to the role of most environmental contaminants in human cancer than the more widely recognized causes such as cigarette smoking, drugs, and occupational airborne particles, a recent Presidential panel concluded that:

> Cancer incitements by so far unrecognized chemicals combine to form a threat to health, that may well be of at least the same general size as the three major threats just described [i.e., cigarette smoking, alcohol abuse and choice of dietary composition]. These chemicals may be natural or synthetic.[6]

Although most of the evidence documenting the effect of carcinogens on man is based on in-

TABLE 9-1. *U. S. Deaths From Various Causes.*

Cancer deaths (1969)	323,000
World War II Battle Deaths	292,000
Auto Accident Deaths (1969)	59,000
Viet Nam War Deaths (6 yr.)	41,000
Korean War Deaths (3 yr.)	34,000
Polio Deaths (1952; worst yr.)	3,300

dustrial exposure to high levels, many cancer experts agree that the lower levels of carcinogens, both natural and man-made, to which the general population is exposed are responsible for the majority of human cancers. Although others espouse the threshold theory for carcinogens or argue that the levels of these chemicals in the environment are too small to be of significance of the general population, the scientific evidence simply does not support this viewpoint. For example, the chemical aflatoxin B_1 is known to cause cancer in man, and in experiments on trout it was shown to produce liver tumors when present in feed in concentrations as low as 400 parts per trillion; even at this low level, its carcinogenic effect was enhanced by addition of various noncarcinogenic oils to the diet.[7] Similarly, dieldrin (a chlorinated hydrocarbon pesticide) has been found to be carcinogenic in several strains of mice at the lowest concentrated tested, 100 parts per billion. Concentrations of dieldrin found in human fatty tissue approximate the range of concentrations of dieldrin found in experimental animals demonstrating metastasizing carcinomas.[8] Therefore, lacking scientific evidence that a threshold existed for *any* chemical carcinogen, the ad hoc Committee on the Evaluation of Low Levels of Environmental Chemical Carcinogens reporting to the Surgeon General in 1970 concluded that: "No level of exposure to a chemical carcinogen should be considered toxicologically insignificant for man."[9]

Carcinogens in Water

Despite considerable interest in the role of air and food in environmental carcinogenesis, hitherto there has been relatively little attention directed to the possibility that carcinogens in drinking water may be causally related to human cancers. Although few studies until recently have analysed for carcinogens in drinking water, the presence of carcinogens in wastes discharged to water supplies has been known for decades. The sources of some of these carcinogens are outlined as follows:[10]

1. *Petroleum Products*—Petroleum refinery wastes containing polycyclic aromatic hydrocarbons, fuel oil, lubricating oils, and cutting oils are being introduced into lakes and rivers from garages, service stations, petrochemical plants, metalworking plants, and ships. Contamination of public water supplies may also result from the use of kerosene, methylated naphthalenes, and similar petroleum products used as vehicles of insecticide sprays, from rain contaminated with air pollutants, or from tarred or asphalted roads.
2. *Coal Tar*—Effluents from gas plants, coke oven operations, tar distilleries, tar-paper plants, and wood-pickling plants all contain carcinogens. Coal tar, pitch, creosote, and anthracene oil are known human carcinogens.
3. *Aromatic Amino- and Nitro-Compounds*—Amino compounds such as beta-naphthylamine, benzidine, and 4-aminodiphenyl are known human carcinogens. These compounds along with their nitroanalogues are released by dye and rubber manufacturing, pharmaceutical factories, textile dying plants, plastic production, and others.
4. *Pesticide, Herbicide, and Soil Sterilants*—Compounds such as DDT, Dieldrin, Aramite, carbon tetrachloride, acetamide, thioacetamide, thiourea, thiouracil, aminotriazole, several urethane derivatives, isopropylchlorophenyl carbamate, and beta-propiolactone are capable of eliciting benign and/or malignant tumors in various organs of experimental animals.

In addition to industrial wastes, discharges from domestic sewage treatment plants may also be responsible for a variety of carcinogenic substances found in water. In a recent study by R. L. Jolley, for example, over fifty chlorinated hydrocarbons were identified in chlorinated domestic sewage effluents.[11] Consequently, Jolley estimated that over one thousand tons of chlorinated organic compounds are discharged by sewage treatment plants into the nation's waterways annually.

While discharges from industry and municipal waste treatment plants represent more or less continuous sources of pollution, spills and accidents resulting from industrial operations, barge traffic, or other transportation accidents near bodies of water may contribute significantly to the level of hazardous substances in public water supplies. For example, in a 240-mile stretch of the lower Mississippi River in Louisiana, over 350 accidental spills were recorded by the U.S. Coast Guard in the twelve-month period, August 1973 to August 1974. Many of these spills resulted in discharges of substantial quantities of carcinogenic substances, such as the 64,000-gallon spill of chloroform on August 19, 1973, and a spill of 1.3 million gallons of crude oil on January 18, 1974. Although the Coast Guard records many of these spills according to the substance discharged, a considerable number are recorded as "unknown" for both substances and amount.

Other sources of carcinogens in drinking water include chlorination of polluted river water at municipal water treatment plants. J. J. Rook was among the first to observe that chlorination of polluted river water produced compounds such as chloroform, dichlorobromomethane, dibromochloromethane, bromoform, and traces of other halomethanes and haloethanes.[12] More recent work by T. A. Belar and coworkers at the Environmental Protection Agency laboratories in Cincinnati has confirmed these results.[13] From analyses of five communities receiving water either from the Ohio or Mississippi Rivers, chloroform concentrations were observed to range from 37 to 152 ppb. Communities receiving well water, which were generally less polluted with organic matter than were surface water, were observed to have less than one-tenth the concentration of chloroform in their drinking water when compared to communities receiving polluted Ohio or Mississippi River water.

From these studies, therefore, it can be concluded that numerous public drinking water supplies are subject to contamination by carcinogenic substances from industrial and municipal discharges, accidental spills, runoff from agricultural and urban areas, and from the chlorination process at water treatment plants. It should therefore not be surprising that recent studies have found widespread contamination of treated municipal drinking water by known or suspected carcinogens.[14]

Subsequent to studies on New Orleans drinking water, in which numerous carcinogens and largely uncharacterized organic chemicals were identified, the Environmental Protection Agency formed an ad hoc Study Group of the Hazardous Materials Advisory Committee to assess the health risk from some of the carcinogens identified in drinking water in New Orleans and elsewhere.[15] Despite numerous constraints to a comprehensive study of all of the potentially toxic chemicals found in drinking water, the Study Group concluded that: "With respect to assessment of health risk associated with exposure to the specific contaminants identified in the charge to the Study Group, it was concluded that some health risk exists."[16]

This assessment was based primarily on the carcinogenicity data for chloroform, carbon tetrachloride, benzene, chloroethers, chloroolefins, and polynuclear hydrocarbons. Chloroform was the only chemical for which the Study Group had sufficient data to extrapolate from laboratory animals to human populations. Based on a chloroform concentration of 311 ppb (the concentration found in Miami drinking water), the study group concluded that "the level of risk . . . might be extrapolated to ac-

TABLE 9-2. *EPA Analysis of Representative Contaminants in 80-City Survey.*

Compound	*Number of Cities in which Detected*	*Range of Concentration (μg/l)*
Chloroform	80	less than 0.1 – 311
Bromodichloromethane	78	0.3 – 116
Dibromochloromethane	72	less than 0.4 – 100
Bromoform	26	less than 0.8 – 92
Carbon Tetrachloride	10	less than 2 – 3
1,2=Dichloroethane	26	less than 0.2 – 6

*EPA, "Preliminary Assessment of Suspected Carcinogens in Drinking Water."

count for up to 40% of the observed liver cancer incidence rate." However, the Study Group emphasized that the data for chloroform was somewhat tenuous, and this estimate could therefore not be considered precise. It was also suggested that for chloroform other types of chronic disease (e.g., cirrhosis) might be of equal or greater concern that possibly cancer production.

In addition to establishing this Study Group, the Environmental Protection Agency initiated a nationwide study of drinking water in eighty communities.[17] For all eighty cities, the EPA monitored for six representatives halogenated (chlorine or bromine containing) organic compounds suspected (bromodichloromethane, dibromochloromethane, 1, 2-dichloroethane, bromoform) or known (chloroform, carbon tetrachloride), to be carcinogenic. Chloroform was found in all eighty cities, while the contaminants 1, 2-dichloroethane and carbon tetrachloride were found in 32.5 and 12.5% of the cities respectively (Table 9-2). Chloroform, bromodichloromethane, dibromochloromethane, and bromoform have been found to arise in the drinking water primarily from the chlorination process, and not from industrial sources. Carbon tetrachloride and 1, 2-dichloroethane appear to be primarily of industrial origin.

In addition to this limited analysis, ten of the eighty water supplies representing five major categories of raw water sources were sampled for a more comprehensive survey of the organic content of their finished water. The cities investigated and their raw water sources were: Miami, Florida, and Tucson, Arizona, (groundwater source); Seattle, Washington, and New York, New York, (uncontaminated upland water); Ottumwa, Iowa, and Grand Forks, North Dakota, (raw water contaminated with agricultural runoff); Philadelphia, Pennsylvania, and Terrebonne Parish, Louisiana, (raw water contaminated with municipal waste); and Cincinnati, Ohio, and Lawrence, Massachusetts, (raw water contaminated with industrial discharges).

A total of 135 organic chemicals were identified in all ten cities. Table 9-3 lists some of the chemicals of known toxicological significance to humans and their presence in the drinking water of the ten cities listed above and in that of New Orleans, the District of Columbia and Evansville, Illinois, that had been surveyed in previous studies.[18] The concentrations detected were usually less than 10 ppb except for the trihalomethanes which ranged up to 311 ppb.

Although it might be expected that the two ground water supplies, Miami and Tucson, would be less contaminated, Miami drinking water was one of the most contaminated in the survey. This is because Miami obtains its drinking water from shallow groundwater aquifers which are contaminated by metropolitan and industrial activities. The obviously superior quantity of Tucson's drinking water when compared to Miami reflects the greater depth of the groundwater aquifers that serve as Tucson's supply. As discussed above, the groundwater sup-

TABLE 9-3. *Selected Potentially Hazardous Chemicals in Drinking Water.*
The symbol + indicates that detectable quantities were present; numbers in parentheses indicate the highest reported concentration in ppb.

	New Orleans	Miami*	Seattle*	Ottuma*	Philadelphia*	Cincinnati*	Tucson*	New York City*	Lawrence*	Grand Forks*	Terrebonne* Parish	Washington, D.C.	Evansville
acetylene dichloride	+	+ (17)			+ (0.1)	+ (0.1)							
benzene†	+	+ (0.1)		+ (0.1)	+ (0.2)	+ (0.3)						+ (0.005)	
bis (2-chloroethyl) ether†**	+				+ (0.5)								+
bromodichloromethane	+	+ (73)	+ (4)	+	+ (20)	+ (15)		+ (17)	+ (1.8)	+ (3.2)	+ (23)	+	+
bromoform	+ (0.57)	+ (1.5)			+	+	+ (3.0)	+ (1.5)				+	
carbon disulfide‡	+	+			+	+		+		+			
carbon tetrachloride†**	+	+		+	+	+		+ (0.13)	+ (0.1)	+ (0.1)	+	+ (0.005)	
chlorobenzene	+	+ (1.0)	+	+	+ (0.1)	+ (0.5)		+ (4.7)	+ (0.12)	+	+ (5.6)	+ (.001)	
chloroform†**	+ (133)	+ (311)	+ (21)	+ (1.0)	+ (65)	+ (38)	+ (0.08)	+ (44)	+ (32)	+ (40)	+ (130)	+	
chloropicrin		+ (0.4)			+ (2)	+ (3)							
cyanogen chloride		+		+	+	+		+	+	+	+		
dibromochloromethane	+ (1.1)	+ (32)	+ (3)		+ (5)	+ (3)	+ (0.01)	+ (0.4)	+ (0.01)	+ (0.1)	+ (1.0)	+	+
dibutyl phthalate‡	+ (0.1)	+ (5)	+ (0.01)	+ (0.1)	+ (0.05)				+ (0.01)		+ (0.02)	+	
diethyl phthalate‡	+ (0.03)	+ (1.0)	+ (0.01)		+ (0.01)	+ (0.1)		+ (0.01)	+ (0.04)				
1,2-dichlorobenzene	+ (0.01§)	+ (1.0)			+	+						+ (0.001)	

*Cities in 1975 EPA study. See reference in Table 2.
†Known cancer-causing chemical (carcinogen).
**Chemical being tested for carcinogenicity by the National Cancer Institute.
‡Chemical currently under study for its possible role in causing birth defects.
§Designated only as "dichlorobenzene isomer".

TABLE 9-3 (cont.)

	New Orleans	Miami*	Seattle*	Ottuma*	Philadelphia*	Cincinnati*	Tucson*	New York City*	Lawrence*	Grand Forks*	Terrebonne* Parish	Washington, D.C.	Evansville
1,3-dichlorobenzene		+ (0.5)			+	+			+				
1,4-dichlorobenzene		+ (0.5)			+	+			+				
1,2-dichloroethane**	+	+			+	+							
hexachloroethane	+ (4.4)	+ (0.5)											+
methylene chloride	+	+	+	+	+	+		+ (0.1)	+ (1.6)	+ (0.1)	+	+	
dieldrin†**	+ (0.05)	+ (0.002)	+ (0.001)	+ (0.002)		+ (0.001)							
lindane†						+ (0.1)							
1,1,1-trichloroethane**				+	+	+							
trichloroethylene†**	+	+ (0.3)		+ (0.1)	+ (0.5)	+ (0.1)			+				
tetrachloroethylene**	+	+ (0.1)		+ (0.2)	+ (0.4)	+ (0.3)	+	+ (0.46)	+ (0.07)	+ (0.2)		+	+
vinyl chloride†		+ (5.6)			+ (0.27)								
vinylidene chloride		+ (0.1)			+ (0.1)	+			+				

†Known cancer-causing chemical (carcinogen).
**Chemicals being tested for carcinogenicity by the National Cancer Institute.

plies in the eighty-city survey were generally significantly less polluted than the surface water supplies.

The drinking water in New York City and in Seattle were generally of superior quality in comparison to the other surface water supplies, probably reflecting the relatively uncontaminated sources of these two supplies. The obviously inferior quality of the drinking water in Philadelphia, New Orleans, Cincinnati, and Lawrence reflect the large industrial and municipal pollution burden of their respective surface water sources. The results for the District of Columbia and Evansville cannot be compared directly with those of the other cities because the analyses employed less sensitive techniques in these two cities.

Although some regulatory officials have

described these analyses as demonstrating that only "very small amounts" or "trace quantities" of contaminants occur in drinking water, the suggestion that concentrations in the parts per billion range are not toxicologically significant simply cannot be supported scientifically. As discussed above, it is precisely in this concentration range that many carcinogens have been shown to cause cancer in laboratory test systems, including dieldrin, aflatoxin and diethylstilbestrol (DES).

Epidemiological Investigations

It has long been recognized by epidemiologists that cancer rates vary according to social, racial, and economic factors, which undoubtedly reflect differential exposures to carcinogens caused by differences in dietary habits, smoking and drinking habits, occupational exposure to carcinogens, urban air pollution, and others. Provided that there are clear differentials in exposure of the general population to specific chemicals, epidemiologic techniques may then be able to correlate certain cancers with certain chemicals or groups of chemicals. However, most environmental exposures to carcinogens defy epidemiological analysis. Epstein discussed some of the reasons for this:

> Human experience generally affords only limited indications of safety or hazard. The reasons for this include the long latent period, in some cases many decades, elapsing between exposure to carcinogenic chemicals and the subsequent development of cancer; demonstration of causality may be obscured by mortality from competing risks during such latent periods . . . Additional difficulties include isolation of the effects of any single chemical from the multitude of others to which populations are concurrently exposed and the lack of base-line data on regional and national incidences of cancer, besides other adverse human effects.[19]

In marked contrast with the large number of epidemiological studies on air pollution and its adverse health effects, including an "urban excess" of lung cancer generally attributed to community air pollution, toxic and carcinogenic effects of chemical pollutants in drinking water have received little attention until recently.[20] Scattered European studies have suggested a possible association between drinking polluted river water and cancer mortalities.[21] Similar studies in the United States observed that counties in Missouri with high cancer incidence clustered around the Missouri River.[22] Although carcinogens in drinking water were suggested as a possible explanation for these observations, this hypothesis was not investigated further.

Because of the historical evidence of high cancer rates in New Orleans[23] and the recent observations of numerous carcinogens in New Orleans drinking water[24], Harris, Page, and Epstein[25] analyzed by multivariant regression analysis the relation between drinking water quality and average cancer mortality rates for the period 1950-1969 by parish (county) in Louisiana (Table 9-4).[26] Exposure variables included the fraction of the parish that is rural; median family income; fractions of the work force in the petroleum and coal products industries, inorganic and organic chemical industries, and mining industries; and the fraction of the parish supplied Mississippi River water. Out of a total 1960 population of 3.3 million, living in 64 parishes, 32% (representing part or all of the population in 11 parishes) were supplied drinking water from the Mississippi River, 66% were supplied groundwater, and 12% were supplied from other surface water supplies. It was considered reasonable to presume that carcinogens in Mississippi River water were in considerably higher concentrations than in other surface and groundwater supplies in Louisiana.

For total cancer mortality rates, the drinking water variable (W) was statistically significant for all groups except white females (Table 9-4). For cancer of the gastrointestinal tract, the drinking water variable was significant in all four population groups. Considered separately, drinking water was significant for cancer of the rectum and stomach, but not large intestine, for nonwhite males and females; large intestine and rectum, but not stomach, for white females; and rectum, but not large intestine and stomach, for white males. For cancer of the urinary organs, drinking water was a significant variable only for white males and nonwhite females. Although not shown in Table 9-4, the drinking water variable was not significant for cancer of the lung, breast, liver, pancreas, or lymphatic system.

Considering the possibility of spurious correlation if cancer patients moved to New Orleans and established residence to take advantage of the availability of specialized and/or free medical care, the regressions were rerun with New Orleans removed from the data. For total cancer the coefficient for the water variable remained positive and strongly significant, making this hypothesis of spurious correlation seem unlikely. Additional problems which might arise from north-south differences in diet or other socioeconomic variables were investigated by restricting the data set to the twenty-nine southern parishes. However, this did not change the basic relationship between the independent variables and cancer mortality rates.

Although no systematic epidemiological study has been conducted in association with the recent eighty-city survey, L. J. McCabe has recently conducted a preliminary study which demonstrates a statistically significant correlation between the chloroform concentration and total cancer mortality ratios (observed/expected) in fifty of the eighty cities for which cancer mortality data were available.[27] A similar correlation was found for total trihalogenated methanes and cancer mortality ratios, but no correlation was found for total (all causes) mortality ratios.

Federal Regulatory Programs

The Federal Water Pollution Control Act Amendments of 1972 (FWPCA) established a comprehensive program for restoring the physical, chemical, and biological integrity of the Nation's waters, including a comprehensive program for regulating the discharge of toxic pollutants into drinking water supplies and providing the necessary funding for research and development to achieve these goals. However, neither regulation nor research and development on the means and techniques of identifying or removing toxic pollutants from industrial or municipal wastes has progressed on anything close to the scale that Congress envisioned.[28]

Although by enacting the FWPCA, Congress established a speedy regulatory program (FWPCA, Sec. 307) to control the discharge of toxic pollutants into water, regulations have yet to be promulgated. As a consequence, such toxic pollutants as PCBs, asbestos, and the numerous other carcinogens identified in drinking water, are discharged virtually unregulated into surface waters that are intended to support sport fishing and other recreational activities, and to serve as drinking water supplies. The Environmental Protection Agency has so far identified over 300 industrial chemicals in treated drinking water, most of which have been shown to arise from industrial discharges, but as yet there are no discharge regulations on any chemicals on this list.[29] Despite the recommendation by the 1972 EPA report on New Orleans drinking water that regulations be adopted on the industries responsible for the discharge of these chemicals, no programs were initiated at that time, nor is there today any systematic effort to identify the sources of those pollutants, let alone attempt to regulate them.[30]

Despite the near standstill of regulation and research and development on toxic pollutants, passage of the Safe Drinking Water Act in 1974 did provide the Environmental Protection Agency with additional regulatory authority to protect the public against contaminated drinking water. However, the Environmental Protection Agency will not be promulgating standards for carcinogens in the Interim Primary Drinking Water Regulations, claiming that more research is needed. Although the widespread occurrence of carcinogens in drinking water suggest the need for such standards, and sufficient data

TABLE 9-4. *Regression Analysis of Cancer Mortality Rates (CMR) by Race and Sex.*
Variable abbreviations: W—fraction of parish supplied with Mississippi River water; R—fraction of parish that is rural; I—median family income; P—fraction of work force in coal products industries; C—fraction of work force in inorganic and organic chemical industries; M—fraction of work force in mining industries. Statistical significance at the 95% level is indicated by the variable abbreviation, and at the 99% level by *. Variables not included in the table were not significant at the 95% level (e.g., for urinary tract cancer, white female, no variable was significant at the 95% confidence level). Numbers in parentheses represent the fraction of variance in cancer mortality rates explained by the indicated variables (coefficient of determination).

$$CMR = a_o + a_1W_i + a_2R_i + a_3I_i + a_4P_i + a_5C_i + a_6M_i + e_i$$

Cancer type	White male	White female	Nonwhite male	Nonwhite female
Total	(0.42)W*R*I*	(0.1)R	(0.37)W*R*	(0.32)W*R
Urinary tract	(0.43)W*R*C*	(0.06)	(0.18)R	(0.12)W
Gastrointestinal tract	(0.39)W*R*I*C	(0.14)W*	(0.33)W*	(0.30)W*

exists to promulgate them, they would not be sufficient to fully protect the public from carcinogens in polluted water supplies. Without adequate controls on the discharge of toxic pollutants from industrial sources, the downstream public will be required to pay the cost of their removal. Such removal techniques, however, will be unable to adequately control all of the potentially toxic organic chemicals discharged by industry.

Conclusions

With the growing recognition that human cancer is an environmental disease, considerable regulatory attention should be focused on the widespread distribution of chemical carcinogens in drinking water. Despite the absence of retrospective case-control epidemiological studies, statistical and comparative studies do suggest that cancer mortality rates are related to drinking water quality in those communities served drinking water from polluted supplies.

Although adequate statutory authority exists under the Federal Water Pollution Control Act to regulate the discharge of toxic pollutants from industrial sources, the Environmental Protection Agency has yet to promulgate such standards. Dwindling research and development funds in the water pollution program has resulted in virtual abandonment of research on techniques of identification and removal of toxic organic pollutants from wastewater. Unless this trend is reversed, the risk of cancer from carcinogens in water can logically be expected to increase.

Notes

1. Natural Health Education Committee, Inc., June 1970. Quoted in *National Program for the Conquest of Cancer.* Senate document no. 9 of the 92d Congress. Ordered printed April 14, 1971.
2. American Cancer Society, *Cancer Facts and Figures* (New York: American Cancer Society, 1974).
3. Department of Health, Education and Welfare, *National Cancer Program,* NIH 74-569 (Washington: U.S. Government Printing Office, 1973).
4. World Health Organization, *Prevention of Cancer,* Tech. Report Series 276 (Geneva: World Health Organization, 1964); S. S. Epstein, "Environmental Determinants of Human Cancer," *Cancer Research* 34 (1974): 2425-35; E. Boyland, "The Correlation of Experimental Carcinogenesis and Cancer in Man," *Progr. Expl. Tumor Res.* 11 (1969): 222-34; H. Higginson, "Present Trends in Cancer Epidemiology," *Canadian Cancer Conference* 8 (1969): 40-75; S. S. Epstein, "Control of Chemical Pollutants," *Nature* 228 (1970): 816-19.
5. Epstein, "Environmental Determinants of Human Cancer."
6. President's Science Advisory Committee, "Chemicals and Health," Report of the Panel on Chemicals and Health, Science and Technology Policy Office, National Science Foundation (Washington, D. C.: U. S. Government Printing Office, 1973).
7. R. O. Sinhuber et al., "Dietary Factors and Hepatomas in Rainbow Trout (*Salmo gairdneri*): I. Aflatoxins in Vegetable Protein Feedstuffs," *J. National Cancer Inst.* 41 (1968): 711—18.

8. S. S. Epstein, Testimony at Hearings on Aldrin/Dieldrin (Environmental Protection Agency and Environmental Defence Fund *vs.* Shell Chemical Company), (EDF ex. 33), in re Shell Chemical Co. FIFRA, docket No. 145, March 1974.

9. Ad Hoc Committee on the Evaluation of Low Levels of Environmental Chemical Carcinogens, "Evaluation of Environmental Carcinogens," report to the Surgeon General, USPHS (Washington, D. C.: U. S. Government Printing Office, 1970).

10. W. C. Hurper and W. D. Conway, *Chemical Carcinogenesis and Cancers* (Springfield: Charles C. Thomas, 1964).

11. R. L. Jolley, "Chlorination Effects on Organic Constituents in Effluents from Domestic Sanitary Sewage Treatment Plants," Publication No. 565, Environmental Science Division (Oakridge: Oakridge National Laboratory [NTIS] 1973).

12. J. J. Rook, "Formation of Haloforms During Chlorination of Natural Waters," *J. Soc. Water Treat. Exam.* 23 (1974): 234-43.

13. T. A. Belar, J. J. Lichtenberg, and R. C. Kroner, "The Occurance of Organohalides in Chlorinated Drinking Water" (Cincinnati: EPA unpublished report, undated).

14. R. D. Kleopfer and B. J. Fairless, "Characterization of Organic Components in a Municipal Water Supply," *Envir. Sci. Tech.* 6 (1972): 1036-37; A. K. Burnham et al., "Identification and Estimation of Neutral Organic Contaminants in Potable Water," *Anal. Chem.* 44 (1972): 139-42; H. J. Svec, J. S. Fritz, and G. V. Calder, "Trace Soluble Organic Compounds in Potable Water Supplies," Office of Water Resource Research, Department of the Interior (Washington, D. C.: U. S. Government Printing Office [NTIS], 1973); G. A. Junk and S. E. Stanley, "Organics in Drinking Water. Part I. Listing of Identified Chemicals," U. S. Energy Research and Development Administration (Washington, D. C.: U. S. Government Printing Office [NTIS], 1975); J. M. Symons et al., "National Organic Reconnaissance Survey for Halogenated Organics in Drinking Water," Water Supply Research Laboratory, EPA (Cincinnati: unpublished, 1975); EPA, "Identification of Organic Compounds in Effluents from Industrial Sources," Office of Toxic Substances (Washington, D. C.: unpublished, 1975); EPA, "Draft Analytical Report: New Orleans Area Water Supply Study" (Dallas: EPA Region VI, unpublished, 1974).

15. EPA, "New Orleans Area Water Supply Study."

16. EPA, "Assessment of Health Risk from Organics in Drinking Water," Science Advisory Board, EPA (Washington: D. C.: unpublished, 1975).

17. EPA, "Preliminary Assessment of Suspected Carcinogens in Drinking Water," December 1975 report to Congress (Washington, D. C.: unpublished, 1975); Symons et al., "National Reconnaissance Survey."

18. Kleopfer and Fairless, "Characterization of Organic Components"; M. A. Scheiman, R. A. Saunders and F. E. Saalfeld, "Organic Contaminants in the District of Columbia Water Supply," *Biomedical Mass Spectrometry* 1 (1974): 209-11; EPA, "New Orleans Area Water Supply Study."

19. Epstein, "Environmental Determinants of Human Cancer."

20. *Ibid.*

21. P. Stocks, "Regional and Local Differences in Cancer Death Rates: Studies on Medical and Population Subjects" (London: General Registry Office, 1974); S. W. Tromp, "Possible Effects of Geographical and Geochemical Factors on Development and Geographical Distribution of Cancer," *Schweiz. Z. Path.* 18 (1955): 929.

22. G. B. Cook and F. R. Watson, "The Geographic Locations of Missourians with Multiple Cancer," *Missouri Medicine* (1966), pp. 997-1002.

23. N. E. Manos, "Comparative Mortality Among Metropolitan Areas of the United States: 1949-51," U. S. Public Health Service Publication No. 562 (Washington, D. C.: U. S. Government Printing Office, 1957); H. F. Dorn and S. J. Cutler, "Morbidity from Cancer in the United States," U. S. Public Health Service Monograph 56 (Washington, D. C.: U. S. Government Printing Office, 1959); T. J. Mason and F. W. McKay, "U. S. Cancer Mortality by County: 1950-1969," National Cancer Institute DHEW Publication No. (NIH) 74-615 (Washington, D. C.: U. S. Government Printing Office, 1974).

24. EPA, "Industrial Pollution on the Lower Mississippi River in Louisiana" (Dallas: EPA Region VI, unpublished, 1972); EPA, "New Orleans Area Water Supply Study"; EPA, "Identification of Organic Compounds from Industrial Sources."

25. R. H. Harris, "The Implications of Cancer-Causing Substances in Mississippi River Water" (Washington: Environmental Defence Fund, unpublished, 1974); T. Page, R. Harris, and S. S. Epstein, "Relation Between Cancer Mortality and Drinking Water in Louisiana" (submitted for publication, 1975).

26. Mason and McKay, "U. S. Cancer Mortality by County: 1950-1969."

27. Symons et al., "National Organic Reconnaissance Survey"; L. J. McCabe, "Association Between Trihalomethanes in Drinking Water (NORS Data) and Mortality," Water Quality Division, EPA (Washington, D. C.: unpublished, 1975).

28. R. H. Harris, Statement to the Subcommittee on the Environment and the Atmosphere of the House Committee on Science and Technology, November 14, 1975 (Washington, D. C.: U. S. Government Printing Office, in press 1976).

29. Junk and Stanley, "Organics in Drinking Water. Part I"; Symons et al., "National Organic Reconnaissance Survey."

30. EPA, "Industrial Pollution of the Lower Mississippi River."

10

Potential Carcinogenic Hazards Due to Contaminated Drinking Water

SAMUEL S. EPSTEIN, M. D.

School of Medicine
Case Western Reserve University

Carcinogenicity as a Public Health Hazard

1. The Scope of Human Cancer

There is now little doubt that many diseases hitherto regarded as spontaneous, particularly cancer, are caused by environmental pollutants. This realization is heightened by the exponential increase in exposure of the general population to currently used and new synthetic chemicals—and their degradation products in water, air, and soil—which, in general, are inadequately characterized toxicologically and ecologically. These considerations apply with even greater emphasis to relatively uncontrolled occupational exposure to a wide range of known chemical carcinogens, in addition to thousands of toxicologically uncharacterized or inadequately characterized synthetic chemicals.

More than 25%, 53 million, of the 200 million people now living in the United States will develop some form of cancer, and approximately 20% of the American population die of cancer. In 1969, 323,000 people died of cancer in the United States. In 1972, approximately 610,000 new cases of cancer were diagnosed, and 1,000,000 were under treatment. Cancer is a leading cause of premature death (Table 1).

The rate of recent increase of cancer deaths appears more rapid than the rate of increase in population and is even more rapid than the increase in the rate of death; this increase in new cancer cases is real and is over and above that due to increases in age alone.

The economic impact of cancer is massive. The direct costs of hospitalization and medical

TABLE 10-1. *U.S. Deaths from Various Causes*

Cancer deaths (1969)	323,000
World War II battle deaths	292,000
Auto accident deaths (1969)	59,600
Viet Nam War deaths (6 yr)	41,000
Korean War deaths (3 yr)	34,000
Polio deaths (1952; worst yr)	3,300

care for cancer in 1969 exceeded $500 million. The direct and indirect costs of cancer, including loss of earnings during illness and during the balance of normal life expectancy, were estimated at a total of $15 billion for 1971; on the basis of analyses of selected cases, it appears that the total direct costs for a particular patient may range from $5,000 to over $20,000. On purely economic grounds, it is clear that control or further limitation in the overall burden of environmental and occupational chemical carcinogens, with anticipated major reductions in the incidence of human cancer, is likely to achieve very significant reductions in total national costs from cancer. These considerations are not appropriately reflected in allocation of federal or state priorities and resources for the prevention, in contrast with the diagnosis and treatment, of human cancer.

2. Chemical Carcinogens as a Cause of Human Cancer

Carcinogens are chemical, physical, or biological agents which, upon exposure of animals or humans to them, increase the probability of tumor induction. This may appear as follows: increase in the tumor incidence in a given population (i.e., an increase in the number of tumor-bearing individuals); increase in the number of tumors in each individual; reduction in the latent period of tumor induction; any combination of the above effects.

Carcinogens may act at the site of initial contact, at the site of selective organ localization or accumulation, at the site of excretion, or at the site of metabolism. Some carcinogens act at single sites only; others act at multiple sites.

There is now growing recognition that the majority of human cancers are due to chemical carcinogens in the environment and that they are hence ultimately preventable. There is also growing interest in the possible role of chemical carcinogens in activating oncogenic viruses.

The World Health Organization has estimated that over 75% of human cancers are influenced by environmental factors. It has also been estimated that approximately 90% of human cancers are chemical in origin; other estimates indicate that between 60 and 80% of human cancers are environmental in origin. The basis for these estimates largely derives from epidemiological studies, in large community populations over extended periods, which have revealed wide geographic variations in the incidence of cancer of various organs. In some of these studies, the role of specific environmental carcinogens has been implicated or identified.

The best documented and most significant data on carcinogenesis due to environmental chemicals are those on tobacco. It has been suspected for several decades that heavy tobacco smoking is directly and causally related to chronic lung disease, especially cancer. Over twenty-nine retrospective epidemiological studies on lung cancer, generally substantiating the causative role of smoking, have been published from 1939 to 1964. One of the most important of these studies demonstrated unambiguously that the incidence of lung cancer could be positively correlated with cigarette smoking. Mortality rates from lung cancer in both men and women have increased exponentially over the last few decades and have now reached epidemic proportions, particularly in the United States.

There are also many studies that demonstrate the contributory role of urban air pollution in lung cancer; additionally, numerous classes of chemical carcinogens have been identified in polluted urban air. These studies have shown that there is an excess of lung cancer deaths in smokers in polluted urban areas in contrast with those living in rural areas. There are marked regional differences in lung cancer mortality patterns in the United States. Increased mortality is clearly related to increased urbanization and to increased levels of organic pollutants in the air. The higher lung cancer rates in urban areas cannot be fully explained by factors such as smoking or occupation. In a survey some twelve years ago, lung cancer rates in the United States—

standardized for smoking besides age—were found to be 39 in 100,000 in rural areas and 52 in 100,000 in cities with populations in excess of 50,000. Similar surveys in England also confirmed the importance of this urban factor and stressed its interaction with smoking. This urban excess of 25% in United States mortality is generally regarded as being due to air pollution. Confirmatory evidence is also afforded by several studies on immigrants who tend to retain the incidence pattern of lung cancer of their country or origin, even though they assume the smoking and other habits of their country of adoption.

In marked contrast with the numerous studies on the relation between urban air pollution and lung cancer, it should be stressed that there have been no systematic and large scale studies relating water pollution to the incidence of total cancers and of specific organ cancers.

Striking regional variations in the occurrence of a wide range of other organ cancers are now well recognized. The high incidence of cancer of the oral cavity in Asia, representing some 35% of all Asiatic cancers, in contrast to 1% of all European cancers, is clearly related to the chewing of betel nuts and tobacco leaves. The high incidence of liver cancer in the Bantu and in Guam may well be due to dietary contamination with aflatoxin, a potent fungal metabolite carcinogen, and to eating cycad plants, containing azoxyglucoside carcinogens respectively. The high incidence of gastric cancer in Japan, Iceland, and Chile has been associated with high dietary intake of fish; suggestions have been made implicating nitrosamines, formed by reactions between secondary amines in fish and nitrite preservatives. The high incidence of cancer of the esophagus in Zambians drinking Kachasu spirits, in the Calvados area of France, and in other clearly defined geographic areas may be related to contamination of alcoholic drinks or food with nitrosamines, some of which produce esophageal cancer in experimental animals.

More restricted data on chemical carcinogenesis in humans derive from follow-up studies on patients treated with carcinogenic drugs. These include immunosuppressive agents, used in transplant therapy; alkylating agents, used in treatment of cancer and to a lesser extent autoimmune diseases; radioactive phosphorus, used in treatment of polycythemia vera; estrogens, used for treatment of prostatic cancer in men and for hormone replacement therapy in women; and diethylstilbestrol, used for treatment of threatened abortions and as a "morning-after" pill.

Data on occupational carcinogenesis extend back to the 18th century, with the discovery by Sir Percival Pott of a high incidence of scrotal cancers in young British chimney sweeps exposed to soot. A very wide range of occupational cancers has since been identified and studied in detail. These include bladder cancer in the aniline dye and rubber industry, induced by such chemicals as 2-naphthylamine, benzidine, 2-aminobiphenyl, and 2-nitrobiphenyl; lung cancer in uranium miners of Colorado, in coke oven workers, in nitrogen mustard factories in Japan, and in United States workers even briefly exposed to bis(chloromethyl)ether; skin cancer in cutting and shale oil workers; nasal sinus cancer in wood workers; lung cancer and pleural mesotheliomas in insulation workers and in others (such as construction workers) exposed to asbestos; cancer of the pancreas and lymphomas in organic chemists; and angiosarcoma of the liver in workers involved in polymerization processes for the manufacture of polyvinyl chloride.

Permanent standards for fourteen occupational carcinogens, recently promulgated by the Occupational Safety and Health Administration, contain no provisions for monitoring of the occupational environment or for a permit system registering and detailing specifics of usage of these carcinogens, in spite of near-unanimous recommendations to this effect by the Standards Advisory Committee to the Department of Labor. In the absence of these provisions, the standards cannot be implemented in practice and could be regarded as cosmetic in intent.

3. *Categorization of Chemical Carcinogens*

From toxicological standpoints, attempts have been made to differentiate "potent" from "weak" chemical carcinogens. Potent carcinogens, such as aflatoxins and nitrosamines, induce a relatively high incidence of cancers in conventional experiments with restricted numbers of test animals, even at the very low levels at which they have been found in foods and elsewhere in the environment. Identification of such carcinogens has encouraged attempts to link their distribution, illustratively in foods, with local geographic patterns of cancer incidence. However, the effects of "weak carcinogens," such as atmospheric pollutants, may easily escape by conventional biological tests. Additionally, as such weak carcinogens are unlikely to be clearly implicated epidemiologically, unless large populations with sharp differentials in exposure levels are followed up for prolonged periods of time, they are likely to pose greater hazards than more obvious and more easily recognizable potent carcinogens.

From societal, legislative, and regulatory standpoints, it is important to differentiate two major categories of carcinogens on the basis of the voluntary or involuntary nature of human exposure. Firstly, there are carcinogens, such as cigarette smoke and certain drugs, to which individuals may voluntarily expose themselves, generally with some degree of cognizance as to the possible benefits besides the potential hazards of such exposure. Secondly, there is the wider range of carcinogens, such as naturally occurring or synthetic pollutants of water, air, and food (e.g., Dieldrin) to which populations are involuntarily exposed, generally in the absence of information on such exposure. The Delaney Amendment to the Federal Food Drug and Cosmetic Act expressly proscribes the deliberate addition to food of additives that have been shown to be carcinogenic in man or experimental animals. Nevertheless, no such protection has yet been developed against widespread environmental contamination of water, food, and air following deliberate or accidental discharge of occupational carcinogens or following use of carcinogenic industrial chemicals or pesticides, such as Dieldrin. Residues of pesticides or other industrial chemicals, such as polychlorinated biphenyls, in food are generally referred to as "accidental food additives," the term "accidental" qualifies their contamination of food and does not, of course, reflect the nature of their deliberate original use or intended application.

Involuntary exposure to such carcinogenic environmental pollutants generally occurs in the absence of individual or collective information; of critical independent analysis of the benefits, alleged or real, besides the potential hazards attendant on such exposure; and finally of the existence of similarly efficacious but less hazardous alternatives. Additional factors that properly influence the degree of concern on involuntary exposure to carcinogens include their persistence, environmental mobility, and biomagnification or accumulation in the food chain.

Problems in Recognition of Chemical Carcinogens

Current toxicological techniques are relatively insensitive and limited in their ability to detect carcinogens individually and in various combinations or mixtures realistically reflecting low or ambient levels and patterns of environmental exposure. Similarly, it is generally considered that epidemiological techniques are unlikely to detect weak carcinogens unless there are sharp differentials in exposure of the general population, as with cigarette smoking. For widely dispersed agents, including unintentional or accidental food additives such as Dieldrin and DDT—to which the population-at-large is generally and ubiquitously exposed in water, food, and air—human experience is unlikely to provide any meaningful indication of safety or hazard.

1. Toxicological Considerations

Chemicals should be tested for carcinogenicity at higher levels than those of general human exposure. Irrespective of the route of administration, Maximally Tolerated Doses (MTD) are recommended as the highest dose in dose-response studies for this purpose; life-long administration of these doses should not induce toxicity, as evidenced by weight loss and increased mortality. Testing at these relatively high doses is now generally regarded as essential to the attempt to reduce the gross insensitivity imposed on tests by the small size of animal groups routinely tested, such as fifty or so rats or mice per dose level per chemical, compared with the millions of humans at presumptive risk. It must be stressed that testing at high dosages does not produce "false-positive" carcinogenic results. There is no basis whatsoever for the contention that all chemicals are carcinogenic when tested at high doses. In fact, the experimental findings of carcinogenicity in synthetic industrial chemicals, such as pesticides, are unusual. To illustrate, in the recent National Cancer Institute-sponsored Bionetics study, about 140 pesticides and industrial chemicals were tested in mice of both sexes and 2 strains by continuous oral administration at MTD levels from the first week of life until sacrifice at 18 months; less than 10% of these agents were found to be carcinogenic.

To illustrate the insensitivity of conventional animal test systems, assume that man is as sensitive to a particular carcinogen as the rat or mouse. Assume further that this particular agent carries a risk of producing cancer in 1 of 10,000 humans exposed; this would result in approximately 20,000 cancers in the United States population. Then the chances of detecting this in groups of 50 rats or mice, tested at ambient human exposure levels, are very low. Indeed, samples of 10,000 rats or mice would be required to yield 1 cancer, over and above any spontaneous occurrences; for statistical significance, perhaps 30,000 rodents would be needed.

Of course, in any particular instance, humans may be less or more sensitive than rodents to the chemical in question. There is consequently no valid basis for the *a priori* quantitative prediction of the relative sensitivities of test animals and man. Illustratively, for thalidomide the lowest effective human teratogenic dose in 0.5 mg/kg/day; the corresponding values for the mouse, rat, dog, and hamster are 30, 50, 100, and 350 mg/kg/day, respectively; thus, humans are 60 times more sensitive than mice, 100 times more sensitive than rats, 200 times more sensitive than dogs, and 700 times more sensitive than hamsters. Moreover, certain aromatic amines, such as 2-naphthylamine, are potent bladder carcinogens for man, monkeys, and dogs, but not for rats, mice, guinea pigs, and rabbits. Hence, it is not feasible to attempt to predict a finite safe level for carcinogens such as 2-naphthylamine from rodent data or from derived mathematical models based on supposed safety factors. Such attempts would clearly expose humans to obvious potential carcinogenic hazards. Thus, threshold levels cannot be predicted for any individual, let alone for populations, comprised as they are of aggregates of genetically heterogeneous individuals with widely varying predispositions and susceptibilities.

Apart from the gross insensitivity of animal test systems and the impossibility of gauging human sensitivity from animal tests, ample data on interactions between individual carcinogens and a wide range of noncarcinogenic agents further confirm that it is not possible to predict safe levels of carcinogens based on an arbitrary fraction of the lowest effective animal dose in a particular experimental situation.

2. Predictive Value to Humans of Carcinogenicity Tests in Animals

There is close parallelism between the pathological and biological characteristics of the very wide range of cancers seen in both rodents and humans. The use of rodent species, particularly the mouse and the rat, has been and still is standard practice in routine carcinogenicity testing in toxicological practice, in addition to basic research. It should also be recognized that lack of carcinogenic effects in animals is generally proferred by industry as evidence supporting claims for safety and in petitions for registrations. Considerations of convenience apart, the key role of the mouse in carcinogenicity testing has been repeatedly and widely endorsed by a wide range of expert committees and authorities, as illustrated by the following statement of the National Academy of Sciences (Food Protection Committee, 1960):

> The rat and mouse have been shown to be susceptible to the carcinogenic action of a large variety of compounds and, indeed, most of our knowledge of chemical carcinogenesis is based on the use of these species. The hamster and guinea pig are relatively resistant to the carcinogenic action of several compounds which produce tumors readily in the rat and mouse. On the other hand, no known carcinogens for the hamster and guinea pig are inactive in the rat and mouse . . . . The mouse has been the most widely used animal in cancer research, and many special strains have been developed that have known spontaneous incidences of different types of tumors.

It should be stressed that many known carcinogens, including those with recognized activity in humans, were first identified in the mouse. These include coal and tobacco tar extracts, polycyclic and heterocyclic aromatic hydrocarbons, estrogens, diethylstilbestrol, vinyl chloride, bis-chloromethylether (and possibly carbon tetrachloride). Mice are sensitive to all major classes of chemical carcinogens, including aromatic amines and azobenzenes, although it must be clearly recognized that the affected target organs are not necessarily the same as in other species, including man. Furthermore, the mouse has been selected by the National Cancer Institute as a standard test animal in large-scale national testing programs on potential carcinogens and on cancer chemotherapeutic agents.

All chemicals known to produce cancer in man, with the possible exception of trivalent inorganic arsenic, which is still under experimental study, also produce cancer in experimental animals, generally in rodents. To test for the reverse, deliberate exposure of selected large human populations, with matching unexposed controls, to graded doses of chemical carcinogens would be necessary.

Prudent public health policy thus demands that any chemical agent found after appropriate testing to be carcinogenic in one or more animal species should also be presumed to be carcinogenic in man in the absence of conclusive evidence to the contrary. The majority of chemical carcinogens produce cancer in various different species, provided that they have been adequately tested. Analysis of the available literature on chemical carcinogens provides little basis to support the concept of "species-specific carcinogens," which has occasionally been adduced by some toxicologists to suggest that carcinogenicity data from a particular test species, generally the mouse, are not relevant to humans.

3. General Epidemiological Considerations

Even with well-planned and well-executed animal testing, it is likely that unexpected carcinogenic effects from toxic chemicals will occur in humans, reflecting the inherent insensitivity and limitations of toxicological test systems. Epidemiological surveys of human and animal populations may provide *post hoc* information on geographical or temporal clusters of unusual types or frequencies of carcinogenic effects after exposure to undetected or untested chemical agents in the environment. Additionally, inadequacies in surveillance systems, in the size of the population studied, and in the duration of study may limit the sensitivity of epidemiological approaches even when temporal or geographical clusters of adverse effects have developed.

Epidemiological techniques serve to detect trends of fluctuations in mortality and morbidity incidence patterns. Provided that there are clear differentials in exposure of the general population levels to specific chemicals or agents, epidemiology may then correlate particular effects with particular agents. A classic example of such a relationship is that between heavy cigarette smoking and lung cancer, although direct causality has only recently been belatedly recognized after decades of human exposure. However, these relationships appear much more difficult to establish when exposure differentials are minimal, as with widely distributed water pollutants, intentional or unintentional food additive residues, or with food additive residues consumed by the general population at similar levels. These considerations seem to apply with particular emphasis to Dieldrin, in view of its widespread environmental contamination in the United States, resulting in its almost ubiquitous presence at relatively high levels in human adipose tissue.

Epidemiological studies may provide retrospective data relating to the apparent safety of chemicals in current use. However, human experience generally affords only limited indications of safety or hazard. The reasons for this include the long latent period, in some cases many decades, elapsing between exposure to carcinogenic chemicals and the subsequent development of cancer; demonstration of causality may be obscured by mortality from competing risks during such latent periods. Illustratively, the latent period for induction of bladder cancer and of lung mesotheliomas following occupational exposure to aromatic amines and asbestos may average 18 and 30 years, respectively; in contrast, latent periods of less than one decade have been noted for carcinogens such as cutting oils and for radiogenic leukemia. Additional difficulties include isolation of the effects of any single chemical from the multitude of others to which populations are concurrently exposed and the lack of base-line data on regional and national incidences of cancer, besides other adverse human effects. These epidemiological limitations are compounded when the population at large is exposed to the carcinogen in question and when there are thus no sharp differentials in levels of human exposure.

Significance of Low Levels of Environmental Chemicals

The significance of low levels of environmental chemical carcinogens stems from the following four major classes of considerations:

a) There is an overwhelming concensus in the scientifically qualified community (see p. 80 below) that there is no valid scientific method for predicting safe levels of exposure to chemical carcinogens. This scientific consensus is also reflected in current regulatory policy, with particular reference to the Delaney Amendment.

b) A wide range of interactions and synergisms between different carcinogenic chemicals and between carcinogenic and noncarcinogenic chemicals may result in carcinogenic effects from exposures to apparently "no-effect" levels of chemical carcinogens.

c) Some chemicals are biologically potent at levels which would be generally regarded as very low, i.e., at the part per billion (ppb) level. Examples include tetrachlorodioxins, which are toxic and teratogenic at the ppb level, Botulinus toxin which is acutely toxic below the ppb level, and aflatoxin, which is carcinogenic even below the ppb level.

d) Certain chemical carcinogens, such as Dieldrin, are persistent, fat soluble, and biomagnify. This can result in the accumulation of relatively high levels in tissues of fish, following exposure to "very low levels" of the carcinogen in water, and relatively high levels in human tissues following exposure to relatively low environmental levels.

The Delaney Concept and its Possible Applicability to Carcinogens in Drinking Water

The toxicological considerations discussed above underlie the 1958 Delaney Amendment, which imposes a zero tolerance for carcinogenic food additives. Although the apparent intent of Congress was clearly to protect the consumer, irrespective of the intent of the manufacturer, it appears that legal precedent generally applies the Amendment to intentional food additives.

As HEW Secretary Flemming stated some fourteen years ago at the Delaney hearings, "Scientifically, there is no way to determine a safe level for a substance known to produce cancer in animals." The same views were recently expressed by HEW Secretary Casper Weinberger, in a letter to Rep. H. O. Staggers, Washington, D.C. (6/18/74), as follows:

> At present, the Department of Health, Education, and Welfare lacks the scientific information necessary to establish no-effect levels for carcinogenic substances in animals in general and in man in particular. In the absence of such information, we do not believe that detectable residues (of carcinogenic animal drugs) should be allowed in the food supply.

The Amendment states that "no additive shall be deemed to be safe if it is found, after tests which are appropriate for the evaluation of the safety of food additives, to induce cancer in man or animals." These conclusions were further emphasized, as follows, by an expert committee of the World Health Organization: "It is agreed that no assuredly safe level for carcinogens in human food can be determined from experimental findings at the present time."

The scientific basis of the Delaney Amendment has been consistently endorsed by qualified independent expert committees, such as the 1966 Symposium of the International Union Against Cancer and the 1970 Ad Hoc Committee Report to the Surgeon General which states:

> The principle of a zero tolerance for carcinogenic exposures should be retained in all areas of legislation presently covered by it and should be extended to cover other exposures as well. Only . . . where contamination of an environmental source by a carcinogen has been proven to be unavoidable should exception be made (and then) only after the most

TABLE 10-2. *EPA Listing of Chemical Carcinogens Found in Treated Drinking Water.*

Carcinogen	Jefferson Parish, La.	New Orleans, La.	Carville, La.	Cape Giraudeau, Mo.	St. Louis, Mo.	Jefferson City, Mo.	Kansas City, Mo.	Burlington, Iowa	Omaha, Neb.	Vicksburg, Miss.	Cincinnati, Ohio	Washington, D.C.
Aldrin		•	•	•	•	•	•	•	•	•		
Dieldrin	•	•	•	•	•	•	•	•	•	•		
DDT		•	•	•	•	•	•	•	•	•		
Benzene		•	•									•
3,4 Benzpyrene											•	
Carbon Tetrachloride	•	•	•									•

TABLE 10-3. *Chemical Carcinogens in U.S. River and Drinking Water*

Source	Reference	Chemical Carcinogens (Concentration ppb)	Comments
Kanawha River, Nitro W. V.	Rosen *et al.*, (FWPCA)* 1963	Bis-chloroethyl ether	Also bis-chloroisopropyl ether
Ohio River, at Wheeling, W. Va. and Cincinnati	Mashni and Booth (FWPCA) 1966	Bis-chloroethyl ether	Part of an analysis on petrochemicals in Ohio river basin.
Carrollton Plant, New Orleans	FWPCA Survey, 1969	Benzene; carbon tetrachloride; bis-chloroethyl ether	36 organics identified, including bis-chloropropyl and isopropyl ethers. Most were toxicologically uncharacterized.
Lower Mississippi River in Louisiana, and municipal drinking water	EPA survey, 1972	Benzene; carbon tetrachloride; chloroform	Essentially a confirmation of the 1969 FWPCA Survey.
Municipal drinking water Evansville, Ind., from Ohio river basin	Kloepfer and Fairless, (EPA) 1972	Bis-chloroethyl ether	46 organics detected, 13 of which identified by mass spectroscopy including bis-chloropropyl and isopropyl ethers.
Literature review of studies on U.S. drinking water	Tardiff and Denzer, (EPA) 1973	Benzene; carbon tetrachloride	60 organic chemicals identified most from EPA Region VI.
Drinking water from various locations	Bellar *et al.*, (EPA) 1974	Chloroform (at maximum levels from 37–150)	Wide range of tri-halogenated methanes, concentrations depending on total organic levels.
Carrollton and two Jefferson water plants	EPA, 1974	Bis-chloroethyl ether (0.07); chloroform; carbon tetrachloride; dieldrin (0.05)	66 organics identified, and highest measured concentrations defined in some instances.

*Federal Water Pollution Control Administration

TABLE 10-4. *Summary of Carcinogenicity Data on Five Selected Chemical Carcinogens Found in Drinking Water*

Carcinogen	Animal Data	Human Data
Dieldrin	Carcinogenic in multiple organs sites in mice and rats, in several studies	Occupational data inadequate to permit inferences
Carbon tetrachloride	Hepatocarcinogenic in mice, hamsters, some rat strains, and trout, by oral, parenteral (and inhalation) routes	Two reports on hepatomas following acute poisoning
Benzene	Not adequately tested	Occupational data on induction of acute and chronic myeloid leukaemias, Hodgkins (and aplastic anaemias)
Bis-chloroethyl ether	Hepatocarcinogenic in two strains of mice	No data
Chloroform	Two reports on hepatocarcinogenicity in mice	No data

extraordinary justification is presented . . . . Periodic review . . . should be made mandatory.

No such instance of "extraordinary justification," necessitating open societal consideration of the benefit-risk calculus, appears to have yet been presented or documented for any carcinogenic industrial chemical or consumer product, such as Dieldrin, the use of which appears to result in widespread environmental contamination in water, air, or food. The scientific basis of the Delaney Amendment and of the absence of threshold or "no-effect" levels for carcinogens was overwhelmingly endorsed at a workshop on this subject convened by the New York Academy of Sciences, January 15–16, 1973. Strong support was also expressed there for extending "Delaney-type" legislation to water, particularly to synthetic chemical carcinogens from point source industrial discharges.

The Federal Food, Drug and Cosmetic Act (27 U.S.C. 301 *et seq.*) defines food to include water. Water may be considered as a food additive, raw agricultural commodity, or a processed food. A chemical discharged into water becomes a "food additive" when the intended use of the chemical "results or *may reasonably be expected* to result, directly or *indirectly,* in its becoming a component or *otherwise affecting the characteristics* of" the water (emphasis added) (§207 (s); 27 U.S.C. 327 (s)).

Thus, the deliberate addition of a chemical carcinogen into drinking water, or point source industrial discharge into water systems from which drinking water is derived appears to be a violation of the Delaney Amendment. In view of the legal basis for regarding water as a food additive, the addition to foods, during processing or cooking, of carcinogen-containing drinking water also appears to be in violation of the Delaney Amendment.

Identification of Chemical Carcinogens in River and Drinking Water

There have been relatively few analytic studies on rivers and drinking water for the presence of toxic organic and carcinogenic chemicals. The impetus for such studies has generally been provided by odor and taste, rather than potential public health problems.

A 1974 EPA report by G. G. Robeck lists 6 carcinogens in treated drinking water from a wide range of cities (Table 10-2).

A more extensive listing of carcinogens found in U.S. river and drinking water is presented in Table 10-3.

These data should be interpreted in relation to the following considerations:

a) A wide range of organic pollutants and chemical carcinogens have been identified in river and municipal drinking waters from several sources. While most of these studies have been made in EPA Region VI, it is clear that contamination of drinking water is a widespread national problem.

b) In spite of the existence of data on contamination of river and drinking waters with organic pollutants and chemical carcinogens for over the last ten years, there has hitherto been little attempt to explore this problem systematically.

c) It is likely that both the numbers of chemical carcinogens, and their concentrations, now recognized in river and drinking waters are underestimates for reasons including the following facts: many organic pollutants identified in water are toxicologically uncharacterized; techniques for recovery of pollutants from water appear to have a relatively low degree of efficiency; and, only a relatively small fraction of total organic pollutants recovered from water have been chemically defined.

The quality of the carcinogenicity data on the various chemicals listed in Tables 10-2 and 10-3 is variable. For illustrative purposes, data for five selected carcinogens are summarized in Table 10-4.

Epidemiological Studies on Cancer Incidence and Rates in New Orleans

Several studies have indicated a relative excess of cancer incidence in New Orleans, as compared with the national average or with rates in other metropolitan areas:

1. *National Cancer Institute (NCI) Cancer Morbidity Studies* (Dorn and Cutler, 1948–1949)

The incidence of cancer in New Orleans was compared with nine other major cities. New Orleans was shown to have the second highest incidence of all cancers and a higher incidence of virtually all types of cancer, except central nervous system and skin.

2. *U.S. PHS Report "Comparative Mortality among Metropolitan Areas of the U.S."* (1949–1951)

Of 163 metropolitan areas studies, New Orleans was shown to have the third and sixth highest mortalities for kidney and bladder cancers, respectively. New Orleans was also shown to have 2.6 and 3.5-fold excesses over the national averages for tongue and buccal cancers, respectively.

3. *NCI Studies on U.S. Cancer Mortality by County* (1950–1969)

New Orleans was shown to be the third highest county for cancer mortality in the U.S.A. and to have a 32% higher mortality than the national average. Nine of the 45 counties in the U.S.A. with the highest mortality rates were in Louisiana.

4. *Environmental Defense Fund (EDF) Preliminary Regression Analysis of NCI County Data in Louisiana* (1974)

Using NCI county cancer mortality data, EDF recently developed preliminary regression analyses on mortality data by county, attempting to standardize for factors including ethnicity, socio-economic level, occupation and source of drinking water (water from the lower Mississippi or from local ground water). The techniques used in these analyses were based on those which have been previously used to analyze cancer mortality from air pollution.[2] The results of this study appear to indicate that drinking water from the Mississippi could be related to an enhanced incidence of total cancers and of bladder and gastro-intestinal tract cancers.

Carcinogenic Hazards from New Orleans Drinking Water

Evidence for the recognition of carcinogenic hazards from New Orleans drinking water rests on the repeated demonstration of chemical carcinogens in the drinking water, coupled with repeated epidemiological demonstrations of relatively high cancer rates in New Orleans. While such evidence cannot be accepted as conclusively establishing causality, it nevertheless raises valid presumptions of causality. It should also be recognized that the evidence relating contaminated drinking water to excess cancer rates is qualitatively analogous to evidence relating urban community air pollution to lung cancer. There is general acceptance of the fact that the "urban excess" of lung cancer in urban and rural communities in populations matched for smoking is due to air pollutants, which have been shown to contain a wide range of chemical carcinogens (see above pp. 74-75).

These considerations should prompt the immediate implementation of appropriate regulatory actions, at both federal and state levels, to prevent or minimize further human exposure to chemical carcinogens in drinking water. As immediate priorities, these actions should include control of point source industrial discharge of carcinogens and the use of activated carbon filtration in water treatment systems. Prudent public health policies dictate that such actions be no longer postponed pending further monitoring or research.

Notes

1. This chapter is based on testimony before the House Committee on Health and Welfare, Louisiana House of Representatives, 2/21/75. The author refers the interested reader to his article "Environmental Determinants of Human Cancer," *Cancer Research* 34 (1975): 2425. This article contains eighty-five references, many of which are pertinent to the issues discussed in this chapter.

2. L. B. Lave and E. P. Seskin, "Air Pollution and Human Health," *Science* 169 (1970): 723.

11

Groundwater: Fact, Fiction, and Future

JAY H. LEHR

National Water Well Association

Columbus, Ohio

WAYNE A. PETTYJOHN

Department of Geology

Ohio State University

Ground Water—Fact and Fiction

Contrary to popular belief, ground water is neither mysterious nor occult, it does not occur in underground lakes, rivers or veins, and spring water is not synonymous with purity. These are but a few of the common misconceptions concerning ground water. Since some misconceptions of ground water are widely held, they should be examined before proceeding with a description of ground water behavior and attributes.

Misconception: Ground water occurs in underground lakes, rivers, and veins.
Fact: Generally, bodies of ground water bear no resemblance to surface water bodies.

In the vast majority of cases, ground water occurs in rock formations that have a sufficient number of interconnected openings (permeable material) for the water to pass through them. Not uncommonly, these aquifers (water-bearing rocks) are connected for hundreds or even thousands of square miles. Along many river valleys, the floodplain deposits consist of permeable sand and gravel. In cases such as these, the ground water commonly flows in the same direction as the surface stream, and there may be a concentrated underground flow adjacent to the stream that is different from the underground flow in nearby areas. This does not mean that the ground water flowing in the stream-side deposits is in any manner similar to water in a surface stream. During the last Ice Age, when glaciers advanced across the northern part of the United States, many of the preexisting stream channels in these areas were filled, and the streams were forced to develop other courses.

Many of these now-buried river valleys contain vast quantities of ground water, but the enclosed water must not be likened to an underground river. This erroneous concept, however, is based on fact, because in areas where limestone forms major aquifers, the water may slowly flow in large underground openings, such as caves and solution channels. In one sense these large openings are similar to underground rivers and lakes.

Misconception: Ground water is mysterious and occult.
Fact: Natural laws control the occurrence and movement of ground water and therefore it is predictable.

In past centuries, before the development of scientific techniques of ground water hydrology, the natural laws controlling water movement were unknown. This led to the concept, preserved in case law, that the occurrence and movement of water in the ground is mysterious and occult—i.e., that the principles of its behavior cannot be known. In fact, using well-established natural laws, the quantity and quality of ground water are predictable both in space and in time.

Misconception: Water rushes so rapidly under ground that its presence can be detected by listening for the noise.
Fact: In most cases ground water flows only a few tens of feet per year.

Some people believe they can detect the presence of large quantities of water underground merely by placing their ear to the earth and listening for the rushing water. This is not possible because ground water moves very slowly through the earth. In fact, in most cases it moves only a few feet per year. Locally, however, it may flow very rapidly in limestone terrain characterized by large openings, such as caves.

In water pollution problems, ground water velocity is of particular importance. Due to its slow rate of movement, an area once contaminated may remain unuseable for years. It is primarily for this reason that disposal of wastes on and in the ground must be closely regulated.

Misconception: Ground water when removed from the earth is never returned.
Fact: Ground water is a renewable resource.

In most parts of the country, ground water when removed from the ground is constantly replaced. Thus, ground water is a renewable resource. Water is constantly seeping into the ground and is continually discharging into streams, lakes, and wells. Areas where water seeps into the ground are called recharge areas; this includes virtually the entire land surface. Some types of soil, however, will accept water much more rapidly than others. For example, the rate of infiltration (the rate at which water seeps into the ground) through sandy soil will be much greater than through a heavy clay soil. Because of this phenomenon, relatively high rates of recharge (infiltration) occur along river flood plains and in sandy or gravelly areas. Areas with high rates of recharge must be carefully managed for two major reasons. First, they provide water to underlying ground water reservoirs (aquifers) and, secondly, water-soluble waste products stored in these areas will also infiltrate and contaminate the underground supply. Consequently, areas of high recharge must be protected in order to maintain the quantity of water in storage and to protect its quality.

Misconception: Ground water migrates thousands of miles through the earth.
Fact: Most ground water withdrawn is replaced in the near vicinity.

Much of the water in a ground water reservoir infiltrates within a radius of a few tens of miles from where it is found. It has not traveled in the ground for thousands or even hundreds of miles as some people tend to believe. There are no huge underground rivers transporting great volumes of high purity water from Canada to Minnesota, Virginia to Florida, or from the Rocky Mountains to Iowa and other Great Plains states.

Misconception: Ground water is not a significant source of supply.
Fact: The amount of ground water in storage dwarfs our present surface supply.

Ground water is commonly considered an insignificant source of supply. The fact is, however, that the quantity of water in underground storage is two to three thousand times larger than the amount in all the lakes, streams, and rivers combined. At the same time, it is cold, of a nearly constant temperature, free of sediment, and of high quality. The major feature of ground water is its omnipresence—it provides a reliable and economical water supply for all kinds of activities for which surface water supplies would uneconomical or even infeasible.

In 1970 the United States was using more than 370 billion gallons of water per day. Of this amount, about 20% was from ground water, and here lies the greatest potential for future development. About a third of all public supplies and about 96% of all rural domestic supplies are derived from wells. About 25% of the water used for irrigation is ground water. Were it not for these huge underground reservoirs, irrigation could not long exist in the arid and semiarid regions of the United States. Ground water makes possible agriculture and industry alike because it occurs at the point of use and does not require transportation over long distances.

Misconception: There is no relationship between ground water and surface water.
Fact: Ground water provides most of the flow of streams and lakes, and swamps are merely reflections of the water table.

It is commonly believed that ground water and surface water represent isolated systems. Consider, however, a stream in late summer. Although it may not have rained for several days or possibly even weeks, there may still be a considerable flow in the stream. Obviously, this water could not have been derived from the surface runoff of rainfall. In large part the stream represents water that has flowed from the ground into the stream channel. Hence, the low flow of a stream may be derived entirely from ground water discharge.

During low-flow conditions, the chemical quality of the stream is similar to that in adjacent aquifers, but during periods of surface runoff and precipitation, the stream quality is considerably different. In many places, the quality of surface water deteriorates during the late summer and fall because adjacent ground water reservoirs are contaminated and these contaminated ground waters provide the stream flow. It is evident that one needs to consider not only waste disposal directly into a stream but also the disposal of water-soluble material on the land surface, particularly in recharge areas. In other words, there is a close interrelationship between surface water and ground water, and one cannot be managed without consideration of the other.

Misconception: The water table is falling throughout the country.
Fact: Although in a few areas the water table has declined significantly, in most places the water table rises and falls with the seasons.

Not uncommonly we read in news media that "the water table is falling." During periods of drought, when water demands increase substantially and withdrawals may exceed the rate of natural recharge, the water table may decline. In most cases the decline is temporary and the water level recovers once the drought is over. A similar problem of declining water levels exists in particular areas, such as a few municipalities and large industrial complexes, and in extensive irrigation districts. This occurs because the rate of ground water pumping over long periods may far exceed the natural rate of replenishment (recharge) within the area of supply. In areas of substantial water level decline, a few shallow wells may go dry or provide only limited quantities of water. Several techniques have been developed to reverse these local water level declines, but what is needed is comprehensive ground water management programs. The water table indeed is falling in a few areas of substantial overdraft, but these areas are generally very small and unmanaged. In the vast regions of the country the water level merely rises and falls with the seasons.

Misconception: The water level in a well remains constant.
Fact: The water level must decline in the vicinity of a pumping well.

The proud owner of a new well may be so enthusiastic that he believes there is no water level lowering or drawdown when the pump is on. However, because water must be forced to flow into the well, there must be a water level decline in the vicinity of a pumping well. In very permeable formations, such as limestone caverns or sand and gravel deposits, the amount of drawdown may be very slight. With increasing yield, however, the drawdown increases and around high-yield wells, water level lowering may extend outward for miles.

Misconception: Spring water is synonymous with exceptional quality.
Fact: Springs are points where ground water is naturally discharging, but they are easily contaminated.

Various product advertisements assert that spring water has exceptional taste, purity, and perhaps medicinal properties. They may also imply that well water is of unacceptable quality. A spring, however, is merely a point where ground water is discharging at the land surface. It has nearly the same quality as nearby wells and streams. On the other hand, springs are easily contaminated, but the quality of well water is predictable and has nearly uniform temperature and chemical properties. Well water is biologically pure except when contaminated by sewage waste, the sources of which lie in the near vicinity.

Misconception: All well water is naturally of drinkable quality.
Fact: The mineralization of ground water generally increases with depth and eventually a point is reached where it is no longer potable.

Not all ground water is drinkable. Lying at depths ranging from a few tens to several hundred feet is the fresh/saltwater interface. Below this interface, the mineral content of the water increases substantially, and at greater depths it may be so mineralized that it is considered a brine. Wells whose total depth approaches the fresh/saltwater interface eventually suffer from deteriorating quality due to the upward migration of salt water as fresh water is removed. In coastal areas, overpumping of fresh water may cause migration of sea water into the aquifer causing its quality to deteriorate. Adequate water management programs, however, can substantially reduce problems caused by migrating highly mineralized water.

Misconception: Since ground water cannot be seen, nothing is happening to it.
Fact: We do not know the extent of ground water contamination, but from available information we must assume that the threat of widespread contamination of ground water is substantial.

If an individual laid a raw sewage discharge line into a lake, it is quite unlikely that he would withdraw his drinking water from a surface intake only twenty-five feet away. The potential health problem would be easily recognized. Since ground water cannot be seen, however, the situation is considerably different. A cess pool or septic tank drain could lie within twenty-five feet of a well, but the potential problem might not be recognized.

J. F. Karuvian examined the petroleum, pulp and paper, and primary metals industries in the United States in an attempt to estimate the effect of these activities on ground water pollution.[1] These industries produce a vast quantity of wastewater, much of which is stored in unlined basins and lagoons. The liquids leak from the containment structures and contaminate ground water. It was estimated that the total leakage during 1973 from these structures was 192,815 acre-feet from primary metals industries, 76,335 acre-feet from petroleum refining, and nearly

134,000 acre-feet from the pulp and paper industry. Thus, the total quantity of leakage from these three industries alone amounted to more than 403,000 acre-feet of contaminated water. (An acre-foot is approximately equal to 326,000 gallons.) The contamination is occurring underground, is hidden from view, and is not easily recognized. Nonetheless, the problem is significant.

Ground Water—Future

Let us examine a few of the ways that ground water becomes contaminated and then consider potential ways in which ground-water contamination can be controlled or at least decreased.

Ground-water pollution is caused by man's activities—excessive pumping from coastal wells, irrigation, manufacturing, mining, and many others, but most importantly ground-water pollution is caused by poor or inadequate waste-disposal practices.

Waste materials are commonly disposed of by placing them in streams, on the land surface, or in the ground.

Wells near streams cause the water to flow from the stream into the ground, and thence to the well. If a stream is grossly polluted, this undesirable water may contaminate nearby ground water and wells—but it may take several years for the polluted water to show up at the well, and by that time perhaps the entire area between the stream and the well is contaminated. This type of ground-water pollution can be reduced by prohibiting the dumping of noxious chemicals and other wastes into streams. Strict enforcement of surface-water quality standards, which already exist, will go a long way towards a solution of this problem.

Man commonly spreads wastes on the land surface and stockpiles raw materials or finished products outside. Many of these substances contain chemicals that dissolve in water—in rain or surface runoff. Once dissolved, the highly mineralized or polluted water may slowly sink into the soil leading to ground-water pollution.

Examples include dumps, manure piles, mineral ores, stockpiles of salt for snow removal, and hundreds of others. Animal feedlots have contaminated ground water with excess amounts of nitrate—a definite health hazard. Salting of roads in winter and salt stockpiles have caused well supplies to taste briny. And ground-water contamination by leachate from dumps has caused foul tastes and odors.

Unquestionably the most serious and widespread causes of ground-water pollution is related to the storage or disposal of wastes in the ground, primarily by septic tanks and secondarily by holding ponds and lagoons. The wastes filter down to the water table and may later be withdrawn from a well. The same is true with outdoor privies and cesspools, waste disposal in excavations, and sanitary landfills. More insidious, and in some ways more dangerous, is the leakage through ruptured and corroded storage tanks and transmission lines of materials such as gasoline, fuel oil, and radioactive wastes.

Wells contaminated by privies, cesspools, or septic tank wastes can lead to waterborne diseases and epidemics at the worst, or the water may have bad tastes and odors and even produce soap suds! Toxic compounds such as chromium and various organic chemicals have appeared in wells because the ground water was contaminated by wastes from holding ponds or lagoons. The disposal of oil-field brines in holding ponds or evaporation pits has led to the pollution of literally thousands of sites with salty water. The contaminated water may be so corrosive that plumbing in pumps literally fall apart, and if the water is used to irrigate a lawn or garden, all of the vegetation dies.

The effects of drinking water contaminated by radioactive wastes are very subtle and may require long periods of time to become evident, but leakage of gasoline into the ground has led to taste and odor problems which are minor compared to the destruction brought about by explosions and fires. In a few cases, gasoline floating on the water table has leaked into base-ments where the fumes were ignited by pilot lights or sparks.

Many of these ground-water pollution problems cannot be easily solved, but controls can be developed so that future problems are minimized. Domestic waste disposal will be less of a problem if state agencies develop strict guidelines on septic tank spacing and prohibit the use of outdoor privies and cesspools. Stockpiles of soluble materials should be covered, or drainage ditches constructed around them, so that runoff can be collected and treated. Industrial and municipal holding ponds and lagoons should, in most situations, be lined with material that will halt the slow infiltration to the water table. The ponds should be protected by dikes and alternate containment facilities, and before a permit is issued, methods should be available for the control and cleanup of ground-water pollution in the event that it should occur. Performance bonds and stiff fines coupled with stringent permit procedures would go a long way towards proper waste disposal management.

There are other causes of ground-water pollution:

a) PROBLEM—The leakage of highly mineralized or polluted water through unplugged abandoned wells and exploratory holes or through improperly constructed wells.
CONTROL—Adequate water well construction standards. State requirement that all abandoned wells and exploration wells be properly plugged.

b) PROBLEM—Drainage wells and collection sumps are designed to drain swampy or water-logged land so that the water flows deeper into the ground or to collect spilled materials. Much of the drainage water and spilled substances are highly mineralized and sink from the wells and sumps into water-bearing strata.
CONTROL—prohibit the use of drainage wells and collection sumps except for domestic use, or require a permit procedure under strict guidelines.

c) PROBLEM—In some instances, water supplies are contaminated by accidental spills, such as truck or train wrecks.
CONTROL—Obviously, it is not possible to prohibit accidents but it is possible to consider if the materials are being handled in the safest manner possible. Furthermore, it is also possible to develop quick cleanup techniques. This should be done.

Summary

There are many sources of ground-water contamination. In order to visualize whether a certain process or activity may cause ground-water pollution, one has to ask only one question—does the product or waste material contain water-soluble substances? If it does, then there is a good chance that ground-water pollution may occur if adequate precautions are not followed.

Notes

1. J. F. Karuvian, "Polluted Ground Water: Estimating the Effects of Man's Activities," Series EPA-680/4-74-002 (Washington, D.C.: U.S. Government Printing Office, 1974).

12

Comments on the History and Economics of Micropollutants in Drinking Water

WALTER A. LYON

Director, Bureau of Water Quality Management
Department of Environmental Resources
Commonwealth of Pennsylvania

Stimulated by the recent municipal water supply findings, in New Orleans and elsewhere, the resultant public concern about the relationship between health and micropollutants in water tends to overshadow the fact that environmental engineers and scientists have for quite some time expressed considerable concern about this problem at many levels.

As early as the 1950s, Mark Hollis, who was then the Chief of the Public Health Service, Sanitary Engineering Program, expressed in public statements a deep concern with organic micropollutants in drinking water. To audiences, Hollis often exhibited flasks containing oil and tar-like substances which had been extracted from carbon filters through which drinking water from Cincinnati and other major cities had been passed.

In 1951 a publication entitled, *Environment and Health,* issued by the Public Health Service, touched on this problem, as did Abel Wolman in a 1960 Conference on the Physiologic Aspects of Water Quality. A group of sanitary engineers, headed by Professor J. E. McKee of California Institute of Technology published in 1961 a list of one hundred problems in environmental health which included the problem of micropollutants.

In 1962 Rachel Carson published her first book, *Silent Spring,* and in it she raised many questions about the exposure of the American people to chemicals "without their knowledge and consent" and raised the question as to whether or not our forefathers, in drafting our Constitution, were able to anticipate this kind of technologic dilemma. During the same year, the "Gross" Committee, a Committee on Environmental Health Problems created by the Department of Health, Education and Welfare, expressed its concerns about this issue to Surgeon General of the Public Health Service, Luther Terry.

In 1966, the so-called "Daddario Committee," a Subcommittee on Science and Research of the House Committee on Science and Astronautics, expressed its concern regarding this issue. In the same year, the "Spilhaus Committee" (Committee on Waste Management and Control) of the National Academy of Sciences and the National Research Council expressed the same concern. The "Linton Committee," a Task Force on Environmental Health and Related Problems, reported the next year to then Secretary of HEW, John Gardner, and recommended that in a few years no new chemicals should be allowed to be used unless they had first been tested. That same year, a Task Force on Environmental Health of the National Commission on Community Health Services, headed by Dr. Gaylord Anderson, in reporting to Marian Folsom, then Secretary of HEW, included consideration of this problem in its recommendations. In 1968, there was a Symposium on Human Ecology which touched on this problem, and in 1970 this author presented a paper on "Water and Health, Are We Concerned Enough?"

So there have been many expressions of concern about micropollutants and health over the last twenty-five years, but only during the past year has there been much serious public attention given to this problem. Broad-based public attention is essential if the nation is going to provide more adequate resources, designed to deal with this problem, for research and programming at the federal and state level.

The Federal Drinking Water Act provides, for fiscal year 1975, $7.5 million to the States to deal with this issue. We estimate that the nation would need $36.8 million in order to have states begin to address themselves to all the issues presented in the Drinking Water Act—approximately five times the amount that has been provided by Congress. This, of course, is the key reason why we today face such serious obstacles in dealing with these new concerns about drinking water.

Around the turn of the century, our main concern with water related to the need for preventing the spread of water-borne communicable diseases, such as typhoid and paratyphoid, caused by bacteria. We have essentially solved that problem by chlorination and treatment of drinking water, although chlorine now emerges as both a blessing and a potential micropollutant; but we have not adequately addressed ourselves to the issues of long-term exposure to low levels of chemicals—particularly now that we are more aware of the presence of these chemicals because of the significant advances that have been made in water chemistry analyses.

If we really intend to deal with this problem, we must be willing to face up to the economic issues that confront us. Let us look at a few of them.

During a recent taste and odor episode in Pittsburgh, there was much discussion concerning the urgency of having the West Penn Water Company provide activated carbon treatment and regeneration in its system. The Company has agreed to do this at an approximate cost of $2.50 per year per customer—certainly not a major expenditure. West Penn and its customers are fortunate because the system serves well over one-half million customers. When we look at a smaller waterworks, the economics of scale raises the customer's cost for activated carbon treatment considerably. A waterworks serving forty thousand customers would have to impose a cost of $12 per year per customer to accomplish this same end. In a small community, a waterworks serving thirteen hundred customers would have to charge $90 per year per customer, and in even smaller water systems serving five hundred customers, the yearly cost would be $300 per year per customer. These estimated costs are in addition to the present cost of water delivery and treatment, which are rising annually due to rising discount and inflation rates. Nearly half of the thirteen hundred public water supplies in Pennsylvania serve less than eight hundred customers and would hardly be able to provide this type of treatment without substantial financial assistance.

One other economic concern relates to the opportunity cost of preventing illness by placing a more preventative medical function on the role of drinking water. At the moment, the carcinogens which are found in our water supplies have directed our attention to cancer prevention. However, there is considerable evidence to show that there is a significant relationship between cardiovascular disease mortality and hardness of drinking water. While the actual casuality has not been determined, it appears that the relationship is significant. We also know that certain viral diseases can be distributed through drinking water and that present conventional treatment methods give us little protection against them. It is, therefore, important that we not only take a broad view about the implications of drinking water and health, but that we also mount a major national research effort in order to give us better and more solid answers about the relationship between water and health.

In summary, concern about micropollutants in drinking water which do not directly produce communicable diseases is at least a quarter of a century old. Thanks to the Environmental Defense Fund and other concerned citizens' organizations, this matter has now been made one of national concern. This is as it should be if the nation is going to concentrate on solutions of this problem. My plea for research should not be misinterpreted to mean that we should not act now. Not every chemical is a carcinogen, but some are. If we find them we should do what we can to reduce or eliminate them from our drinking water and from sources of supply. My main concern is that we have no illusions about the impact of the new Federal Drinking Water Act on the health protection of many of our public water supplies. Unless we fund research and government and public works programs much more adequately, the Act will turn out to be an empty gesture as far as many Americans are concerned.

13

Sudanese Native Methods for the Purification of Nile Water During the Flood Season

SAMIA AL AZHARIA JAHN

Department of Physiology
University of Khartoum
Sudan

The flood season of the Nile is the result of abundant summer rains in the highlands of Ethiopia, around Lake Tana, where the Blue Nile and its main tributaries, the Dinder and the Rahad, originate. The main tributary of the Nile, the Atbara River, also comes from this area. When the water rushes down from the Ethiopian plateau, it carries off great amounts of solids due to the heavy erosion of soils and rocks.

More solid material is added to the water of the Blue Nile by drainage from the Ingassana Hills, located on the western side of the river, through many water-filled wadis. Although the frequency of rainfall decreases in the North, sudden, very heavy rains can occur in the Northern Province, which, in a few hours, can transform a dry valley into a vast stream with dangerous whirlpools. These water masses block the roads and exert a great erosive power on the bed and shores of the wadi, carrying even more suspended solids to the river.

Where there are lake-like basins on the Nile, such as above the dams of Roseires and Sennar, conditions are favorable for a slow, spontaneous sedimentation of the suspended matters. Thus, the amount and composition of the suspended solids undergo many variations along the course of the Blue Nile and the River Nile.

The flood season begins in late June or July, reaches its maximum in August, and lasts until the end of September or beginning of October. The highest water level does not coincide with the highest amount of suspended solids. Table 13-1 shows the maximum concentrations in total solids during different flood seasons and at different points in the riparian Sudan.

TABLE 13-1. *Fluctuation in the Maximum Amount of Suspended Solids During Flood Seasons.*

Site of measurement	Year	Day in August	Suspended matter in mg/l	Reference
Blue Nile, *Sennar* (before construction of Roseires dam)	1959	26	4875	District Engineer, Egyptian Irrigation Dept. see El Badri, 1972
	1960	10	4055	
	1961	24	3800	
Blue Nile, *Khartoum*	1905	12	1166	Beam, W., 1906
	1967	2	3850	El Badri, 1972
	1968	17	1888	
	1969	4	3640	
River Nile, *Dongola*	1971	22	7832	Communication from Ministry of Irrigation, Hydrology Dept., Khartoum, 1975 (Measurements from Irrigation Dept. Dongola)
	1974	22	6823*	

*In 1974 the Blue Nile at Khartoum contained 2290 mg/l suspended solids on 4 August. According to the monthly examination carried out in the Chemical Laboratories of the Ministry of Health, the annual maximum was probably still higher.

According to the International Standards for Drinking Water compiled by the World Health Organization (1963), the total solids should have an acceptable maximum of 500 mg/l, and an allowable maximum of 1500 mg/l.[1] The maximum concentration of solid material in the River Nile is, therefore, 4.5 to 5 times higher than the allowable maximum. Farmers, travelers, and families in distant villages still drink untreated water from the river or from an adjacent irrigation canal. Towns such as Damazin, Wad Medani, Khartoum, Omdurman, Shendi, Atbara, parts of Karima, and Dongola are supplied with clean drinking water. In some towns—for example, Ed Damer and Berber—many people use wells. The majority of villages under the Gezira Board Scheme are provided with good well water. In other parts of the Nile Valley, however, well water, if present, has a very bitter taste and is regarded as hard water; therefore, Nile water is preferred. People living in rural communities, with schools and small dispensaries, are now showing increased interest in improving the quality of their drinking water.

I personally experienced problems with drinking water during the flood season of 1973, in parts of Karima (located near the River Nile), and in Nuri (where the water is taken only from irrigation canals). Subsequently, I began to develop a research program on native attempts of water purification. The investigation was started during the flood season of 1974, in collaboration with Sudanese chemists, microbiologists, and botanists, and a German expert on soil research. During five study trips to Gezira, the Northern Province, and the Blue Nile Province, I collected information and obtained samples of untreated water from various sources discussed below.

Water Purification with Clay Soil

The Nile shores show great variation in geological origin and mineral composition.[2] People have found empirically that a type of clay soil from certain sites along the river flocculates the high concentration of solid material in the Nile water of the flood season. This type of soil is called *rauwāq* (clarifier) in Gezira and the Northern Province. In the Blue Nile Province at Roseires, it is known as *turāb al terwīq* (clarifying earth). A small amount of water is added to a handful of rauwāq, producing a kind of dough or thick broth which is stirred well for at least ten minutes and later poured on the turbid Nile water kept in large storage jars. Water is usually stored in the jars for no longer than twenty to twenty-four hours. Nile water from the flood season left in containers without rauwāq, becomes clean by spontaneous sedimentation in fifteen to twenty days.

The rauwāq method for water purification has spread from Gezira to other parts of the riparian Sudan. In Wad al Said, a village near the sugar factory of Guned, rauwāq has been used for more than fifty years. Previously, merchants from the Northern Province bought rauwāq from the next town, Rufa'a, until a good quality rauwāq was discovered ten to twelve years ago in their own area, near Karima, in Uqda. In the area between Sennar and Roseires, rauwāq is said to have been in use for only twenty years.

According to the Roentgenographic analysis by Dr. Heinrich Rösch, Federal Geological Survey (Bundesanstalt für Geowissenschaften und Rohstoffe), Hannover, the main components of rauwāq are montmorillonite, also called bentonite, and quartz (Table 13-2). Thus, it seems that the flocculating effect of rauwāq is due to the presence of montmorillonite, which has been utilized to some extent to flocculate suspended solids in water.[3] A detailed analysis of suspended matter in the Blue Nile at Khartoum showed that suspended matter in water samples taken at 0.5 meters from the river bed contained 75% quartz and no montmorillonite, but samples taken at 0.5 meters from the water surface contained 70% montmorillonite and no quartz.[4] We have tried to determine the optimal concentration for the effect of different types of rauwāq. The laboratory investigation was conducted by Mohammed Osman al Obeid (analyst in the Chemical Laboratories of the Ministry of Health), under the guidance of Dr. Joseph Zaki Boutros of the same laboratories.

Rauwāq samples collected at different times from Gezira and Uqda were not of the same quality. The lowest optimal concentration for rauwāq from Gezira tested on water from the Blue Nile at Khartoum, containing 2040 mg/l suspended solids (8/29/74), was 1.5 g/l; the corresponding lowest optimal concentration for rauwāq from Uqda was 2 g/l. Concentrations above and below the optimal level leave turbid water.

Water Purification with Plants

The majority of plants used for flocculation of suspended solids in the Nile water, or in water

TABLE 13-2. *Roentgenographic Analysis of Rauwāq Samples.*

Type of Rauwāq	Main component	Additional component	Traces
Uqda (Karima) Northern Province	Montmorillonite (= Bentonite) Quartz	Feldspar Calcite Dolomite(?) Chlorite	Amphibole Mica Kaolinite
Wad al Said Nile shore (Gezira)	Montmorillonite Quartz	Feldspar	Chlorite Dolomite Mica Palygorskite(?)
Wad al Said khūr (= Wadi or pond carrying water only during the rain season)	Montmorillonite Quartz	Feldspar Calcite	Dolomite Chlorite Kaolinite Mica

TABLE 13-3. *Indigeneous Plants Used for Flocculation of Suspended Matters.*

Family	Species	Arabic name in the Sudan	English name	Part of the plant
		Monocotyledons		
Palmae	Phoenix dactylifera L.	tamr (nakhla)	date-palm	pounded datestones
	Hyphaene thebaica (L.) Mart.	dōm	dom-palm	epicarp—mainly for cleaning of leather water bags (qirba)
Cyperaceae	Cyperus rotundus L.	se'da or tīb al khadim	sedge-plants	pounded rhizomes
Gramineae	Cymbopogon nervatus (Hochst.) Chiov.	nāl		stalkes and leaves
	Sorghum sp.	dhura	millet	leaves or as surface layer of flat bread dough (rauwāq al ajīn)
		Dicotyledons		
Capparidaceae	Caparis decidua (Forsk.) Edgew.	tundub		crushed branches
	Boscia senegalensis (Pers.) Lam. ex Poir	shagara al mukhēt or Shagara Alkursān		crushed branches; or branches twirled in water for 20 minutes or more
Moringaceae	Moringa oleifera Lam.	shagara al rauwāq or figl al kilāb	horseradish tree	pounded seeds
	Moringa peregrina (Forsk.) Fiori	shagara al rauwāq		pounded seeds
Elatineae	Bergia suffruticosa (Del.) Fenzl.	rimit		fresh leaves
Ficoidaceae	Trianthema salsoloides Fenzl. ex Oliv.	dāle		stalks and leaves
Tamaricaceae	Tamarix nilotica (Ehrenb.) Bunge	shagara al tarfa "mūr" in the Nubian "Fadidja" of Wadi Halfa area		branches
Cucurbitaceae	Cucurbita sp.	qara'	gourd	pounded dry seeds (libb)
Bombacaceae	Adansonia digitata L.	tibaldi	baobab tree	pounded seeds (qonqulīz)
Malvaceae	Abelmoschus esculentus (L.) Moench. (before: Hibiscus esculentus)	bāmia	okra (ladies fingers)	slices of the fresh fruits

TABLE 13-3. (cont.)

Family	Species	Arabic name in the Sudan	English name	Part of the plant
Papiliona-ceae	Faba fona Medic. (before: Vicia faba)	fūl masri*	broad bean	pounded dry seeds
	Pisum sativum L.	bisila*	garden pea	pounded dry seeds
	Lupinus termis Korsk.	turmus*	lupin	pounded dry seeds
	Lablab niger Medic. (before: Dolichos lablab)	lūbia 'afin		pounded dry seeds
	Arachis hypogea L.	fūl sudani	ground or monkey nut	pounded dry seeds
	Trigonella foenum graecum L.	helba*	fenugreek	pounded dry seeds
Salicaceae	Salix subserrata Willd.	safsāf	Nile willow	leaves
Rutaceae	Citrus auranti-folia (Christm.) Swingle	shagara al limūn*	lime	leaves
Balanitaceae	Balanites aegypti-aca Del.	shagara al higlig or shagara al lalūb		pounded bark
		Imported plants		
Acanthaceae	Hypoestes cancel-lata Nees.	mahlab		pounded seeds
Rosaceae	Prunus amygdalus	lūz	almond	pounded stones

*cultivated

from irrigation canals during the flood season, are indigenous plants. Table 13-3 shows plants known to date. Two specialists in the taxonomy of Sudanese plants, Dr. Hassan Mustafa Hassan and his research assistant, Dr. Zainab Abdel Halim, from the Department of Botany, University of Khartoum, assisted in the identification of the plants according to their vernacular names and determined the identity of some dried specimens collected during my travels.

Geographical distribution of the plants

The choice of plants in the valley of the Blue Nile and the River Nile depends on the local occurrence of indigenous plants and on the characteristics of soil and climate for cultivated plants. At present, the use of clay soil (rauwāq) is more popular along the rivers of the Sudan than the use of plant material. The names I collected concern both plants widely known as "clarifiers" ten to thirty years ago and those which are still in use today. The following details of geographical distributions are not complete, but do give a general outline.

In the area of Wadi-Halfa, *Tamarix nilotica* was known in villages close to the Nile, since the tree grows on islands in the Nile. In the villages far from the Nile, purification of water is done only with cultivated plants of this area, particularly with broad beans and peas. Around Dongola, people have tried date stones, Moringa seeds, millet, broad beans, and *Lablab niger.*

In the area of Karima and Merowi, including the villages Nuri and Al Qureir, the water supply is drawn from Nile irrigation canals. Here, millet and *Cyperus rotundus* are known. Elderly people in Merowi remember gathering moringa seeds during the flood season from a private garden. Among the cultivated plants, broad beans are still used in many houses and some families also take fenugreek and *Lablab niger.*

Between Shendi and Ed Damer, the following plants are known: millet, *Cyperus rotundus, Hyphaene thebaica,* and moringa. Okra slices were mainly used for clarification of washing water. Among cultivated plants, broad beans and lupin are used. In earlier periods, the people of Omdurman used *Balanites aegyptiaca.* On the Nile island Tuti, situated between Khartoum, Khartoum North, and Omdurman, the turbid Nile water is still treated with lime leaves.

Moringa oleifera is well known by older people around Rufa'a (Gezira). In villages far from the river, the water is also treated with *Trianthema salsoloides* and *Cymbopogon nervatus.* The most important cultivated plants are ground nuts.

In the area of Sennar, people have tried *Bergia suffruticosa, Tamarix nilotica, Caparis decidua,* and *Salix subserrata.* The most important indigenous plant is still *Boscia senegalensis,* and the most important cultivated plant is ground nuts. Around Roseires, people use *Boscia senegalensis, Adansonia digitata, Moringa peregrina,* and the cultivated ground nut.

In the past, some wealthy families treated the Nile water during the flood season with pounded almond stones or pounded mahlab seeds *(Hypoestes cancellata).* Mahlab is an important ingredient in several famous homemade perfumes and cosmetics in the Sudan. Dried gourd seeds are mainly known in the western Sudan but seldom used in the Nile valley.

Controversy remains whether *Moringa oleifera* should be considered an indigeneous or an imported plant. In *Sudanese Flora,* by F. W. Andrews, *Moringa oleifera* is described as a tree introduced from India.[5] Sudanese botanists who have travelled in Darfur claim, however, that the tree is at least indigeneous to southern Darfur. I have not yet been able to determine if the flocculating properties of moringa seeds are known there or not. On the other hand, it is quite likely that *Moringa oleifera* trees found in public gardens in Wad Medani, Shendi, or in private gardens anywhere in Gezira or the Northern Province were planted by foreign government officials who brought the seeds from abroad.

Moringa peregrina trees, which I found in Roseires, were probably introduced there by people who knew the effect of moringa seeds on suspended solids. The vernacular name for both moringa species is *shagara al rauwāq* (clarifier tree). Strangely enough, this name is not listed at present in *Sudanese Flora.* The other vernacular name for *Moringa oleifera* is not widely known: *Figl alkilāb* means "dog radish," because the root resembles the radish in shape and smell. This name is reminiscent of the English "horseradish tree." *Moringa peregrina* is regarded as indigeneous to the Eastern Sudan (Red Sea Hills) and to Kordofan. It is also known as a tree of Sinai under the name of *libān* or *bān* (Table 13-3).[6]

Choice of plants suitable for the treatment of turbid water

We do not know when the use of plants for the purpose of water purification was discovered in the Sudan; people seventy years and older remember it from childhood. Plants are also used for the same purpose outside the Nile Valley, especially in the western Sudan.

Looking at Table 13-3, one may wonder if suitable plant material was found only by chance, or if certain groups of plants were more likely to be tested. Women preparing food, drink, and native medicine probably had good opportunities to observe that some plant material put into the muddy dark-brown Nile water of the flood season improved its appearance. Men living in primitive huts in the fields or close to the riverside had, on the other hand, a better chance to discover that stalks, leaves, bark, or seeds from wild shrubs and trees purified water.

Plants providing foodstuffs: Fruits or seeds of the following plants are either foodstuffs or are used to flavor certain dishes: *Sorghum* sp., *Cucurbita* sp., *Abelmoschus esculentus (Hibiscus esculentus), Faba fona, Pisum sativum, Lupinus termis, Lablab niger, Arachis hypogea,* and *Trigonella foenum graecum.* In areas where these plants are grown, they are usually stored in the home; the women often take what is readily available, and not necessarily what yields the best results. Broad beans are most popular in the Northern Province, as are ground nuts in parts of Southern Gezira and Blue Nile Province.

Plants flavouring water: *Cyperus rotundus* is known as *tīb al khadim* (perfume of the servant or perfume of the poor). Its rhizomes, the fragrant part of the plant, are used for flocculation. As was mentioned above, the wealthy could afford to treat their water with mahlab. The stalks and leaves of *Cymbopogon nervatus,* used for roofs of the round straw houses in Southern Gezira and Blue Nile Province, are also fragrant. Lime leaves and pounded almond stones also improve the taste of water.

Plants with effects on gastro-intestinal diseases: In Sudanese native medicine, fresh roots of *Cyperus rotundus* are recommended against indigestion in children.[7] The Nubians of Wadi Halfa mix the sun-dried pounded rhizomes with sugar and administer the powder for stomach disorders. In other parts of the Northern Province and in Halfa, powdered Cyperus rhizomes are also taken as "hot tea."

The seeds of the baobab tree, called *qonqulīz,* are a well known drug against diarrhoea. They are chewed, served in cold or hot water, or in a porridge-like dish called *medīda.*[8] According to R. G. Anderson, a decoction of mahlab seeds *(Hypoestes cancellata)* was used in Kordofan for colic in children.[9]

The seeds of fenugreek are also still widely used as a local remedy against dyspepsia and diarrhoea. They are taken in cold or hot water, hot milk, or in medīda. The fruits of *Caparis decidua,* called "humbuq," are said to cure dysentery; the bark of *Balanites aegyptiaca* is considered a purgative and antihelminthic.[10]

Plants growing at the riverside: Plants which are found along the Nile shores are *Cyperus rotundus, Tamarix nilotica* and *Salix subserrata.*

Different ways of using plant material in native homes

Around Merowi, where water purification with pounded broad beans is still rather common, I went to some villages and asked the women in five or six neighboring houses to show me their jar water in a drinking glass. Although the turbid water from the river or canal had arrived at approximately the same time, and the type of plant material was the same, the appearance of the water with respect to color and turbidity varied greatly. Apart from difficulties in obtaining an optimal concentration of the clarifier adjusted to variations in the amount of suspended solids, the manner of cleaning and pounding the beans and the careful daily cleaning of the jar are essential. Women who throw "handfulls" of pounded beans in the jar find an offensive smell in the water five or six hours after the beginning of treatment. Women also complain about the bad smell of water treated with fenugreek, *Lablab niger,* and groundnuts. The odor is due to the presence of putrefying material in the jars. A few families in Merowi treated the turbid water only with an extract from *Faba fona* seeds, obtained by pouring the pounded material through a sieve. If this method is combined with proper choice of concentration and great cleanliness in handling, the drinking water from the jar does not acquire any odor. I have tasted water of this kind after more than twenty hours and it was still in good condition.

In the Blue Nile Province, some women complained about deterioration of the water after use of Moringa seeds. During my visit to Roseires, the flood season was almost over, but I tried seeds from *Moringa oleifera,* during September in Merowi, according to a method whereby the plant material is only extracted and not allowed to enter the jar. Pounded seeds were put in a small bag of thin clean material. By means of a thread tied to this bag, the clarifier can be immersed in the water and moved about slowly for 20 to 30 minutes; it is then left for 2 more hours. Satisfactory purification of about 20 l of water could be obtained with only five seeds.

Preliminary laboratory observations on plant material

In the beginning of September, water from the Blue Nile at Khartoum was sampled and poured into two-liter beakers. Plant material was prepared according to native methods and added directly to the turbid water. Initial stirring was done for five to ten minutes with glass rods.

Moringa oleifera seeds gave the best results with respect to time of reaction and final appearance of the purified water. They were used in concentrations of 0.5 and 1.0 seed per liter. After two hours the water already had a whitish appearance. After three hours, it was as clear as tap water; there was no odor after twenty-four hours. In the experiment done during field work, the concentration would be equal to only 0.25 seed per liter.

Faba fona and *Trigonella foenum graecum* seeds were tested in concentrations of 0.2 and 2.0 grams per liter. The effects were similar. The higher concentration acted faster, but after three hours the water was not yet clear, as in the case of moringa. Even after twenty-four hours, the water was still slightly whitish opaque in the case of *Faba fona* and yellow-whitish opaque in the case of *Trigonella foenum graecum.* All samples had acquired an unpleasant odor.

Adansonia digitata, tested in a concentration of 0.25 grams per liter, and *Boscia senegalensis,* tested by stirring the water for 30 minutes with two thin branches, improved the appearance of the water to a lesser extent. After 24 hours the water was still discolored and opaque (Fig. 13-1). Planned bacteriological investigation of water after treatment with different plants could not be carried out at this point.

In Germany, flocculation of suspended matters was observed when certain aquatic plants, particularly *Scirpus lacustris,* were adapted to exist in contaminated waters.[11]

Bacteriological studies on untreated Nile water and water samples from jars after treatment with clarifiers

During the flood season, the Blue Nile and the River Nile become very dangerous due to the high water level and to great erosion at many places along the banks. At that time, people try to descend to the river at easily accessible places and where the current is weaker. In villages close to the Nile, girls and women fetch the water. For homes some distance from the river, water is usually bought from professional water sellers

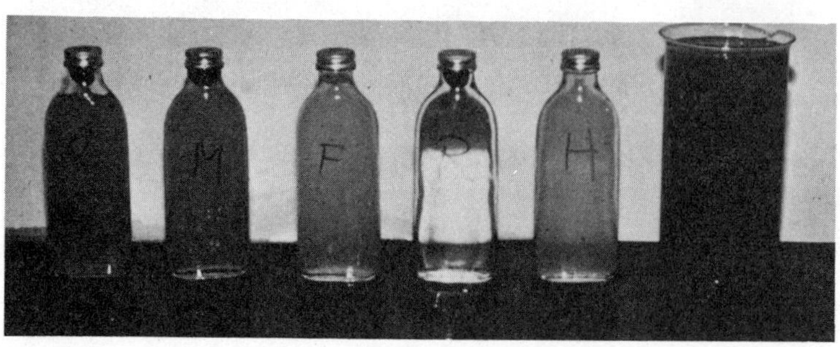

FIG. 13-1. Water purification with plant material. Samples of water after 24-hour treatment with (Q) *Adansonia digitata,* (M) *Boscia senegalensis,* (F) *Faba fona,* (R) *Moringa oleifera,* and (H) *Trigonella foenum graecum.* The beaker on the far right contains untreated water from the Blue Nile at Khartoum (9/8/74).

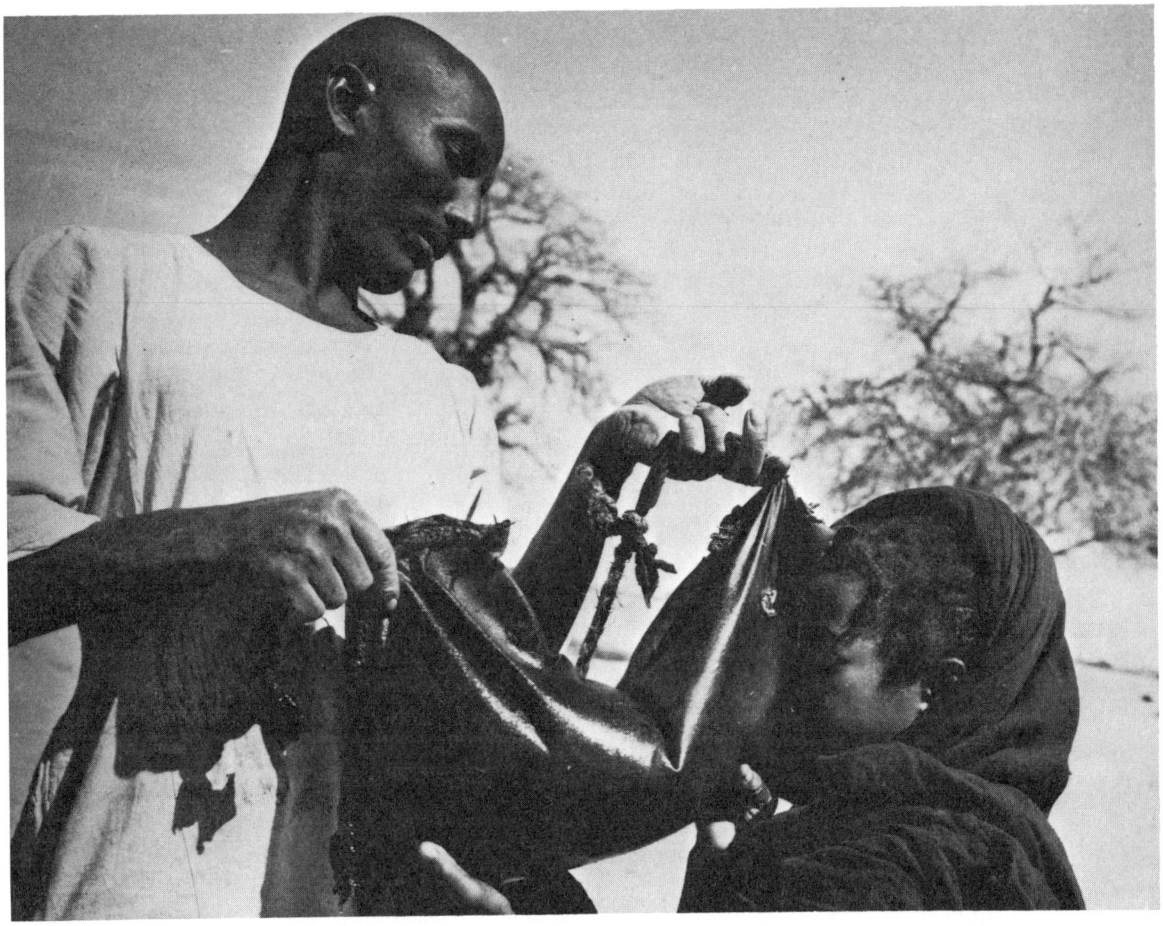

FIG. 13-2. Qirba, a large water bag, can be used for many years, if kept full, and regularly treated with tar (qutrān) from coloquintes (root crop) or sesame oil.

who transport the water by donkey in bags of leather or other strong material.

In Merowi, water was primarily obtained from a site at the ferry-boat dock. The donkeys enter the river, and the water is poured into the bags. At the same place, cattle wait for transport to the far shore of the Nile, and small children play along the water's edge. Under such circumstances, human and animal wastes certainly contaminate the water. In the villages of Gezira, between Rufa'a and Guned, water is taken from the same places where cattle bathe or drink. The dangers of pollution with wastes are still greater in villages like Nuri and Al Qureir, where the only source of water is irrigation canals containing semi-stagnant water.

In the Northern Province, where rainfall is slight, day temperatures during the flood season are usually above 40° C. The donkey boys are busy from sunrise to sunset supplying homes with water; little time remains for cleaning water bags (Fig. 13-2). In the native homes, water is poured into earthenware jars *(zīr)* for storage and cooling. If these jars are not handled carefully, the drinking water becomes even more polluted. Studies carried out by a World Health Organiza-

TABLE 13-4. *Bacteriological Studies on Drinking Water During the Flood Season 1974 in the Northern Province.*

Untreated Water			
Place of sampling	Date		Viable count *E. coli*/ml at 37° C
Nile at ferry-boat landing-place in Merowi	8/14		98,000
Nile at ferry-boat landing-place in Merowi (new site due to increase in water level)	9/18		17,600
Irrigation canal Nuri, Gereif*	8/14		74,300
Irrigation canal Nuri, Hilla al Gabal	8/14		85,200
Irrigation canal Nuri, Haluf	9/18		uncountable
Irrigation canal Nuri, Hilla al Gabal	9/18		uncountable
Water from jars after treatment with clay soil (rauwāq)			
Locality, profession of owner of the house	Date	Type of Rauwāq	*E. coli*/ml at 37° C
	Nile water		
Merowi, cook in boarding school	8/14	Uqda	46,400
	Water from irrigation canals		
Nuri, Hilla al Gabal, builder	8/14	Uqda	41,000
Nuri, Haluf, kindergarten	8/14	Uqda	21,600
Nuri, Haluf, chief gardener, government gardens Nuri	8/14	Basa	8,500
Nuri, Hilla al Gabal, date palm cultivator	9/18	Uqda	uncountable
Nuri, Haluf, teacher	9/18	Uqda	12,300
Nuri, Haluf, date palm cultivator	9/18	Uqda	15,400
Water from jars after treatment with seeds of Faba fona			
Locality, profession of owner of the house	Date		*E. coli*/ml at 37°C
	Nile water		
Merowi, senior nurse	8/14		96,000
Merowi, government official	9/18		uncountable
Merowi, tailor	9/18		uncountable

*Gereif, Hilla al Gabal, and Haluf are quarters of Nuri

tion team in Banat, southwest of Omdurman, show that even high quality water kept in jars was fecally contaminated.[12]

Our own bacteriological investigation was carried out in collaboration with Dr. El Sheikh Mahgoub, Head of the Department of Microbiology, University of Khartoum, and with his research assistant, Dr. Samia Ahmed Guma'a. Samples of Nile water from the river and from irrigation canals and samples of jar water from homes were collected early in the morning in sterilized bottles and brought in ice boxes by car or air to Khartoum, where laboratory examination was done the same day. Treatment with clarifiers had taken place 14 to 17 hours before sampling.

The viable count test was always preceded by a presumptive coliform count carried out according to McCrady's tables. Besides *Escherichia coli,* the pathogen *Pseudomonas pyocyaneus*

FIG. 13-3. Rainwater storage in trees. In Kordofan and Darfur (Western Sudan), rain water is stored in Tibaldi trees *(Adansonia digitata).* A hole is made in the trunk at the level of the first branches, and the inside of the tree is dug out for the water storage area. After filling the tree with rain water, the opening is closed with a mixture of soil and animal dung. The water can be stored in this way for long periods, and is still potable even up to ten years. It is thought that a substance in the bark leaches into the water to preserve the water quality. The tree is not damaged by this use. In the picture, water is being drawn with a leather bucket (dēlu).

was isolated. According to the method used for the *E. coli* count, amounts above 90,000/ml are indicated as "uncountable."

Table 13-4 shows the results from the Northern Province. All samples of untreated water were fecally contaminated. There is evidence that rauwāq treatment reduces *E. coli* to a certain extent. The effect appears to depend on the "quality" of rauwāq. Type "Basa," from the village Basa, located between Merowi and Nuri, was discovered only one year ago. A possible entrapping of microbes in the floc could be due to the high content of montmorillonite in rauwāq, since montmorillonite is able to adsorb organic matter.[13]

Treatment with *Faba fona* did not improve the bacterial count. In samples taken in September 1974 the number of *E. coli* had even increased. This could be due to the presence of decaying plant material in the jars.

Discussion

Our experience gained during the 1974 flood season calls for continued research with the aim of improving the drinking water conditions in the riparian Sudan. Existing health risks from poor-quality water are even higher in a country with great water consumption. In the Sudan, the average daily consumption per person was estimated at 5 liters, whereas in central Europe it is only 1 liter. The corresponding figures for total water ingested, including water taken in food and drink are 6 liters for the Sudan and 6.85 litres for central Europe. This means, also, that in the Sudan most of the water is ingested without previous boiling.[14]

The dangers for fecal pollution in water from the river could be minimized by encouraging public health authorities to establish strict rules concerning suitable places for the collection of water. There will always be the risk, however, of contamination occuring in subsequent handling.

The ideal clarifier should combine flocculating and bactericide properties. Water purification according to methods applied in public drinking water treatment facilities are not fit for native households. In Khartoum, the maximum allowable amount of aluminium sulphate, according to international rules (200 mg/liter) is not exceeded during the flood season, even if the optimal concentration for perfect floc formation is higher. People in rural areas who try to use aluminium sulphate are usually working with an overdose, with the result that the water causes stomach disorders or is only suitable for washing purposes.

With respect to native methods, preference should be given to clarifier material which can be clearly identified and measured quantitatively. Proper application rates need to be established, and the future use of plant material should be guided by systematic laboratory experiments. There is already evidence that "extracts" give better results than admixtures of plant material. Extracts of already known plant clarifiers must be tested, first from a toxicological point of view, and, second, for possible bactericide action. In this regard the Tibaldi tree is of potential interest. In the western Sudan, water is stored in live, hollowed out Tibaldi trees. This tree is thought to produce a bactericidal substance that leaches into the stored water (Fig. 13-3).

To improve the quality of water supplied from irrigation channels in the Northern Province, one might consider designing experiments with Sudanese aquatic plants similar to the successful application of aquatic plants for water improvement in German water treatment facilities. It should also be considered whether or not the most powerful water clarifier, *Scirpus lacustris,* which metabolizes inorganic and organic contamination and reduces significantly coliform bacteria, could not be grown in the Sudan.

Studies of this kind would be of great value for other parts of the Sudan, where water purification is practiced by similar native methods as in the Nile valley. Such research and pilot installations might be of great benefit, not only for the Sudan, but also for other African countries.

Notes

1. C. R. Cox, "Operation and Control of Water Treatment Processes," *WHO* (Geneva, 1964).
2. Omar El Badri, *Sediment Transport and Deposition in the Blue Nile at Khartoum Flood Seasons 1967, 1968, 1969,* Master's Thesis University of Khartoum, 1972.
3. Cox, "Operation and Control."
4. El Badri, "Sediment Transport."
5. F. W. Andrews, *The Flowering Plants of the Anglo-Egyptian Sudan* (Arbroath, published for the Sudan Gov't ed. T. Buncle, 1950).
6. V. Tackholm, *Alfred Kaiser's Sinai-Herbarium,* Publications from the Cairo University Herbarium No. 2, Cairo, 1969.
7. A. F. Broun, *Catalogue of Sudan Flowering Plants* (Khartoum, 1906).
8. Ahmed Abdel Halim, "Native Medicine and Ways of Treatment in the Northern Sudan," *Sudan Notes and Records* 22 (1939): 27-48.
9. R. G. Anderson, "Medical Practices and Superstitions of Kordofan,"*3rd Report of Wellcome Res. Lab,* Khartoum (1908): 281-322.
10. Broun, *Sudan Flowering Plants.*
11. K. Seidel, "Macrophytes as Functional Elements in the Environment of Man," *Hidrobiologia (Bucuresti)* 12 (1971): 121-30.
12. W. J. van Zijl, N. Blioumel, and D. G. Schliessmann, *Report of Studies on Diarrhoeal Diseases.* The WHO Diarrhoeal Advisory Team in Co-operation with the Ministry of Health, Khartoum, 1961.
13. L. A. Pinck, "Adsorption of Proteins, Enzymes and Antibiotics by Montmorillonite," *Clay and Clay Minerals* 9 (1962): 520-29.
14. Figures according to a note from a course for Public Health Inspectors, Khartoum, 1967.

Part IV

Biological Treatment of Wastewater

14

Macrophytes and Water Purification

KÄTHE SEIDEL

Max Planck Institute
Krefeld, West Germany

Alternatives

In 1953, I had the opportunity to first present methods for the alleviation of over-fertilization, pollution, and siltation of inland waterways and for their revitalization through the use of appropriate species of plants in a report published by the Max Planck Institute.[1] At that time, the following view predominated among water experts:

1. Water pollution alleviation considerations were limited to:
   a) Physical: the withholding of coarse particles such as sand, paper, cardboard, metals, plant residuals, etc.
   b) Chemical: the introduction of oxygen and precipitants for flocculation.
   c) Biological: the stimulation of bacterial activity to separate organic compounds and to flocculate organic and inorganic substances and transform them into sludge.
2. The controlled use of macrophytes for water purification was not taken into consideration.

At the time, it was believed to be an established fact that, for example, bulrushes (*Schoenoplectus*[2] *lacustris* Palla) and other higher plants were healthy only in clean water. The ability of higher plants to eliminate toxic substances in water was not recognized. Opinions were based on the analyses of stems and leaves of bulrushes only. Plants as a link in an ecological chain were not considered.

We began our work by investigating the growth efficiency of certain higher plants in contaminated waters. Our fundamental question was whether plants, generally speaking, have a positive effect on contaminated waters and sludge, and if so, which plants had such an effect. Today, the number of known plant species is roughly estimated to be 200,000, which indicates the magnitude of our task. Furthermore, the rigid division of plant species into terrestrial and aquatic species can be misleading: it is easy to overlook, for example, the fact that water-lilies can be cultivated on land, as can marsh marigolds, and that even the common duckweed can be grown outside of water.[3]

FIG. 14-1. Gigantic growth of *Alisma plantago* in municipal sewage (right) as compared to the same species grown in its natural habitat (left).

In setting out on a research program, we made use of my doctoral dissertation on *Schoenoplectus lacustris.*[4] This plant proved a worthwhile subject for us in further research and practice.

Biological Facts and Verification

Field of Activity

Three different research media were considered: the polluted surface waters from streams, rivers, and lakes; the brackish and salty water from the Baltic and North Seas, and the sewage from villages and industries. Initially, there were four basic questions:

a) Which types of plants could endure varying water conditions?
b) Could certain plants eliminate water comtaminants?
c) Do plants have an influence on the condition of water, subsoil, and sludge?
d) Do plants themselves change when confronted with new environments, i.e., have plants latent qualities of which we are unaware because they have not been challenged until now?

Morphological, Anatomical, and Physiological Changes in Plants Grown in Polluted Water

We know only a little about genetic changes in higher plants brought about by environmental factors in water; we are, however, surprised at the adaptability of plants when confronted by the types of effluents generated today. Until recently, it was believed that everything about a particular species' form, size, and habitat requirements, its rhythm of life, and its reproduction, was irrevocably determined. However, it was overlooked that our knowledge of the characteristics of many species is based only on observations of plants growing in a normal environment. If certain species are planted in water containing "modern" pollutants, new characteristics may emerge that had been latent. This refers to morphological as well as to anatomical and physiological changes. For example, *Alisma plantago,* a plant with a small storage bulb, grew to gigantic size in fecal sewage, and formed a very large storage bulb which was replaced by a new one each year (Fig. 14-1).

With *Schoenoplectus lacustris* the reverse was observed: in uncontaminated water, its roots are discarded each year, but if grown in certain types of sewage, they are not. In unsettled fecal sewage it is completely free of roots, while the rhizome loses its scale-like leaves and takes over

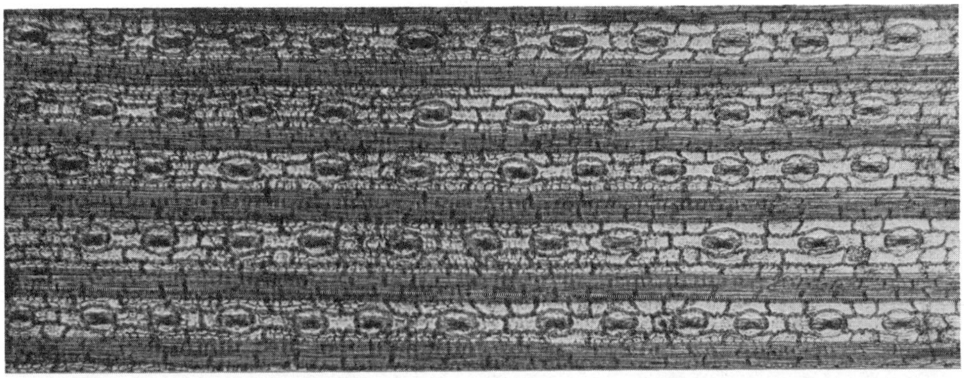

FIG. 14-2. Uniform arrangement of stomates *(Schoenoplectus lacustris)* in uncontaminated water.

nutritional absorption. When a planted bed is floated in sewage, there are no alterations in the characteristics of the stems of the plants.

On the other hand, *Schoenoplectus lacustris* planted in the unsettled effluent of a fabric printing plant showed an unusually abundant growth: plants developed 10 to 20 times more plentifully than in uncontaminated water.

We also observed closely the anatomical changes in *Schoenoplectus lacustris* in oxygen-poor mediums, where the aerenchyma expanded greatly while the number of fibres remained constant. In uncontaminated water, the stomates of this species are uniform both in arrangement and size (Fig. 14-2). When the plants had to feed from phenol-containing water, however, the stomates became irregular in arrangement and form (Fig. 14-3). When plants were sunk into six meters of such water, they formed a large number of stomates. In each of these cases, growth efficiency was preserved.

When *Schoenoplectus lacustris* was confronted with water containing phenol, a physiological change was noted as well. Coming into contact for the first time with phenol-containing water (particularly in concentrations over 50 mg/l), specimens from a healthy biotype cast off their stems, but retained their roots. The newly-formed stems were green, growing well, and did not exhibit any new physiological reactions. At first a lag phase set in, during which the plants did not take in any phenol. In lower concentrations (40 to 60 mg/l), this phase could last one to two days, while with higher concentrations, it lasted two to three days.

Another example of the arousal of latent

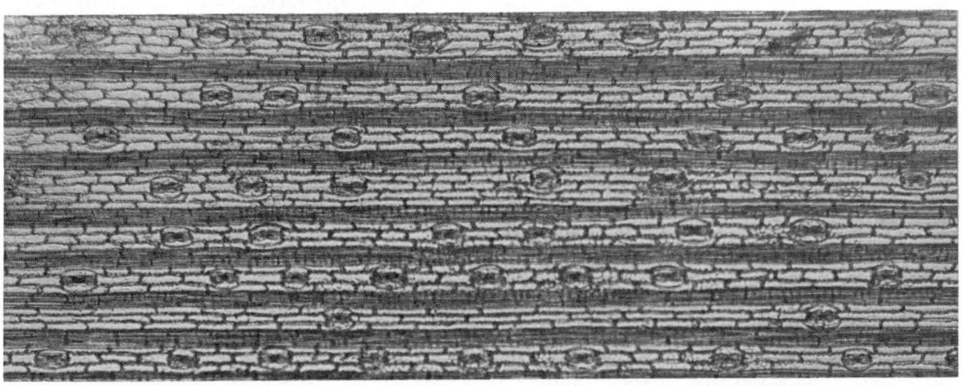

FIG. 14-3. Irregular arrangement of stomates *(Schoenoplectus lacustris)* under the influence of phenol.

FIG. 14-4. Formation of water cells on the epidermis of *Schoenoplectus lacustris* cultivated in phenol solution.

characteristics is the formation of hydropods (water cells) on the epidermis of *Schoenoplectus lacustris* when cultivated in phenol solutions (Fig. 14-4).

Nature offers an abundance of plant species; man produces many types of sewage. How plants will react to new chemical wastes produced by man, what latent qualities will emerge enabling them to survive, and what arrangements will have to be made to lighten their way into an encumbered future are questions of vital importance.

Elimination of Inorganic and Organic Wastes from Water

a) Inorganic Matter

Table 14-1 shows that the relative quantity of inorganic matter absorbed by species of the same biotype in a given time period varies widely when described in terms of g/kg or mg/kg of dry plant substance. But because individual plant species form varying quantities of biomass and thus varying quantities of dry substance, these numbers become significant only when converted to yield per square meter of surface. Numbers are only of value in this tested, healthy biotype.

Because the ability of a plant to absorb inorganic matter varies in accordance with the characteristics of its environment, the figures in Table 14-1 are of little use in calculating the elimination of inorganic substances in industrial waste water. Table 14-2, showing the absorption capacity of *Schoenoplectus lacustris* in two different environments, illustrates how the absorption capacity of a species can be increased many times by the characteristics of a specific environment.

b) Organic Matter

We have worked extensively in the field of organic pollution, and have eliminated monovalent phenol and its derivatives such as xyol, pyrocatechol, chlorophenol, pentachlorophenol, and cyanide. These substances, found in industrial wastes, are a grave concern in the field of water treatment.

We tested a wide range of plants for their compatibility with phenol. Many of these plants could not bear the exposure to phenol and died off completely. Finally, a group of plants suited to our purpose emerged. These plants all have similar structures: they have a large aerenchyma and a basal meristem, i.e., they grow from the

TABLE 14-1. *Absorption of Chemicals by Various Macrophytes.*
Uptake figures given in mg/kg dry weight and mg/m² growth areas. (K. Seidel, "Reinigung von Gewässern durch höhere Pflanzen," *Naturwissenschaften* 53(1966): 289-97.)

	Cu		Co		Zn		Ni		Mo		Mn		B	
	mg/kg	mg/m²	mg/kg	mg/m²	mg/kg	mg/m²	mg/kg	mg/m²	mg/kg	mg/m²	mg/kg	mg/m²	mg/kg	mg/m²
Scirpus lacustris	4.8	16.13	5.63	2.28	50	168.00	1.71	5.75	0.55	1.85	1200	4032.00	14.6	49.6
Carex stricta	5.6	15.23	6.72	1.96	63	171.36	2.46	6.69	0.29	0.79	970	2638.4	21.4	58.21
Iris pseudacorus	5.7	14.14	1.11	2.75	50	124.0	1.75	4.34	0.33	0.93	382	947.36	10.3	25.41
Typha angustifolia	4.7	6.77	0.44	0.63	43	62.92	1.86	2.68	0.30	0.43	779	1121.76	24.5	35.28
Glyceria aquatica	5.6	11.65	0.48	0.99	73	151.84	1.99	4.14	0.24	0.49	586	1218.88	15.0	31.20
Phragmites communis	4.2	18.82	0.62	2.78	37	165.76	1.53	6.85	0.26	1.16	166	743.68	8.2	36.74
Acorus calamus	4.1	6.56	0.53	0.85	38	60.8	1.08	1.73	0.30	8.48	383	612.8	56.9	91.04
Sparganium erectum	5.6	7.17	1.07	1.37	76	97.28	2.27	2.91	0.24	0.31	604	773.12	40.2	51.46
Myosotis palustris	12.2	3.90	1.44	0.46	104	33.28	3.19	1.02	0.53	0.17	2000	640.0	30.8	9.86
Mentha aquatica	8.5	14.28	0.54	0.91	78	131.04	2.03	3.41	0.47	0.79	381	640.1	38.3	64.34

	Ca		Mg		P		K		Na		Fe		Si	
	g/kg	g/m²	g/kg	g/m²	g/kg	g/m²	g/kg	g/m²	g/kg	g/m²	g/kg	g/m²	g/kg	g/m²
Scirpus lacustris	3.95	12.17	0.98	3.29	2.00	6.72	10.30	34.61	6.30	21.17	0.78	2.62	12.60	43.34
Garex stricta	4.84	13.16	2.06	5.60	2.20	5.93	13.90	38.31	2.20	5.58	3.80	10.34	17.60	47.87
Iris pseudacorus	16.97	42.09	2.58	6.39	2.50	6.20	35.40	87.79	1.70	4.22	1.30	3.22	4.30	10.46
Typha angustifolia	14.36	22.48	1.49	2.29	2.20	3.17	14.10	20.30	12.20	17.57	1.10	1.58	3.70	5.33
Glyceria aquatica	4.79	9.96	1.46	3.04	2.50	5.20	26.00	54.00	1.20	2.50	1.30	2.70	14.80	30.78
Phragmites communis	1.70	7.62	0.82	3.90	1.40	6.27	8.10	36.29	1.10	4.93	0.92	4.12	21.70	97.22
Acorus calamus	12.33	19.73	1.84	2.94	2.90	4.64	20.00	32.00	4.10	6.56	0.97	1.55	3.30	5.28
Sparganium crectum	10.11	12.94	2.10	2.69	3.90	4.99	24.80	31.74	6.80	8.70	3.60	4.61	6.80	8.70
Myosotis palustis	9.13	2.92	1.50	0.48	2.00	0.64	42.50	13.60	5.00	1.60	10.20	3.26	8.30	2.66
Mentha aquatica	16.95	28.48	1.58	2.75	2.20	3.70	10.50	17.64	6.90	11.59	3.10	5.21	4.10	6.89

TABLE 14-2. *The Metal Absorption Capacity of* S. lacustris *in Two Environments.*

	Cu	Co	Mn	Cr	Ni	V
Healthy Lake	18	3	260	2.5	3.5	6
Sewage	50	15	2500	115.0	30.0	115

stem base. A few plants of the *Juncus* species fit this category, as do a few members of the *Schoenoplectus* species, especially *Schoenoplectus lacustris,* which had been our test plant for many years. We cultivated them in water containing up to 1000 mg/l phenol; we fed them, in a long series of tests, from 10 to 100 mg/l phenol for three years. Fig. 14-5 shows the elimination of phenol.

It was surprising that these plants, supplied only with tap water and phenol, increased in biomass every year. Furthermore, those fed smaller amounts of phenol increased the least.

After the first phenol feeding, we found that many stems had crystallized inclusions of phenol. After a short time, however, we found no more phenol in the stems. It would be generally assumed that bacteria of the root system had cleaved the aromatic ring of the phenol and that the plant was served only this cleavage product as food. To test this possibility, a phenol solution was injected into the plant rhizome. All precautions for proper control were observed. This test demonstrated that, at least in *Schoenoplectus lacustris,* this poisonous material was cleaved by the plant and afterward metabolized to amino acids for the production of protein.

We had misgivings when requested by industry to find plants for the elimination of chlorophenol and pentachlorophenol. These poisons are a dangerous burden for water, because they destroy all animal and plant life.

We tried a combination of what we call a cascade system (MPI System), which will be referred to later on, and a tidal inundation system. A surprisingly large amount of pentachlorophenol was absorbed by using this system.

Cyanogen, often used in industry, is also one of the most dangerous of pollutants. After primary tests at the Max Planck Institute, we built, with the assistance of a Rumanian study team, a pilot plant to eliminate cyanide from the effluent of a Rumanian steel combine by means of plants. As a control, a similar but unplanted pilot installation was supplied with the same effluent in the same quantity and allowed identical detention time. According to Rumanian reports, the planting beds caused considerable improvement in effluent quality.

c) Pathogenic bacteria

Not only unsettled sewage laden with feces but also clarified sewage from conventional sewage plants almost always contains quantities

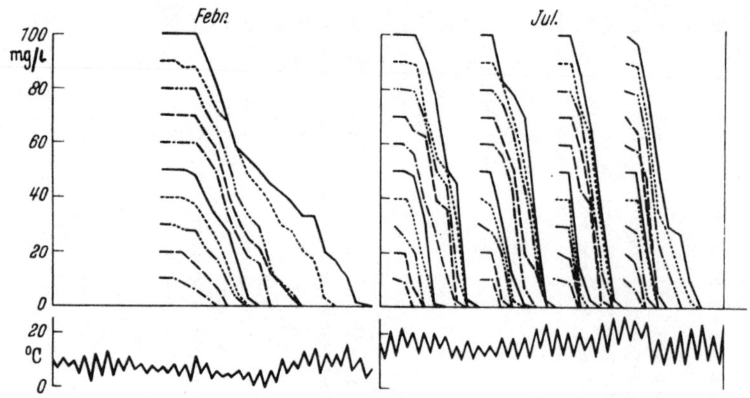

FIG. 14-5. Reduction of phenol in containers of 5 liter of well water planted with *Schoenoplectus lacustris* (300 gram of biomass each). Vertical axis: phenol concentration. Horizontal axis: absorption time. Horizontal bar: temperature readings in centigrade. Note: a lag phase in plant adaption to high phenol concentrations is particularly evident in February experiments. (K. Seidel, "Reinigung von Gewässern durch höhere Pflanzen," *Naturwissenschaften* 53 (1966): 289-97.)

FIG. 14-6. Root nodules on *Alnus glutinosa.*

of pathogenic bacteria, e.g. *Salmonella* and *Enterococci.* Because these pathogenic bacteria can be transmitted easily from surface to ground waters, they pose an especially serious threat to public health. In conventional treatment plants, such bacteria are eliminated by heavy chlorination. This is, however, exchanging one evil for another, because chlorination can create carcinogenic compounds, as shown by U.S. Environmental Protection Agency testing in New Orleans.[5]

It has been shown that root excretions of *Mentha aquatica, Acorus calamas, Juncus effusus, Phragmites communis,* and the root bulbs of *Alnus glutinosa* (Fig. 14-6) can either partially or completely kill disease bacteria in contaminated water. Benign bacteria, however, are left intact.

We propose the following hypothesis in accordance with results from our investigations: first, that excretions from plant roots protect them from decay caused by certain bacteria and fungi, maintaining the entire plant in a healthy condition, especially during the winter months. Second, that a "root tent" (rhizosphere) provides a protective space for benign bacteria in times when contaminants such as acids, phenols, cyanides, or mercury threaten to destroy them. After such contaminants have passed by, surviving bacteria can recolonize sites outside their "root tent" and continue their vital work.

In conventional sewage plants, this protection for benign bacteria does not exist. They are often destroyed by poison or acid and must either be reintroduced artificially through bacterial sludge or must re-establish themselves naturally. This can take six to eight weeks, and during this period sewage passes through the system without the essential bacterial breakdown of organic forms. Therefore, our process involving planting beds can serve a most important function, if it comes after conventional sewage treatment; fecal indicator and pathogenic bacteria (Table 14-3) can be destroyed possibly by excretions

TABLE 14-3. *Average Percent Reduction of Fecal Indicator and Pathogenic Bacteria After Two Hours Contact with Various Macrophytes.*

	Plant tested						
Bacteria	Alisma plant.	Mentha aquat.	Juncus effusus	Scirpus lacustr.	Phragm. comm.	Iris pseud.	Phragm. + Scirpus + Mentha
E. coli	80	90	80	70	70	50	70
Enterococci	80	90	80	80	40	20	50
Salmonella	90	90	50	60	40	10	50

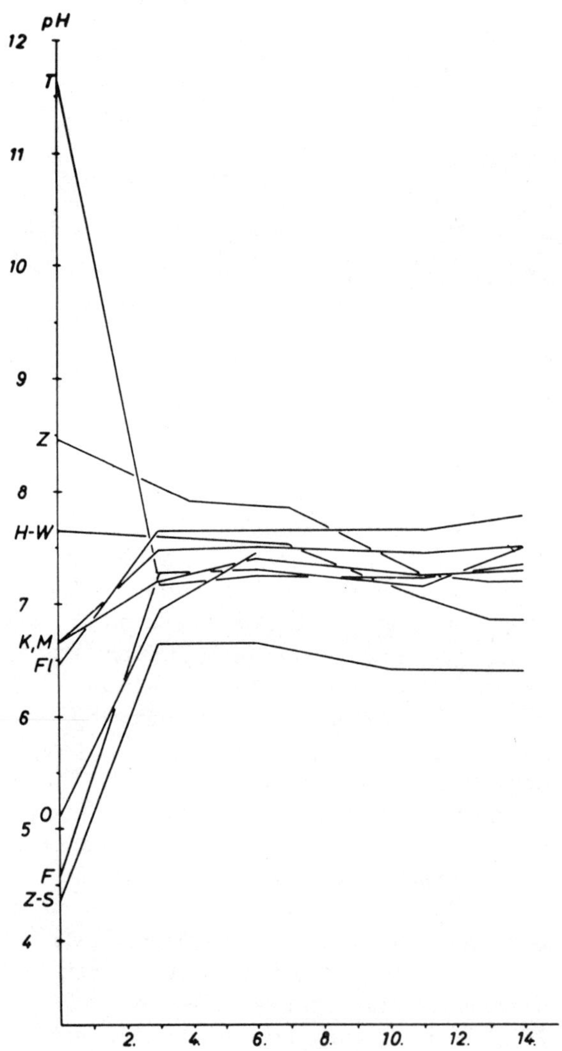

from plant roots without the use of chlorination.

d) pH Regulation

The pH of water is affected by the action of certain plants. If acid or alkaline sewage flows through the root zone of *Schoenoplectus lacustris,* for example, the water is neutralized (i.e. altered to pH 7). This results in optimum bacterial and chemical reactions, which in turn optimizes the quality of lake, river, and ground waters. Thus, the use of expensive and problematic chemical remedies can be discarded. The rapidity of the neutralization reactions caused by certain plants remains, however, unexplained. Fig. 14-7 shows the neutralization of industrial effluents in controlled environments utilizing macrophytes. A further example of pH neutralization from a fabric printing works in South Carolina is shown in Table 14-4.

Sludge

a) Sludge and Silt Stabilization

It is equally important to use plants for the purification and reclamation of sludge from coastal areas and from sewage. Several years of experi-

FIG. 14-7. Alteration of pH value in industrial effluents exposed to *Schoenoplectus lacustris* during a 0 to 14 day time period shown on the abscissa. pH values are shown on the ordinate. Symbols: O = fruit processing effluent; M = margarine production; FL = meat processing effluent; K = bakery plant effluent; F = special foods manufacturing; T = dough cannery; Z = sugar refinery effluent; H-W = yeast and spice processing effluents; Z-S = effluent of sugar refinery mixed with municipal effluents.

ments were stimulated by a request for assistance from a nuclear research institute. Environmental law in Germany prohibits the transport of nuclear laboratory sludge. Our proposed solution was based on my observation approximately twenty years ago that drifting silt at the sea coast turned hard in summer, lacking oxygen and almost all life. We used to call it "summer cement." However, where *Spartina Townsendii* and *Schoenoplectus lacustris* grew, the silt was crumbly, abundant in oxygen, and vital; it had been transformed into genuine soil. The roots had made this possible.

To solve this problem posed by the nuclear research center, we had to find a plant with a large root zone that grows rapidly and transpires large volumes of water and whose stem nodes would direct adventitious roots into the continuously shifting and draining sludge and initiate its mineralization. *Phragmites communis* met these requirements. The necessary plant trenches were prepared and planted in a special way. They will be referred to later on. When the plants were 60–100 cm high, 30–50 cm of sludge was placed upon them. This sludge separated itself from the water, which drained off. The sludge mass dried out and its volume was considerably reduced. (Fig. 14-8). After a few days a new layer

TABLE 14-4. *pH Neutralization by Macrophytes.*

	Time	*Influent pH*	*Discharge pH*
25 Jan. '72	morning	4.0	6.5
25 Jan. '72	evening	4.9	6.5
10 Feb. '72	evening	12.0	6.7
11 Feb. '72	morning	11.8	6.8

of sludge was applied. In this manner we were able to stabilize and reduce the volume of sludges in the nuclear research center, resulting in 80-90% dried matter and clear water. Plants grew repeatedly in this converted sludge and regenerated the soil, and, in inhabited areas, this soil was even cultivated.

In other work we have been an advisor to a municipality that had a serious problem. Up to 90% of its wastes consisted of brewery effluents, resulting in sludge that could only be disposed of on land. After treatment of the sludge by *Phragmites* beds there was a dramatic improvement in effluent quality as shown in Table 14-5.

In yet another application, municipal sludge was dewatered by a *Phragmites* bed. The sludge of this industrial city consists mainly of feces, dyeworks refuse, and heavy metals from steel production. During a period from August to February, layers of this sludge totaling 8.5 m in

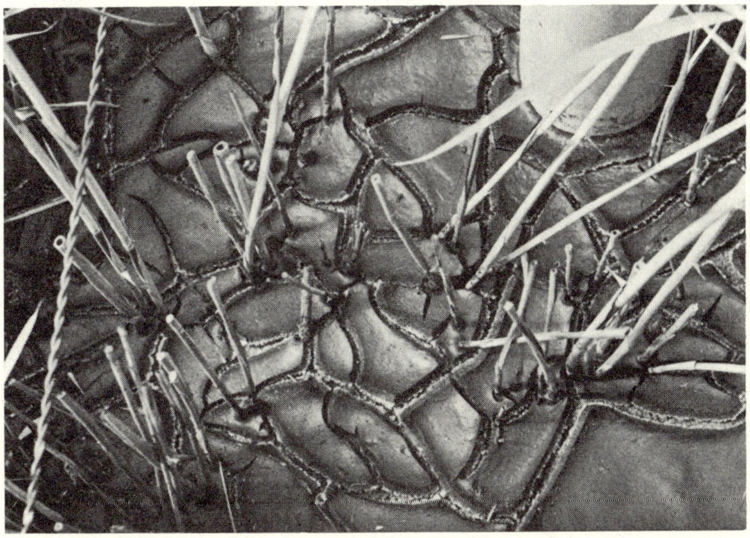

FIG. 14-8. Sludge drying process in planted beds of *Phragmites communis* at nuclear research center.

TABLE 14-5. *Sludge Water Quality Improvement Using* Phragmites *Beds.*

	Influent Sludge Water	*Discharge Water*
pH	7.4	7.7
Transparency (cm)	0	53.0
BOD_5 (mg/ l)	1300	5.5
Settling solids (cc/l)	825	0

height were applied to a *Phragmites* planting bed. The water draining from this bed was always clear, and complete analyses were constantly made. At the end of this period, sludge was reduced to the height of only 9 cm.

b) Sludge Sterilization

The elimination of disease-causing bacteria from sludge is a grave concern throughout the world. The only solution at present is to burn the sludge in an energy-intensive process. This is a costly method and is not always feasible. Recently, a considerable number of experiments were undertaken to eliminate disease germs in sludge by radiation. At our *Phragmites* station we observed that nature can also manage this task; an example is shown in Table 14-6.

Ecological Reflections

The Max Plank Institute (MPI) System relies on purposefully used plant cultures. This requires professional preparation and precise technical and biological execution. The following considerations guide installations:

a) Plants should not be allowed to feed from nutritious soils, but should rely only on nutrients available in the water applied to the planting bed. For this reason, plants must stand in inert material, e.g., gravel or sand, which must be washed.

b) The system requires oxygen, both for the chemical reactions induced and for maintenance of bacteria efficiency. Therefore, the water in the system should be enriched with oxygen by means of small waterfalls (cascade construction) and by drainage pipes in the sludge beds. A considerable improvement in the supply of oxygen is obtained by the use of a tidal inundation principle: repeated inundation followed by drainage. By this method, air is pulled into the root zone.

c) In most eutrophic sewage, large amounts of algae form. When they decompose, they cause secondary pollution and loss of oxygen. This accounts for further reduction in the quality of effluents. Therefore, provisions for algae control must be included.

d) Effluents must be spread evenly over the treatment beds. In addition, planting beds must be narrow enough to be controlled from the banks. It must be possible to periodically harvest the stems (biomass) easily. For this reason the installation should not have the shape of a pond but rather take the form of a trench.

e) Because the stem mass must be periodically harvested from the trenches so that it does not

TABLE 14-6. *Reduction of Bacteria in Sludge Treated by* Phragmites *Beds.*

		Influent	*Discharge*
Krefeld	E. coli	650	7
after 24 hrs.	Salmonella	450	15
0.1 ml	Enterococci	800	15
Ustersbach	E. coli	800	0
after 5 min.	Salmonella	3200	6
0.1 ml	Enterococci	3200	9
Fritzlar	E. coli	uncountable	80
after 2 hrs.	Salmonella	"	0
0.1 ml	Enterococci	"	60

return trapped material to the water or decompose, it is important that the timing and methods used for harvesting follow biological considerations, i.e., during the growing season harvesting should be done at night, and care must be taken not to damage the planting beds.

f) For certain types of effluent (e.g., dyes from the synthetic fiber industry), a plantation system alone cannot provide the desired success, and an intermediary stage of chemical treatment is of great value. The sludge produced and the treated sewage are then passed through the MPI system. The chemicals used in the intermediary stage are low in cost. We have been quite successful in the precipitation of latex and zinc.

The stems of *Schoenoplectus lacustris* have been found useful as a fodder which is rich in protein and minor elements. Thorough research and feeding experiments have been conducted at the University of Kiel. In one case, the effluent of a sugar factory was treated, and the ground up stems of bulrushes were then used to feed 10,000 to 20,000 ducks per year. However, if one does not want to use stems for fodder, they can also be used as a compost material, of great value in arid and semi-arid areas. In Europe, the stems of *Schoenoplectus lacustris* can be used as wicker, often in expensive furniture.

Sun and Gravity as Sources of Energy

Conventional sewage plants have not only high operating costs, but also high maintenance and energy costs (Fig. 14-9). The pumps used for aeration and transport of water use oil and electricity; the energy source for the system we have developed is primarily solar energy and gravity. The sun is still the most economic and most reliable source of energy we have. Today, we make the most complicated efforts to capture and use the energy of the sun. The green plant is an age-old model that we cannot respect and value too highly. The yet to be explained process of assimilation (photo- and biochemical

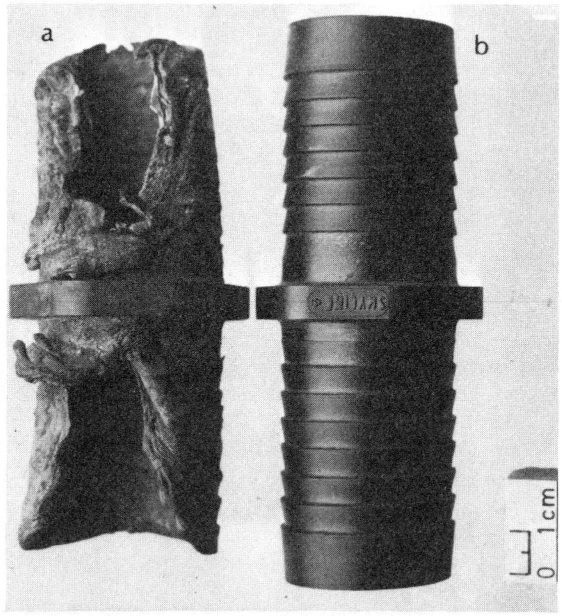

FIG. 14-9. *Schoenoplectus lacustris* (a) grown under conditions that led to the structural collapse of plastic pipe (b).

syntheses) constantly utilize nutrients and contaminants in the environment. Gravity can be used in the transport and aeration of liquid wastes through the design of sloping trenches arranged in sequence on varying levels.

Concepts for the Future

The biological MPI System described in this paper consists of two parts:

1. Sludge retention and decomposition and reduction of certain bacteria counts by beds of *Phragmites* plants.
2. Extensive elimination of dissolved organic and inorganic substances and of other kinds of bacteria by means of a cascade plantation consisting of *Schoenoplectus lacustris* or other appropriate plants.

I have no doubts that this system will prove particularly valuable at campgrounds and park sites. It offers a vital contribution to nature conservation. Exemplary work has already been carried out in the Netherlands. At eight camping

FIG. 14-10. *Schoenoplectus lacustris* in an artificial aquifer recharge basin at a drinking water treatment plant in Haltern, Germany. Single stems without nodes are easily separated to be transplanted from indigenous stands.

sites this principle has been used and the results are very promising. Further, at a winter campground an added precaution was taken by building a conventional sewage plant in addition to our system; the sewage plant has not yet been needed.

Other potential sites are farms, hotels, schools, factories, lakes, and ponds that have pollution problems. It is also possible to reduce sludge loads, ballast substances, and disease germs in polluted streams. This can be accomplished through small check dams and by planting beds in flood plains subject to frequent inundation. Likewise, an aquifer recharge area can be improved in its function by improving the quality of the water that percolates (Fig. 14-10). This has been demonstrated in Berlin where the heavily polluted Spree River water was treated in plantation ditches prior to artificial recharge and use as drinking water. A similar system was used in Krefeld, Germany.[6]

Only through many years of experience are we beginning to be able to deal with the difficult problems of heavily polluted industrial effluents. We are still amazed by the strength of plants. In spite of our emphasis on *Schoenoplectus lacustris,* we feel that we still have not explored all of its latent characteristics. We found an improbably abundant growth in the roots of this plant when it was placed in the heavy effluents produced by a printing press. From one single plant alone we could make twenty new plants by dividing it after a year of growth. This could not have been possible with a plant grown in uncontaminated water.

Nature has an immensely large catalogue of plants available for us. With the proper concern, training, "green thumbs," and patience, we may choose the best alternative for preserving water as the basis of life. I regard finding this alternative as worthy of our efforts and our sacrifices.

Notes

1. K. Seidel, "Pflanzungen Zwischen Gewässern und Land," *Mitteilungen aus der Max-Planck-Gessellschaft* ["Plantings between Water and Land," *Journal of the Max Planck Institute* (Germany)] (1953), pp. 17-20.

2. Formerly *Scirpus.* Although the species is renamed in this chapter, the *Scirpus* designation is used in other chapters.

3. K Seidel, "Uberraschende Möglichkeiten der Nutzung von Höherer Wasserpflanzen," *Mitteilungen aus der Max-Planck-Gessellschraft* ["New and Surprising Opportunities for the Use of Higher Aquatic Plants," *Journal of the Max Planck Institute* (Germany)] (1974), pp. 479-94.

4. K. Seidel, "Die Flechtbinse *Schoenoplectus lacustris,* Ökologie, Morphologie und Entwirkung Bedeutung" ["The Bulrush, *Schoenplectus lacustris,* Its Ecology, Development of Use by Different Populations and Its Economic Value"] (thesis, Germany, 1955).

5. For a discussion of carcinogens in drinking water, see Chapters 9 and 10.

6. The water treatment system used in Krefeld is described in Chapter 33.

15

The Potential of Submersed Vascular Plants for Reclamation of Wastewater in Temperate Zone Ponds[1]

CLARENCE D. McNABB, JR.

Department of Fisheries and Wildlife
Michigan State University

Introduction

The State of Michigan occupies a portion of mid-continental North America that drains by regional rivers to one or the other of the surrounding St. Lawrence Great Lakes. The immense value of these lakes as a freshwater resource and their gradual degradation by the input of wastes from the urban and agricultural activities in the drainage basin has been well publicized. There are sound political, economic, and ecological reasons for protecting this resource from contamination by human pathogens, toxic substances, and the processes of eutrophication. To do so, wastes of Michigan need to be contained and recycled on the watersheds of the state. A facility has recently been established (1974) on the campus of Michigan State University at East Lansing for the purpose of research applied to this goal.

This system consists of a primary treatment unit, an activated sludge facility, four ponds with an aggregate area of 40 acres and an average depth of 6 feet (Fig. 15-1), and 143 acres of land that are equipped for spray irrigation (Fig. 15-2). Three one-acre marshes have been incorporated in the design so that the potential for treatment by that ecological community may be evaluated. Wastewater for this system originates in the university of forty thousand students and a residential community of similar size. The portion of the waste flow shunted to the research-demonstration facility can amount to a maximum of 2 million gallons per day.

Interest and involvement in the research being carried out at this facility are shared by a number of elements in the university; these are coordinated by the Institute of Water Research on the campus. This paper reports on a single biological component of the many being used in the treatment and recycling system—submersed aquatic vegetation.

FIG. 15-1. The system of ponds receiving the effluent of primary and secondary (activated sludge) treatment of wastes of the Michigan State University community.

Adapted Species and Their Requirements for Growth

Submersed vascular plants have the habit of growing attached to the bottom sediments and producing upright stems and leaves that can fill the underwater space with vegetation. Unlike the microscopic free-floating algae of wastewater ponds, their growth form lends itself to efficient harvest with low energy expenditure. Since they are known to concentrate deleterious compounds and other elements of wastewater from relatively dilute solutions, their potential for management in water reclamation has long been suspected.

In Michigan, populations commonly develop in those municipal waste stabilization ponds that are continually aerobic, or nearly so, through the growing season. In anticipation of operation of the Michigan State University treatment system, certain of these ponds have been the subject of our studies over recent years (Fig. 15-3). The species found to occur represent a very restricted segment of the submersed plant flora of the region. In particular, the clay seals of newly constructed aerobic facilities tend to become occupied by thin-leafed species of the genus *Potamogeton (P. foliosus, P. berchtoldii,* and *P. pectinatus).* As ponds supporting these species become progressively enriched over the years with organic particles and compounds, this assemblage is replaced by certain plants of the hypereutrophic communities of regional lakes, namely, *Ceratophyllum demersum* or *Elodea canadensis.* This progression of plant species is suggested in Fig. 15-4. While seed production is greater in the potamogetons, all of these species rely heavily on an over-wintering vegetative carpet along the pond bottom for regeneration in the subsequent spring. A substantial rate of growth occurs only at water temperatures greater than 10°C, which are normally confined to the period from approximately May 15 to the first week in October in Michigan ponds.

In addition to these limitations imposed by oxygen and temperature, the size of the annual harvestable crop of submersed vascular plants in wastewater ponds is ordinarily limited by light, to a lesser degree by the physical character of the sediments and excessive wave action, and conceivably by an overabundance of essential or nonessential elements that would be functionally toxic.

Light limitation on submersed plant growth in wastewater ponds can be imposed by suspended inorganic or organic sediment particles or by planktonic algae. Establishment of the plant community is dependent upon a relatively low density of such materials, which retard light penetration to the bottom stratum from which germination occurs. The free-floating microscopic algae are particularly important in this regard. The excessive nutrient supply of the wastewater system weakens the competition for nutrients that occurs in lakes between these algae and the submersed aquatic plants; consequently the algae tend to be abundant and cloud the water.

In Michigan ponds, the water clarity necessary for the development of a high-density harvestable crop of rooted plants also depends on algal (and detrital) grazing activity of relatively stable zooplankton populations commonly dominated by the waterfleas, *Daphnia pulex* and *Daphnia*

FIG. 15-2. Old fields, woodlots and cultivated fields for land disposal of wastewater from the Michigan State University community. Pond portion of the water reclamation system in right-background.

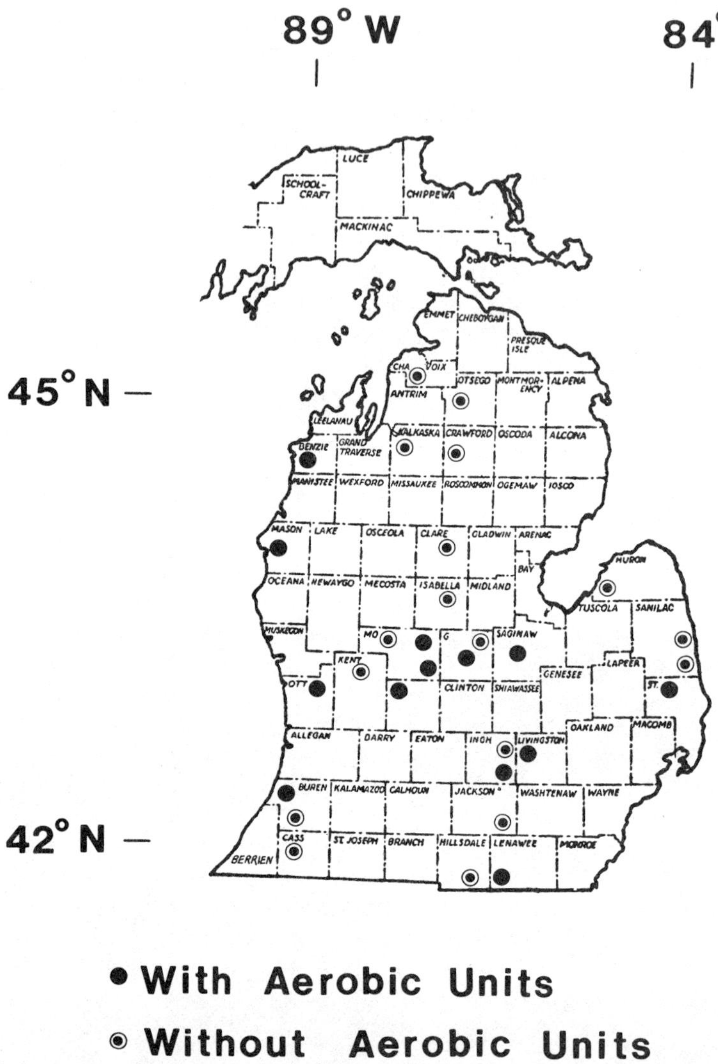

FIG. 15-3. Municipal stabilization pond systems in Michigan sampled for aquatic vegetation 1968-1974. Systems with ponds that are aerobic in the growing season have submersed vascular plants; those without such ponds do not.

magna. Because the isolation of these ponds slows the immigration of fish that prey on these filter-feeding zooplankters, and because the tendency toward annual winter-kill of predators (in the absence of oxygen), zooplankton populations are unusually effective at grazing. That they play a major role in maintaining growing-season water clarity is suggested in Fig. 15-5. The particular system for which the figure has been drawn receives untreated sewage from a community of approximately five thousand. Pond 2 in this system is a facultative cell without submersed vascular plants. It supports a stable, high-density community of typical waste-pond free-floating algae. Light penetration in this pond over the growing season is very poor as shown. Ponds 3, 4, and 5 of this system are ordinarily aerobic cells in which submersed plants flourish in good light and in the presence of high-density populations of grazing zooplankters. In August of

	Placed in Operation	No. of Ponds	No. Ponds Occupied	P. foliosus	P. berchtoldii	P. pectinatus	C. demersum	P. zosteriformis	E. canadensis	H. reticulatum	Cladophora fracta	P. filiformis	Najas marina	Lemna trisulca
St. Charles	1972	2	1	▲										
Coopersville	71	2	1	▲	▲									
Beulah	70	4	1	▲	▲	▲								
Carson City	69	3	1	▲									▲	▲
Ithaca	68	3	2		▲	▲						▲		
Belding	66	5	3	▲	▲	▲	▲	▲			▲			▲
Scottville	65	3	1	▲	▲		▲	▲						
Stockbridge	65	3	1	▲		▲	▲							
Yale	65	2	1	▲		▲	▲							
Morenci	65	2	1		▲	▲								
Bangor	65	3	2			▲			▲					
Fowlerville	64	3	2				▲	▲	▲					
Edmore	1965	3	1							▲				

FIG. 15-4. The dominant submersed vascular plant vegetation (species of *Potamogeton, Ceratophyllum* and *Elodea*) of municipal waste stabilization ponds in Michigan arranged by years in operation. Epipelic algae (species of *Hydrodictyon* and *Cladophora*) and submersed vascular plants *(P. filiformis, N. marina* and *L. trisulca)* that occur sporadically as co-dominants are positioned to the right.

the year of the data, 0.5–1.0 cm fathead minnows *(Pimephales promelas* Rafinesque*)* were planted to Pond 3 at a rate of 0.4 kg fresh weight per hectare by the Fish Division of the Michigan Department of Natural Resources. Three months later, a harvest of 424 kg fresh weight per hectare of these fish was made. This was some fraction of the total fish crop in the pond. During the months of impressive fish growth, the zooplankton populations of Pond 3 were decimated. In the two weeks following the planting of fish the water of Pond 3 turned green with planktonic algae and the light regime shifted dramatically to the range of penetration shown for late season in Fig. 15-5. Poor conditions of light persisted through the remainder of the growing season in this pond, while the penetration of light in Ponds 4 and 5 remained essentially unaltered through the June-October interval of measurement. An appreciation for the degree of clarity maintained by the zooplankters of Ponds 4 and 5 in the face of a continuous flow of algae from upstream ponds can be obtained by noting that if these ponds were deep enough there would be adequate light at 6 meters (approximately 19.5 feet) for the growth of rooted aquatic plants. This degree of water clarity would be considered excellent for recreational lakes of the region.

From these comments concerning light, there are two important observations for management to be made: zooplankton populations are a key to maintaining water clarity in nutrient-rich aerobic wastewater ponds in the growing season, and a *maximum* harvest of submersed vascular plants and a *maximum* harvest of animals that control the planktonic algae either directly (zooplankton) or indirectly (zooplanktivorous fish) are not compatible goals for the same pond in that season.

Limitations to growth of submersed plants due to sediment type and the scouring action of waves at the shoreline of ponds appear to be of little importance as these units are presently constructed in Michigan. That is not to say that these features can be overlooked in design, if growth and harvest of these plants are principal goals of water reclamation systems. Regarding chronic toxicity to the plants from a buildup of ions or compounds in the sediments, it may be instructive to note that growth rates and annual maximum standing crops that we have observed are as high as those reported for submersed species in various lake and pond environments around the world. Many of the substances likely

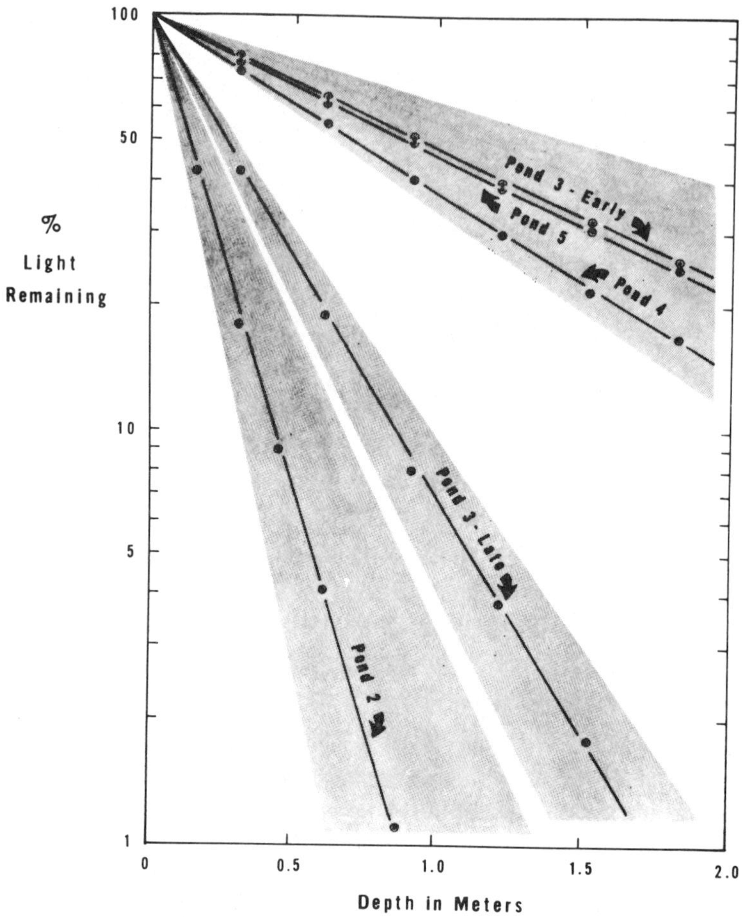

FIG. 15-5. Summertime light penetration in a municipal wastewater pond series. Penetration changed in Pond 3 following the introduction of fathead minnows *(Pimephales promelas)* in August. Pond 2 was a phytoplankton dominated facultative cell; ponds 3, 4, and 5 were aerobic cells with high density vascular plant crops. Lines are drawn for the mean condition; shaded areas show the range.

to be toxic to these plants are sedimented or decomposed in earlier portions of the treatment process. Additionally, as with plants in general, the living portion of the individual cell is considerably buffered by the capturing capacity in the chemical structure of the outlying, nonliving cell wall. Toxic concentrations of various materials (particularly heavy metals) may be trapped in the wall without disrupting the vital functions of the enclosed cell.

The Potential for Harvest

It is apparent from the above discussion that crops of submersed vascular plants are restricted in their occurrence to that portion of a treatment system that is tertiary in nature. They can participate in the reclamation of wastewater principally by trapping soluble compounds and complexes that pass from preceeding stages of the treatment process. In particular, phosphorus and nitrogen removal from secondary effluents is important for protecting the St. Lawrence Great Lakes and smaller lakes of the region from the degradation associated with eutrophication. While growing or at maximum standing crop, the submersed plants contain a concentration of these elements on the order of $0.5-1.0 \times 10^4$ greater than the concentration of total phosphorus or total nitrogen in the wastewater in which they have developed. While the question needs further definitive work, it appears that this

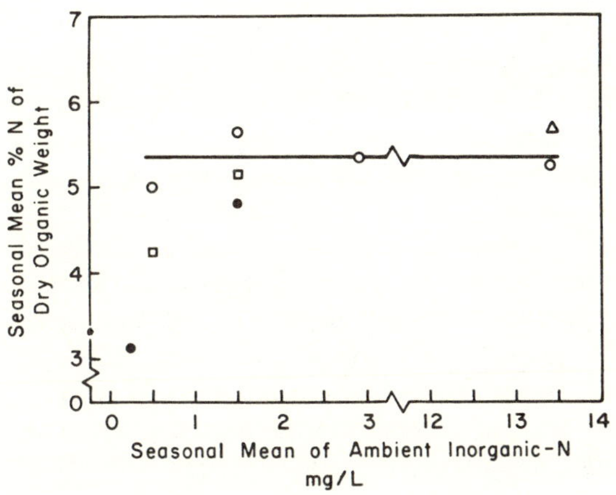

FIG. 15-6. Tissue concentrations of nitrogen in submersed vascular plants growing at different concentrations of inorganic-N in the ambient water of ponds. Line of best fit shown for the data of *Ceratophyllum demersum* (O). That a generalized curve for submersed plants may break downward near 1.0 mg/l inorganic-N is suggested by significant differences at the 5% level between concentrations in tissues of *Elodea canadensis* (□) and *Elodea nuttallii* (●). △ is for *Egeria densa.* Ordinates for % dry weight can be estimated by reducing the values here by 20%.

accumulation in wastewater ponds is made largely from the water through the foliage, rather than from the sediments via the roots. Where land irrigation is an important feature of the water reclamation process (as at Michigan State University), the submersed aquatic vegetation can be used as one of the tools for protecting the productivity and longevity of the soil system. The need to protect crops and the groundwater from excessive accumulations of nitrates is well documented in the literature. Harvestable submersed vegetation from ponds of our region will contain 50–90 mg of boron per kg dry weight. The small, free-floating vascular plant *Lemna minor* (common duckweed) is known from our work to be an accumulator of boron relative to other aquatic plants. It holds quantities in excess of one order of magnitude greater than the range given above while growing on the same wastewater with the submersed plants. Boron in our waters originates primarily in household products and their manufacture. It has been implicated in crop toxicity on sites in our region irrigated with municipal wastewater.

The processes of accumulation of nitrogen and phosphorus in submersed vegetation differ for these elements in ways shown by a comparison of Fig. 15-6 and 15-7. The nitrogen concentration in the tissue in hypereutrophic ponds is constant—near 4.2% of the dry weight of the plants (5.3% of the organic dry weight)—and is independent of the mean concentration of the available forms of nitrogen (inorganic-N) in the water during growth (Fig. 15-6). A significant linear regression regarding the relationship between tissue and ambient phosphorus is shown in Fig. 15-7. The curves presented have been drawn for *Ceratophyllum demersum.* That these relationships are not specific for *C. demersum* is suggested on the figures by the inclusion of data points for *Egeria densa, Elodea canadensis,* and *Elodea nuttallii.* Other dominants in the pond vegetation discussed

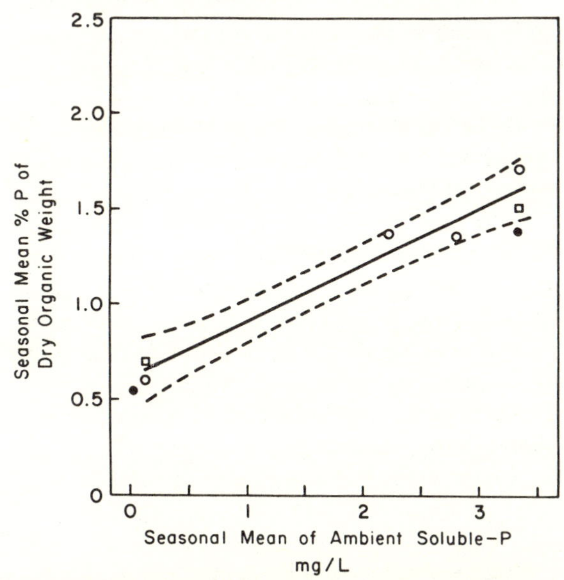

FIG. 15-7. Regression of tissue-P on ambient concentration of soluble-P for the period of growth of submersed vascular plants. $\hat{y} = 0.62 + 0.29x \pm 0.06$ is for *Ceratophyllum* demersum (O); 95% confidence limits are shown. *Elodea canadensis* (□) and *Elodea nuttallii* (●) are shown for comparison. Ordinates for % dry weight can be estimated by reducing the values here by 20%.

above (e.g., species of *Potamogeton*) have been shown to behave similarly as well. Knowing the mean concentration in the water in a particular pond during the period of growth, values from these figures can be used to predict the amount of each element to be removed by the harvest of a particular weight of plants. We know that the relationships expressed here are essentially correct; the accuracy of the predictions regarding phosphorus can be somewhat improved by additional work.

Predictions regarding the pattern of growth of species adapted to the newly constructed ponds at Michigan State University have been made from measurements taken in recent years on other municipal systems. These have been made on units that were aerobic in the growing season, and had a favorable light regime for the reasons discussed earlier. The assumptions important in these predictions of growth are implied in the concepts of Fig. 15-8.

Development of a crop normally begins in the spring from a low weight of reproductive tissues. This begins to grow near mid-May. By early June (June 5 has been chosen as representative), the plant population reaches a rate of growth at which it doubles each 11 days (on the average) for a period extending to between 60 and 70 days. At that time, a maximum standing crop exists. Vegetation fills the available lighted space in a depth of 4–5 feet of water and further growth is thus limited. The dry weight of vegetation to be expected per square meter at this point in the season is on the order of 225–250 gm. This is present in early August. Maximum quantities of harvestable nitrogen and phosphorus are contained in the vegetation at this time. These can be estimated with the data of Fig. 15-6 and 15-7.

Various strategies for timing the harvest of the plant crop are possible. The strategy of Fig. 15-8 calls for a first harvest in early August as plant weight per unit area approaches the maximum. This harvest is to consist of approximately one-half of the weight of the vegetation, to be removed after cutting an appropriate portion of the tops of the plants. The remaining plant tissue is expected to generate new growing parts. After a lag time of 3 days, this tissue will double its weight in an average of 11 days, reaching the surface and a maximum crop density approximately 14 days after the first harvest.

This procedure can be repeated an additional three times in the growing season with a total yield from this depth of water of approximately 400 gm dry weight per square meter. Following the fourth harvest the vegetation is allowed to expand to the maximum density, reaching that point as temperature limitation to growth occurs in early October. In the pond systems that have been studied, 98–99% of the vegetative biomass degenerates in the winter months, leaving 2–4 gm of tissue per square meter to begin development in the subsequent spring. A fifth harvest of the crop in the fall will have the effect of reducing the weight of viable tissue that over-winters. The size of the crop in the following year will be drastically reduced if this is done.

The submersed plants differ dramatically in the last regard from emergent species that are considered elsewhere in this volume. These tend to over-winter a weight of regenerative tissue 2-3 orders of magnitude greater than the submersed plants. This is contained in underground rhizomal tissue that is lacking in submersed species adapted to wastewater systems. The annual production of harvestable biomass in emergent plants from year to year may be greater for this reason. It is intended that the marshes of the Michigan State University facility will be managed to demonstrate the potential of the emergent community.

The harvest strategy described above for submersed plants will provide a maximum annual yield; any other will produce a smaller yield. While studies on the submersed plant species are not complete, there is the inference in our work that the off-season loss of viable regenerative parts may be reduced in ponds where the bottom stratum of water and the sediment surface are in a less reduced (more oxidized) state than those for which the observations for Fig. 15-8

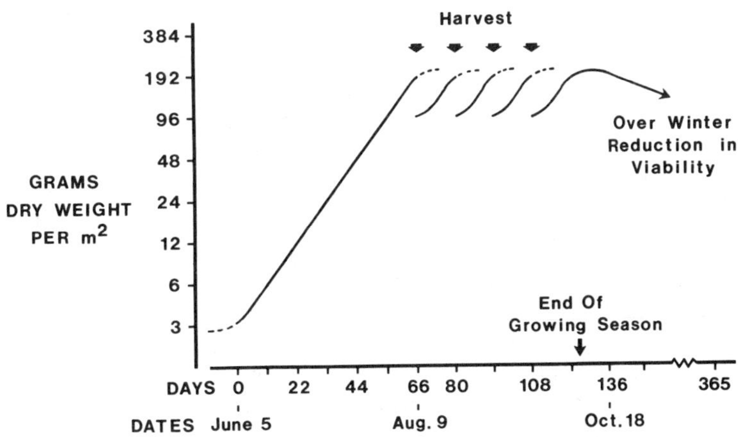

FIG. 15-8. Generalized pattern of growth and over-winter degeneration of submersed vascular plant tissue in municipal wastewater ponds in Michigan, with strategy suggested for maximum annual yield by harvest.

were made. The ponds on our campus may be managed over the years in ways that promote this condition. The capability to improve the over-winter survival of those plants reproducing principally by vegetative means (e.g., *Elodea canadensis)* to 20–30% of the weight of the fall crop appears within reach. Having accomplished this, the first maximum crop would be obtained approximately one month earlier than shown in Fig. 15-8, and at least two additional harvests could be obtained using the format of that particular figure. One should bear in mind that these predictions concerning an anticipated yield have come from systems that are essentially unmanaged for the goals discussed here. Management research will likely improve this yield for systems in which such a yield is considered important in the total reclamation scheme.

By our best estimate, a harvest like that of Fig. 15-8 would remove 20–25% of the phosphorus and 50-70% of the nitrogen that is influent (as total-P and total-N) to an aerobic pond in the growing season operated at a retention time on the order of 28 days. Increasing or decreasing the retention time (thus changing the rate of supply to the absorbing plant tissues) would be expected to have a relatively small effect on the efficiency of removal for phosphorus. The absolute quantities of this element in each harvestable crop are influenced by water concentration in the manner of Fig. 15-7. Since the nitrogen concentration in the plant tissues does not change over the range of water concentrations noted in wastewater ponds of previous studies (Fig. 15-6), increasing or decreasing the retention time would tend to have an effect in the same direction on plant efficiency. Similarly derived estimates of efficiency for other elements harvested with the crop appear to be on the order of 80-100% for manganese, 20-30% for iron, 5-10% for copper and zinc, and 1-3% for cadmium, chromium, cobalt, and nickel.

The degree of benefit to the water reclamation process from harvesting and recycling submersed aquatic plants will be demonstrated for the temperate zone facilities at Michigan State University in the years ahead. The limits of certain of the anticipated benefits can be ascertained from this discussion.

Summary

The crop-producing species of submersed vascular plants of municipal wastewater ponds in Michigan are restricted to a few representatives of the regional flora. *Potamogeton foliosus, Elodea canadensis,* and *Ceratophyllum demersum* are principal among them. The first of these invades the clay seal of newly constructed ponds, while the others replace this species over time.

The size of the annual harvest of these plants

tends to be limited in such ponds by temperature, oxygen, and light. Appreciable growth occurs above 10°C for the interval from mid-May to early October. Persistent growth of vegetative tissues requires oxygen; crops do not develop in facultative or anaerobic ponds of a system. Planktonic algae tend to cloud the nutrient rich water of aerobic ponds. They are ordinarily held in check by the grazing of the zooplankters *Daphnia pulex* and *Daphnia magna* that are themselves exposed to an unusually low predation pressure in ponds of municipal systems. These zooplankters are the key to maintenance of a light regime in which the submersed vascular plants flourish.

A harvestable crop typically begins development from a low weight (2–4 gm/m^2), over-winterizing biomass of vegetative and, to a lesser degree, seed or specialized dormant tissues. In the 60–70 day interval of early June to early August the plants double in dry weight per unit area on an average of each 10–12 days. The annual maximum dry weight of the standing crop produced is on the order of 225–250 gm/m^2. A harvest strategy with maximum annual yield that will be tested in newly constructed ponds at Michigan State University has been presented. It will yield 400 gm dry weight per square meter. A severe constraint on this yield is the requirement to leave a maximum density of vegetation in ponds in the fall in order to insure an adequate amount of tissue for regeneration of the crop in the following spring. In municipal systems that have been studied, only 1–2% of the biomass of the annual maximum crop over-winters. This value is from ponds in series that receive untreated municipal sewage. The inference exists in our data that this can be improved to 20–30% for species over-wintering vegetatively (particularly *Elodea canadensis*) by management programs aimed at preventing the development of winter conditions associated with low oxidation-reduction potentials in the sediments that accumulate over the clay seals. If this can be achieved, the annual yield can be substantially increased and the sediments thereby further protected from organic accumulations.

Graphs have been presented that can be used to make close estimates of the amounts of phosphorus and nitrogen that can be harvested with the submersed plants from ponds for which data exist regarding concentrations of soluble-P and inorganic-N in the water during growth. By our best estimates from field conditions, a harvest strategy with the yield mentioned above would remove a quantity of phosphorus equivalent to 20–25% of the total phosphorus influent in the growing season to a pond having a retention time on the order of 28 days. Similarly calculated estimates are 50–70% for nitrogen, 80–100% for manganese, 20–30% for iron, 5–10% for copper and zinc, and 1–3% for cadmium, chromium, cobalt, and nickel. These efficiencies for the submersed vascular plants should be evaluated from the perspectives of energy use and resource economics as compared to efficiencies achieved by other strategies applicable in the temperate zone to the water reclamation process as a whole.

Notes

1. D. A. Bulthuis, B. T. Coffey, J. R. Craig, R. P. Glandon, S. R. Kosek, J. B. Lisiecki, D. C. Mahan, and D. P. Tierney have been participants in the evaluation presented here while progressing through their graduate degree programs. Laboratory facilities have been kindly made available by B. D. Knezek and F. M. D'Itri. J. L. Kotenko provided technical assistance. Financial support has been given by the Michigan Agriculture Experiment Station under Project No. 1157, the Rockefeller Foundation, and the U. S. Department of Interior Office of Water Resources Research (Grant A-073-MICH) under funds administered by the Institute of Water Research at Michigan State University and by the U. S. Environmental Protection Agency Office of Water Programs under Training Grants WP-264 and T-90031.

16

The Purification of Wastewater with the Aid of Rush or Reed Ponds

JOOST DE JONG

Federal Commission for the IJsselmeerpolders
Netherlands

Introduction

After a large flood in the Zuiderzee area in the beginning of this century the Dutch government decided to seperate the Zuiderzee from the North sea by means of a barrier dam and to create five polders in the fresh water lake, the IJsselmeer (Lake Yssel), that was going to be formed (Fig. 16-1). The aim for making these polders was an extension of the area of land suited for agricultural use. During the last fifteen to twenty years there has been a change from an agricultural development of the polder into a multipurpose development in which urbanization, recreation, preservation of areas of high biological interest, and agricultural exploitation are considered.

Two polders have been completed; one is nearly finished; the fourth is being developed now; and the fifth (The Markerwaard) still has to be made. In the third and the fourth (Flevoland), polder townplanning, nature preservation, and recreation are becoming important factors (Fig. 16-2). In these polders two new towns, Lelystad and Almere, are being built that will have populations respectively of about 100,000 and 250,000 inhabitants in the year 2000.

In Flevoland many recreation sites are situated along the lakes between the polders and the older areas of the Netherlands. Here beaches and camping sites have been made. On both beaches and camping sites the wastewater is collected with a system of sewers and is pumped into the polder where it has to be purified. Where the distance between these beaches and camping sites and the existing sewage treatment plants in the neighboring villages was too big one had to consider alternatives that would enable treatment of rather small amounts of sewage, only existing in summer. Any treatment method for such a situation should fulfill the following requirements:

a) ease of operation
b) low installation and building costs

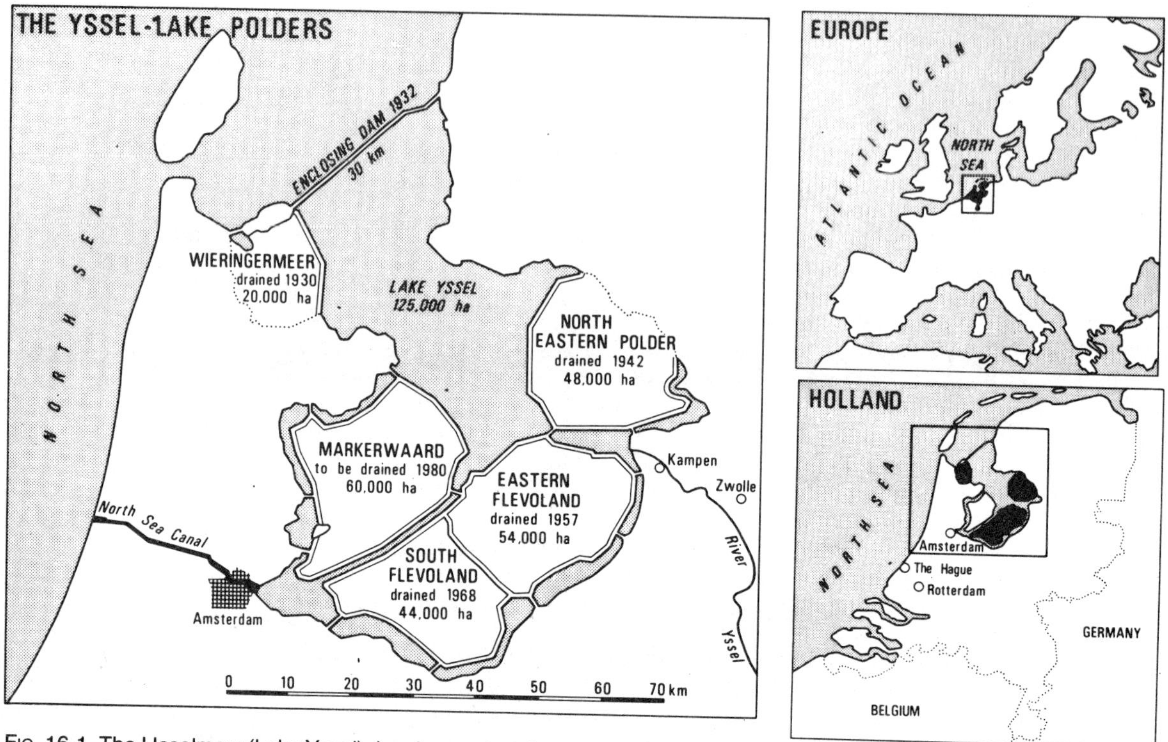

FIG. 16-1. The IJsselmeer (Lake Yssel) development project.

c) low operation costs
d) insensitivity to fluctuating loads
e) good purification results

This last requirement should hold not just for the biological oxygen demand (BOD) or chemical oxygen demand (COD) but also for the major nutrients in the water, phosphorus and nitrogen, which might otherwise cause secondary pollution by eutrophication.

Waste water treatment with the aid of ponds with rushes or reeds

The work of Käthe Seidel describes the possibility of wastewater treatment with ponds in which aquatic plants are grown.[1] In particular, the rush *Scirpus lacustris* was said to possess high wastewater purifying qualities. Based on this information it was decided in 1967 to start experimental treatment of the sewage of a camping site with a rush pond. In 1969 a number of experimental ditches for sewage treatment were dug that enabled a comparison of the properties of rushes, reeds *(Phragmites australis),* and polypropene fibres in purifying sewage. From this comparison and laboratory research it was found that the aquatic plants function mainly by providing attachment sites for microorganisms purifying the sewage. Secondly the aquatic plants consume part of the nitrogen and phosphorus supplied to the pond.[2] The first experimental pond (Fig. 16-3) had the size of one hectare (2.5 acres) and a depth of 0.4 meter. Since the sewage supply from the camping site was varied each day, additional water was supplied from a neighboring lake.[3]

To realize an optimum utilization of the available space a star-shaped layout was chosen for this pond. In practice, however, the maintenance of this arrangement proved very difficult, so subsequent ponds were constructed as long ditches that could be mechanically maintained.

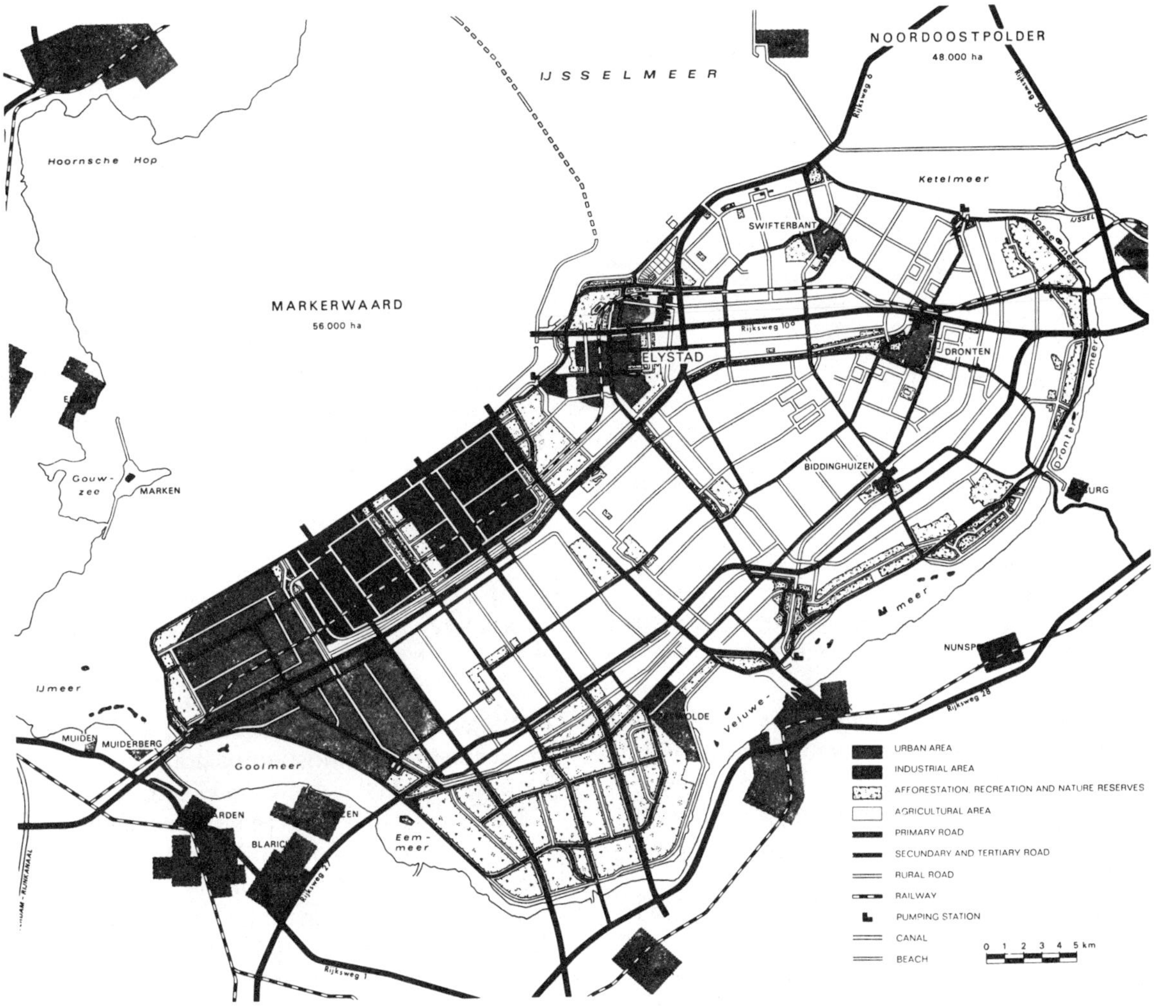

FIG. 16-2. The land-use plan of Flevoland. This plan is already realized in the north-eastern part of the polder and is being realized in the other parts of the polder.

For a calculation of the purification obtained with the pond it is necessary to know the amount and composition of the water flowing in and out. The following flows have to be considered:

a) the sewage	BOD_5^{20}	200–400 mg/l
	COD	400–1000 mg/l
b) the supply water	BOD_5^{20}	10–20 mg/l
	COD	60–70 mg/l
c) the rain	BOD	less than 5 mg/l

d) the evaporation
e) the infiltration into the soil below the lagoon
f) the effluent

From the measurements it was apparent that there was a significant amount of infiltration into the bottom. It decreased from 20 mm/day in 1967 to about 7 mm/day in 1973/74. The evaporation from the water surface and the plants reached an average value of 5 mm/day. For 1969 the measured amounts of water during

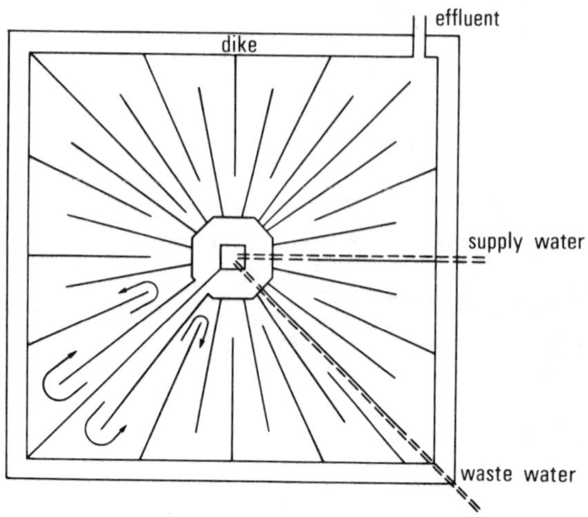

FIG. 16-3. The first experimental rush pond. Area 1 hectare (2.5 acres) depth 0.4 m. A star shape layout was chosen in order to obtain optimum utilization of the available area, which, however, complicated mechanical maintenance.

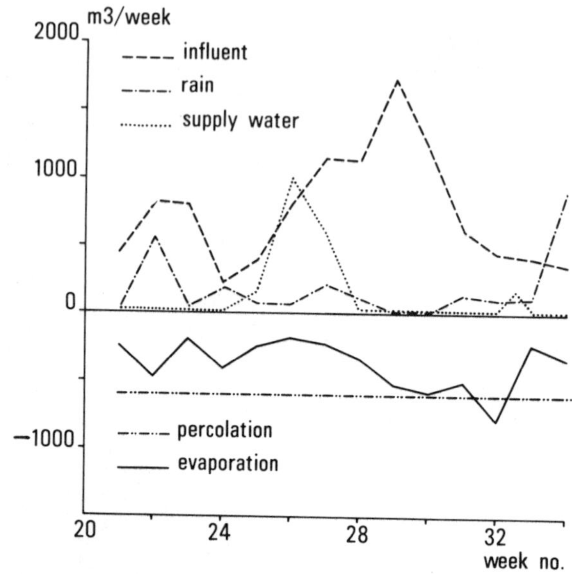

FIG. 16-4. The amount of water, according to route, entering and leaving the rush pond during operation in 1969 (June, July, August).

the height of the summer are given in Fig. 16-4. In 1969, the third year of operation of the pond, when we considered the results to be free of starting effects, both influent and effluent were sampled and analyzed for BOD, COD, and bacteriological quality (most probable number: MPN). Some of the results are show in Table 16-1.

It can be seen that in all circumstances, even in the weeks with peak loads, a good purification of the water is obtained. Also the bacteriological contamination is highly reduced. The obtained reduction in total-phosphorus (total-P) and total-nitrogen (total-N) is shown in Fig. 16-5.

It appears that at first the removal of phosphorus and nitrogen is considerable, but the removal efficiency decreases to almost zero in the last weeks. The reason for this is that the supply of phosphorus and nitrogen exceeds the uptake capacity of the vegetation. This can be seen too from Table 16-2 in which for rush and reed the uptake of phosphorus and nitrogen is compared

TABLE 16-1. *The Purification Results with the Rush Pond in 1969. BOD_5^{20} and COD in mg/l. Bacteriological results as MPN/ml.*

mg/l		week number 26	28	30	32	34	mean week 26-33
BOD_5^{20}	influent	285	331	347	276	127	257
	effluent	12	8	18	17	7	11
COD	influent	661	734	900	590	285	530
	effluent	48	54	94	83	66	70
MPN/ml	influent	43×10^4	40×10^4	38×10^4	52×10^4	41×10^4	36×10^4
	effluent	1	64	2670	8	14	313

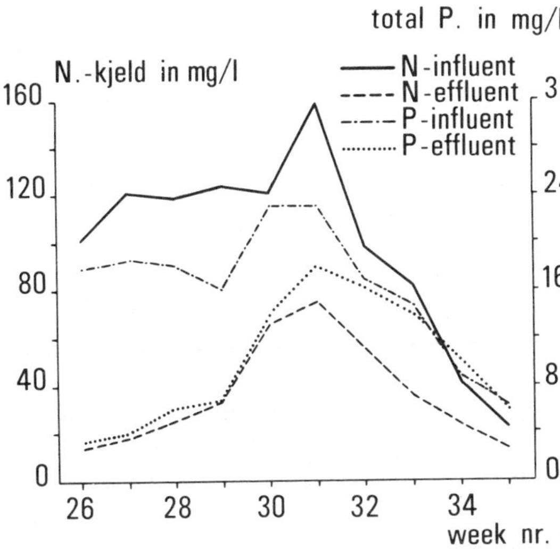

FIG. 16-5. The amount of N-Kjeldahl and total-P in the influent and effluent of the experimental rush pond during operation in 1969 (June, July, August).

with the amount supplied to the pond. When we want to achieve a satisfactory elimination of phosphorus and nitrogen the load on the pond should not exceed the uptake capacity of the vegetation.

The purification of the sewage is based on two phenomena. First, part of the water (about 9 mm/day in 1969) infiltrates into the soil below the pond. Investigation of the groundwater below the pond (1–2 m depth below the bottom of the pond) showed almost no increase of BOD, COD, and phosphorus content when compared to reference samples. A slight increase of the ammonium-nitrogen content was noticed. From this it was concluded that the water that infiltrated into the soil was highly purified in the soil just below the pond. Secondly, the water that flows through the pond is purified by microorganisms in the pond and by uptake of nutrients by these organisms and the plants. In the center of the lagoon where the sewage and supply water enter, some sedimentation of coarse and heavy materials occurs. This sediment is removed once a year.

It is clear that the retention time of the water in

TABLE 16-2. *The Annual Production of Reed and Rush Expressed in kg/ha for P and N and Compared with the Load of the Pond in 1969.*

Vegetation	Annual production (kg/ha)			
	Above ground		Below ground	
	P	N	P	N
rush	50	260	55	320
reed	35	270	20	160
total load of the pond	167	1004		

the pond affects the purification results. So the influence of this factor on the results was determined. The overall purification as a function of the retention time is given in Fig. 16-6.

As can be seen, for retention times of about 10 days or more good purification results can be obtained. If the purification that occurs in the soil below the pond (assuming this water will be purified 100%) is subtracted from the overall results a relation between retention time and the purification obtained in the water layer can be found. The result is shown in Fig. 16-7. The obtained overall purification results for retention times of 10 days or more are good and more complete with respect to phosphorus, nitrogen, and bacteria removal than most biological sewage treat-

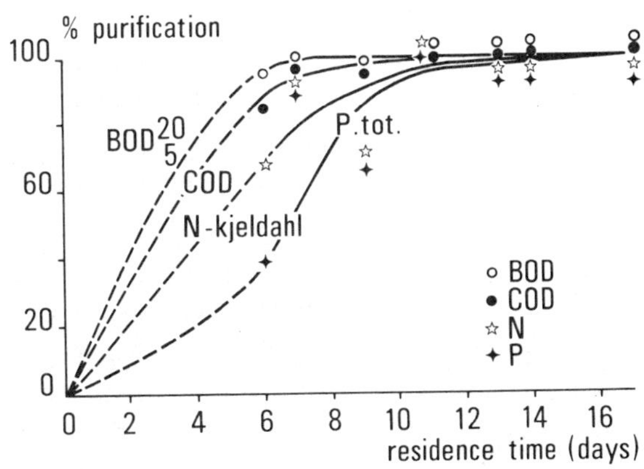

FIG. 16-6. The overall sewage purification results with the rush pond in 1969 as a function of the retention time of the water.

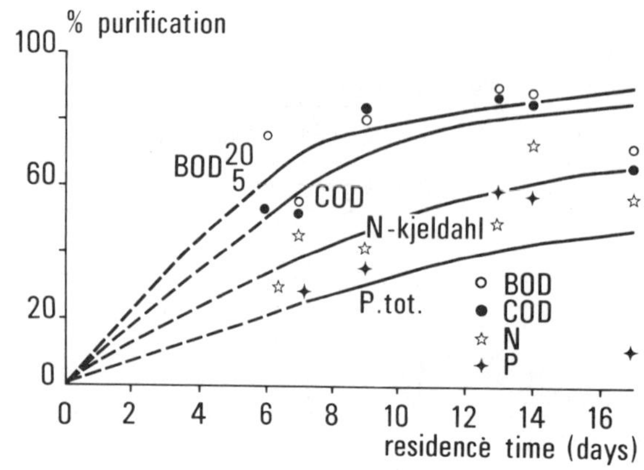

FIG. 16-7. The sewage purification results with the rush pond in 1969 as a function of the retention time after correction for the purification in the soil below the pond.

ment plants. For Dutch situations the comparison of purification results shown in Table 16-3 can be made.

Financial aspects

A comparison of both investment costs and annual costs has been made for an activated sludge type treatment plant (compact type with a capacity of 2000 population equivalents [p.e.]) and a rush lagoon (also designed for 2000 population equivalent). It is noted that one population equivalent corresponding to 100 liters sewage/day, 54 g BOD_5^{20}/day, about 10 g N-Kjeldahl/day and about 3.5 g P/day. From this comparison we learned that for these small size treatment units, treatment of sewage with the rush pond is considerably cheaper than with activated sludge type plants. The results are summarized in Table 16-4.

TABLE 16-3. *A Comparison of Purification Results for Dutch Situations.*

Type of treatment plant	% reduction			
	BOD	total-N	total-P	MPN
trickling filter	60-95	< 50	25-30*	70-95
activated sludge	50-95	< 50†	25-30*	70-98
pond with rush (10 days retention time)	> 98	95	93	> 98

*With phosphorus removal (Fe, Ca) up to 90%
†Oxi-denitro process enables up to 80–90% nitrogen removal.

Future work

Since the rush or reed ponds are used for the treatment of sewage from camping sites, we have no results for operation during the winter. In 1972 a pond was designed to treat the effluent of a biological treatment plant in order to obtain an overall polishing of the effluent. The results were good, but the required area was regarded as being too large to be used for the treatment of the effluent of the two rather large planned cities. Therefore, it was decided to use the purification properties of the soil as far as we can in order to decrease the area of land needed. Experiments on a laboratory scale were started in 1974. Soil columns of 0.5 m with different filtration properties were percolated with effluent of a biological sewage treatment plant. All experiments were done at a temperature of 20°C. It appeared that sand with a permeability of about 1.5 m/day removed 35% of the total nitrogen (N-Kjeld. + $N\text{-}NO_3$) and over 99% of total phosphorus. A sandy clay with a permeability of 2 cm/day removed 80% of the total nitrogen and over 99% of the total phosphorus from the effluent (Table 16-5).

We are now designing a first experimental infiltration field for the purification of part of the sewage effluent of a treatment plant of 40,000 population equivalents. Also the first infiltration

TABLE 16-4. *A Comparison of Investment and Annual Costs for a Rush Pond and an Activated Sludge Type Plant.*

Rush pond	investment/p.e.	Dfl. 30.—*
2000 p.e.	annual costs/p.e.	Dfl. 4.50
Activated Sludge Plant	investment/p.e.	Dfl. 183.—
2000 p.e.	annual costs	Dfl. 18.50

*1 U.S. dollar equals about 2.5 Dfl.

TABLE 16-5. *Sewage Effluent Purification Results of 0.5 m Soil Columns with Different Permeability for Water (laboratory results).*

		N–Kjeld.	$N–NH_4$	mg/l $N–NO_3$	total–P	$P–PO_4$	Filter percolation speed (m/day)
	Effluent composition	3.1	0	25.0	13.3	13.0	
Sand	Filtrate composition	1.5	0	16.5	0.06	0.04	1.65
Sandy clay	Filtrate composition	4.7	0	0.9	0.08	0.01	0.02

field for sewage treatment with reed as the vegetation was put in operation in the summer of 1975. We believe that the results obtained with these fields will be as good as with the ponds while consuming less space.

Notes

1. Käthe Seidel, "Abbau von Bacterium Coli durch Höhere Wasserpflanzen" [The Removal of Coli Bacteria by Higher Aquatic Plants], *Naturwissenschaften* 51 (1964); 395; Käthe Seidel, "Reinigung von Gewässern durch Höhere Pflanzen" [Purification of Water by Higher Plants], *Naturwissenschaften* 53 (1966): 289-97; Walter Czerwenka and Käthe Seidel, "Neue Wege einer Grundwasseranreicherung in Krefeld" [New Techniques of Groundwater Recharge in Krefeld], *Das Gas—und Wasserfach* 106 (1965): 828-33.

2. A. H. Koridon, "De invloed van biezen (*Scirpus lacustris* L. ssp. L. ac.) op het afsterven van *Escherichia coli* en van biezen en microorganismen op de ofbraak van fenol," H_2O (Rotterdam) 4 (1971): 296-98. These findings were in agreement with those of K. Viehl in *Gerässerschutzwäsen-Abwässer* 2 (1968): 12-13.

3. A detailed report on the results with this pond has been made by T. Kok, "De reiniging van afvalwatter van een kampeerterrein met behulp van een biezenveld," H_2O (Rotterdam) 7 (1964): 537-44.

17

Application of Vascular Aquatic Plants for Pollution Removal, Energy, and Food Production in a Biological System

B. C. WOLVERTON
R. M. BARLOW
R. C. McDONALD

National Space Technology Laboratories
National Aeronautics and Space Administration

Vascular aquatic plants such as water hyacinths (*Eichhornia crassipes* [Mart.] Solms) and alligator weeds (*Alternanthera philoxeroides* [Mart.] Griesb.), common to tropical and subtropical regions of the world, appear to be among the most promising candidates for solving many of man's problems, including pollution removal and increased food, energy, and fertilizer requirements. Vascular aquatic plants, when utilized in a controlled biological system including a regular program of harvesting to achieve maximum growth and pollution removal efficiency, may represent a remarkably efficient and inexpensive filtration and disposal system for toxic materials and sewage released into waters near urban and industrial areas. The National Aeronautics and Space Administration/National *Space Technology Laboratories* (NASA/NSTL), as a result of searching for economical solutions to upgrading the effluent quality of its sewage and chemical waste treatment lagoons, began investigating the potentials of vascular aquatic plants for pollution control.

The fact that vascular plants can absorb, translocate, and metabolize or concentrate various chemicals has been known for forty years.[1] This ability, for example, has been used to great advantage by entomologists using systemic insecticides in controlling plant-eating insects.[2] Also, the capability of vascular aquatic plants to assimilate nutrients and remove excess nitrates and phosphates from sewage effluents has been recognized for several years.[3] The phenomenon involved in systemic uptake,

FIG. 17-1. Water hyacinths (left) and alligator weeds (right) in NSTL greenhouse.

translocation, concentration, and/or metabolic breakdown of pesticides, and the vast potential vascular aquatic plants have for removing chemical pollutants from water systems are just beginning to be fully appreciated by environmental scientists. Interestingly, vascular aquatic plants, such as the water hyacinth in particular, have been the subject of considerable research concerning mineral and nutrient uptake, growth rates, and mechanical harvesting practices. Most of the research effort has been associated with control and eradication, since, in the natural state, water hyacinths are considered a major pest due to their tremendous growth rate and extreme hardiness. These characteristics become desirable attributes when the plants are utilized in a controlled biological system for pollution removal. Research has also been done on developing useful products from plants harvested to clear waterways, including evaluation as animal feeds and human food and assessment of their nutrient content.[4] No sustaining efforts to process and utilize the plants on a commercial basis are presently known. Owing to the problems encountered in harvesting in the wild state and transportation to processing sites, economical utilization could not be achieved. Using a well-designed system to increase harvesting efficiency and locating the processing equipment on-site to eliminate logistics problems, an economical operation is entirely feasible.

Laboratory Investigations

As a result of preliminary studies on utilization of vascular aquatic plants for pollution control, laboratory investigations were undertaken at NSTL to determine the pollution removal capabilities for various pollutants and nutrients (Fig. 17-1).

It has been reported that under favorable conditions one acre (0.40 hectare) of water hyacinths can produce over 534 pounds (240 kg) of dry plant material per day, which is one of the greatest yields of organic matter ever reported.[5] This same surface area of water hyacinths has the potential for removing over 3,500 pounds (1,591 kg) of nitrogen and over 800 pounds (364 kg) of phosphorus annually from sewage effluent, absorbing and metabolizing over 150 pounds (68 kg) of phenol every seventy-two hours from water polluted with this chemical, and removing over 120 grams of trace heavy metal contaminates every twenty-four hours.[6]

To effectively remove nutrients and other chemicals from waste effluents, water hyacinths must be harvested at intervals that allow for maximum biomass production. One acre (0.40 hectare) of water hyacinths has the potential of producing over 70 tons (63,640 kg) of dry plant material annually when grown in a desirable nutrient media such as domestic sewage effluent under proper climatic conditions.[7] This large volume of biomass has the potential of producing over one million cubic feet of biogas through anaerobic decomposition with 70 tons (63.5 metric tons) of residual high grade fertilizer being produced as a by-product.[8] Water hyacinths also have a dry weight nutrient content similar to that

TABLE 17-1. *Capability of Water Hyacinths to Remove Various Pollutants from Waters Polluted with these Substances.*

	LABORATORY EXPERIMENTS	FIELD POTENTIAL
Chemical and Metal Pollutants	Quantity-Removed, Absorbed, or Metabolized per Gram Dry Plant Weight per Day	Quantity-Removed, Absorbed or Metabolized (kg) per Acre per Day†
Cadmium*	0.67 mg	0.161 kg
Lead*	0.176mg	0.042 kg
Mercury*	0.150mg	0.036 kg
Nickel*	0.50 mg	0.120 kg
Silver*	0.65 mg	0.156 kg
Cobalt*	0.57 mg	0.137 kg
Strontium*	0.54 mg	0.130 kg
Phenols	36 mg	8.640 kg

*Ionized form
†Based on removal of mature plants every 24 hours

TABLE 17-2. *Capability of Alligator Weeds to Remove Various Heavy Metals from Waters Polluted with these Metals.*

	LABORATORY EXPERIMENTS	FIELD POTENTIAL
Metal Pollutants (ionized form)	Quantity-Removed, Absorbed, or Metabolized per Gram Dry Plant Weight per Day	Quantity-Removed, Absorbed, or Metabolized (kg) per Acre per Day
Lead	0.10 mg	0.024 kg
Mercury	0.15 mg	0.036 kg
Silver	0.44 mg	0.106 kg
Cobalt	0.13 mg	0.031 kg
Strontium	0.16 mg	0.038 kg

TABLE 17-3. *Final Filtration of Sewage Utilizing Water Hyacinths and Alligator Weeds*

	Percent Reduction in Wastewater Characteristics							
	Raw Sewage				Secondary Effluent			
	7-Day Exposure		14-Day Exposure		7-Day Exposure		14-Day Exposure	
TYPE OF PLANT/ MEASUREMENTS TAKEN	with Plants	Control (Free of Plants)	with Plants	Control (Free of Plants)	with Plants	Control (Free of Plants)	with Plants	Control (Free of Plants)
WATER HYACINTHS								
Total Kjeldahl Nitrogen	92%	18%	—	—	75%	13%	89%	15%
Total Phosphorus	60%	13%	—	—	87%	11%	99%	25%
Total Suspended Solids	—	—	—	—	75%	15%	77%	12%
BOD_5	97%	61%	—	—	77%	6%	—	—
pH	Increased From 7.05 to 7.35	Increased From 7.05 to 7.75	—	—	—	—	Decreased From 8.80 to 7.20	Decreased From 8.80 to 8.20
ALLIGATOR WEEDS								
Total Kjeldahl Nitrogen	97%	18%	97%	14%	61%	10%	76%	14%
Total Phosphorus	50%	13%	78%	35%	44%	15%	62%	41%
Total Suspended Solids	—	—	—	—	94%	48%	98%	60%
BOD_5	92%	68%	97%	65%	—	—	—	—
pH	—	—	Increased From 7.1 to 7.4	Increased From 7.1 to 8.25	—	—	Decreased From 8.9 to 7.2	Decreased From 8.9 to 8.35

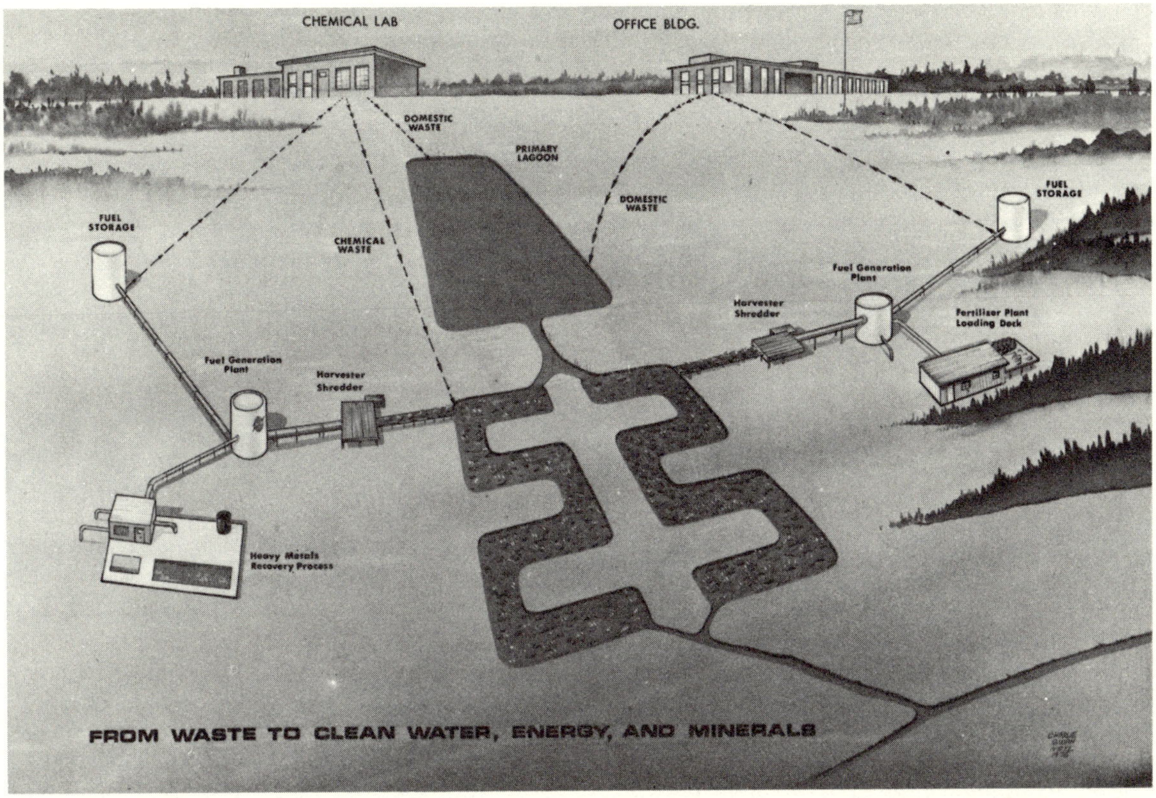

FIG. 17-2. NASA aquatic plant filtration system.

of many agricultural crops.[9] The nutrient content varies with water fertility and stage of plant growth.

Tables 17-1, 17-2, and 17-3 summarize the laboratory investigations on the pollution removal capabilities of water hyacinths and alligator weeds. Experiments were conducted using plant controls free of pollutants and pollutant controls free of plants. Aqueous samples were taken at 1, 3, 6, and 24-hour intervals for heavy metals and at 7-day and 14-day intervals for nutrient removal analyses.[10]

Applications

Two of the most pressing problems facing the United States and other industrial nations today are rapid depletion of vital natural resources and environmental pollution. One important factor in the rise of the United States to its present high industrial level has been an abundance of fossil fuel resources. Presently, coal, oil, and large reservoirs of underground natural gas are all produced through natural decomposition of prehistoric forms of life. Modern society is depleting these resources at an alarming rate. Renewable sources must be developed within the near future. As we deplete our natural resources, we are also polluting and contaminating our environment at ever increasing rates. Fortunately, the minerals and nutrients contaminating and polluting our water systems can potentially be recovered, utilizing natural biological processes.

Based on laboratory results with vascular

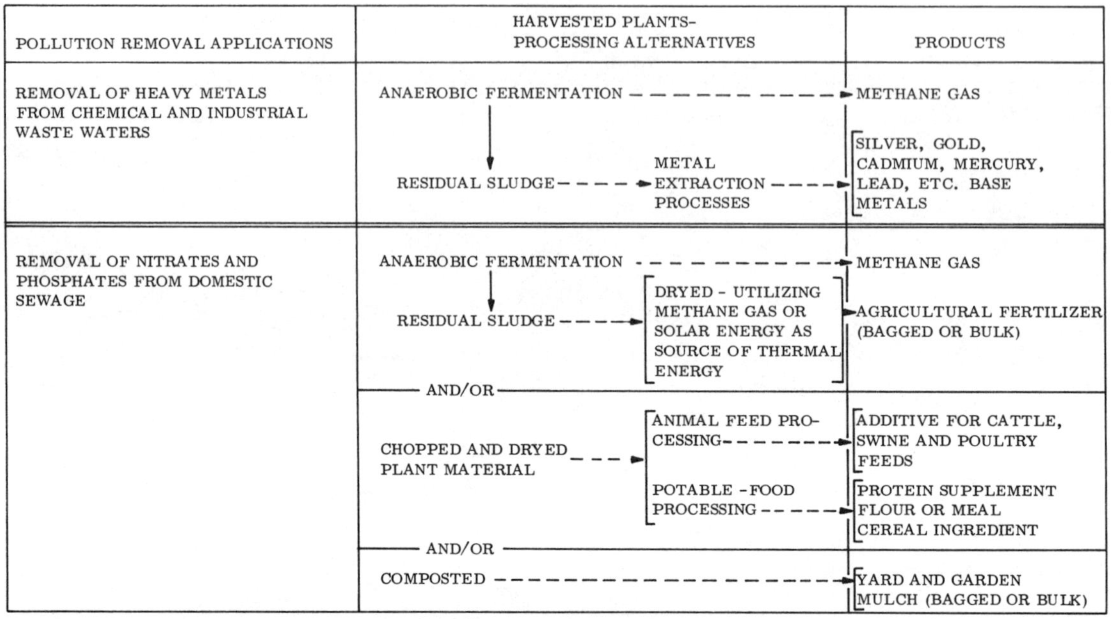

FIG. 17-3. Conversion of vascular aquatic plants to useful products.

aquatic plants and the preliminary results of field tests being conducted under a NASA Office of Applications-sponsored program at the National Space Technology Laboratories, Bay St. Louis, Mississippi, some innovative applications now appear possible. A system of lagoons filled with vascular aquatic plants appears possible for removal of chemical pollutants with the harvested plants being converted to bio-gas and heavy metals being extracted from the sludge. Since no toxic levels of heavy metals have been found in plants grown in domestic sewage samples taken from the City of Bay St. Louis, Mississippi, sewage lagoon, the harvested plants are potential sources of animal feed, human food, and fertilizer.

Fig. 17-2 conceptually illustrates a biological system for removal of chemical and sewage pollutants from waste waters, utilizing either water hyacinths or alligator weeds as a biological filtration system in a zig-zag canal-type lagoon. Mature plants harvested to promote optimum growth rate and removal of pollutants and contaminants become a valuable source of raw material for conversion to useful products. Fig. 17-3 displays some of the processing alternatives and products that may be derived from the harvested biomass.

Fig. 17-4 conceives a potentially self-sufficient agricultural homestead through the installation of a vascular aquatic plant filtration system and ancillary processing equipment to produce energy, fertilizer, and feeds.

Field Demonstrations and Continuing Efforts

Field demonstrations utilizing a chemical waste lagoon and a municipal sewage lagoon are in process. The primary objective of the field tests is to demonstrate the pollution removal effectiveness of vascular aquatic plants. Fig. 17-5 shows the hyacinth-filled, zig-zag canal type lagoon at NSTL being utilized to evaluate the removal of pollutants from chemical wastes.

Efforts are continuing toward the develop-

FIG. 17-4. Applications of a vascular aquatic plant filtration system to an agricultural homestead.

ment and demonstration of harvesting equipment. The planned harvesting scheme is to gather the water hyacinths, remove them from the water, and chop them into approximately one inch (2.5 cm) pieces. From this point the plant material will be delivered to the selected processing equipment. The harvesting equipment is designed to process at a rate of 15 to 20 tons (13.6 to 18 metric tons) per hour.

Processing equipment is being developed, including scaled-up laboratory models of biogas generating units. This equipment will be operated to gather data for sizing a pilot plant for field test and evaluation. Investigations of methods for processing the residual sludge from the biogas units into fertilizers will be conducted, including evaluation of the chemical and nutritive content. Animal feed processing will be accomplished by reducing the moisture content of the freshly chopped plant material to approximately that of well cured forage. A solar dryer will be evaluated for curing the plant material. Several feed formulations will be produced and evaluated in a beef cattle feeding program. Several other processing possibilities are also being investigated on a limited scale. Laboratory studies are being initiated to develop methods for metal extraction from sludges containing heavy metals. Laboratory processes for converting harvested plant material (free of

FIG. 17-5. Water hyacinths in NSTL zig-zag lagoon.

toxic substances) into human food will be undertaken to produce protein supplements, cereals, or flour and meals.

Based on field tests to demonstrate pollution removal and to demonstrate the processing of harvested plants into usable products, a comprehensive economic assessment will be conducted including capital investment requirements, operating costs, and potential sales of products to offset operating costs.

Notes

1. A. M. Hurd-Karrer and F. W. Poos, *Science* 84 (1936):252.

2. S. H. Bennett, "The Behavior of Systemic Insecticides Applied to Plants," *Annotated Review of Entomology* 2 (1957):279-96.

3. H. H. Rogers and D. E. Davis, "Nutrient Removal by Water Hyacinth," *Weed Science* 20 (1972):423-27; C. W. Sheffield, "Water Hyacinth for Nutrient Removal," *Hyacinth Control Journal* 6 (1976):27-30; B. C. Wolverton and R. C. McDonald, "Water Hyacinths and Alligator Weeds for Final Filtration of Sewage," *NASA Technical Memorandum*, TM-X-72724 (1975); W. T. Haller and D. L. Sutton, "Effect of pH and High Phosphorus Concentrations on Growth of Water Hyacinth," *Hyacinth Control Journal* 7 (1968):59-61 C. E. Boyd, "Vascular Aquatic Plants for Mineral Nutrient Removal from Polluted Waters," *Economic Botany* 24 (1970):95-103.

4. C. E. Boyd, "Evaluation of Some Common Aquatic Weeds as Possible Feedstuffs," *Hyacinth Control Journal* 7 (1968):26-27; K. G. Taylor, "Extraction of Protein from Water Hyacinth," *Hyacinth Control Journal* 9 (1971):20-22; C. E. Boyd, "The Nutritive Value of Three Species of Water Weeds," *Economic Botany* 23 (1969):123-27; J. F. Easley, "Nutrient Elements for Livestock in Aquatic Plants," *Hyacinth Control Journal* 12 (1974): 82-84; K. G. Taylor and R. C. Robbins, "The Amino Acid Composition of Water Hyacinth and its Value as a Protein Supplement," *Hyacinth Control Journal* 8 (1968):24-25.

5. J. L. Yount, "Report of the 35th Annual Meeting" (Florida Anti-Mosquito Association, 1964), p. 83.

6. Rogers and Davis, "Nutrient Removal"; Wolverton and McDonald, "Final Filtration"; Haller and Sutton, "Effect of pH"; B. C. Wolverton, "Water Hyacinths for Removal of Phenols from Polluted Waters," *NASA Technical Memorandum*, TM-X-72722 (1975); B. C. Wolverton and R. C.

McDonald, "Water Hyacinths and Alligator Weeds for Removal of Lead and Mercury from Polluted Waters," *NASA Technical Memorandum,* TM-X-72723 (1975); B. C. Wolverton and R. C. McDonald, "Water Hyacinths for Removal of Cadmium and Nickel from Polluted Waters," *NASA Technical Memorandum,* TM-X-72721 (1975).

7. Yount, "Annual Meeting."

8. "Methane Digesters for Fuel Gas and Fertilizer," New Alchemy Institute Newsletter No. 3 (1973); R. B. Singh, "Bio-Gas Plant Generating Methane from Organic Wastes" (Ajitmal, Etawah [U.P.], India: Gobar Gas Research Station, 1971); B. C. Wolverton, C. M. Ladner, and J. Gordon, "Bio-Conversion of Water Hyacinths into Methane Gas and Fertilizer," *NASA Technical Memorandum,* TM-X-72725 (1975).

9. Boyd, "Some Common Aquatic Weeds."

10. Wolverton and McDonald, "Final Filtration"; Wolverton, "Removal of Phenols"; Wolverton and McDonald, "Removal of Lead and Mercury"; Wolverton and McDonald, "Removal of Cadmium and Nickel."

18

Land Treatment of Wastewater by Overland Flow for Improved Water Quality

P. G. HUNT
C. R. LEE

Environmental Effects Laboratory
U.S. Army Engineer Waterways Experiment Station

Introduction

The United States and numerous other industrialized countries face the apparently conflicting needs of environmental quality and energy requirements. The neglect of either would certainly result in catastrophic problems. The impacts of allowing pollution of the environment or of needless use of power supplies are staggering, and careful assessment leads to the conclusion that total regard for the environment and satisfaction of energy needs is an impossible goal. As true as this conclusion may be, both problems should and can be approached with a similar philosophy, for both problems can be classed with others as problems of finite resources. The world has acutely felt the reality of the finite supply of fossil fuels so vital for production of energy and synthetic materials. It has also realized the finite nature of food production. To a somewhat lesser degree, the industrialized nations have realized the finite nature of the diluting capacity of clean air and water or the resiliency of numerous ecological processes.

It is our hope, however, that some of the present adversity experienced by the industrialized world in food and fuel will point to the more subtle problems of environmental quality. One such area in which progress, although slow, is being made in the United States is the recycling of nutrients from wastewater into soil via land treatment.

This paper presents an overview of land treatment of wastewater; a somewhat detailed development of the overland flow system of treatment, which is probably the least understood mode of land treatment of wastewater; and possible modifications of current concepts of overland flow treatment for improved water quality.

Land Treatment Systems

What is Land Treatment of Wastewater?

Land treatment as opposed to land disposal of wastewater is a method whereby wastewater is applied to land in a controlled quantitative manner to achieve the removal of various polluting fractions of the wastewater.[1] The source of wastewater can be industrial or municipal, and treatment before land application can vary from simple screening to secondary treatment and disinfection. The particular combination is dictated by factors such as the type of waste, facility location, and degree of treatment desired.

As with conventional wastewater treatment systems, design and operation of land treatment systems depend upon a number of factors such as source of wastewater, facility location, regularity of flow, and construction cost. Common systems familiar to the sanitary engineering profession are trickling filter, extended air, and anaerobic lagoons. Land treatment systems that are becoming equally familiar to the environmental engineer are slow infiltration, rapid infiltration, and overland flow.[2]

In the rapid and slow infiltration systems, wastewater is renovated by the soil, plants, and microorganisms as it moves through the soil profile. The slow infiltration system is normally an integral component of an agricultural operation.[3] Rapid infiltration systems are usually operated separate from agriculture and are on thick deposits of sandy or coarse gravelly soils.[4] Overland flow systems are somewhat different in that most of the water flows over a relatively impermeable soil surface and the renovative action is more dependent upon microbial and plant activity.[5] Classification and operation of land treatment systems in these categories are based upon their different hydraulic characteristics. Such an approach has been an improvement in land treatment concepts; in the past there was a tendency to believe that only a medium-textured soil that would allow sufficient but not rapid infiltration while maintaining aerobic soil conditions was necessary for a successful land treatment system. The development of rapid infiltration and overland flow treatment systems has enabled both very porous and rather impermeable soils, as well as medium-textured, permeable soils, to be of value in properly designed and operated wastewater treatment systems.

Mechanisms for Pollutant Removal

Regardless of which land treatment system is chosen, the major pollutants normally are nitrogen, phosphorus, trace elements, and oxygen-demanding materials. These water pollutants are, of course, the major components of fertilizers and soil amendments that are necessary for food production. Recycling them into the soil can save on fertilizer consumption as well as improve water quality. In general, the pollutant materials are removed by soil, plants, or microorganisms. The particular mechanisms for removal vary with the system. A summary of the mechanisms involved follows, but these highlights of treatment are very brief. The actual treatment of wastewater by any one of the three methods requires management based on known operational principles to maximize the renovation of wastewater. Many variables, such as the cover crop, wastewater characteristics, and weather, cause variation in the operation on an unscheduled basis. The treatment of wastewater via a recycling mode is no less complicated than any other. Its advantage is that it works and it recycles rather than consumes resources.

Nitrogen

Of the major pollutants, nitrogen occurs in the most forms. It can be removed in major quantities by the soil, plants, or microorganisms (Fig. 18-1); but a removal mode may be important in one system but not in another. For instance, denitrification is of paramount importance in a rapid infiltration system, because while great quantities of water are passed into the soil, the plant system is only capable of removing a very

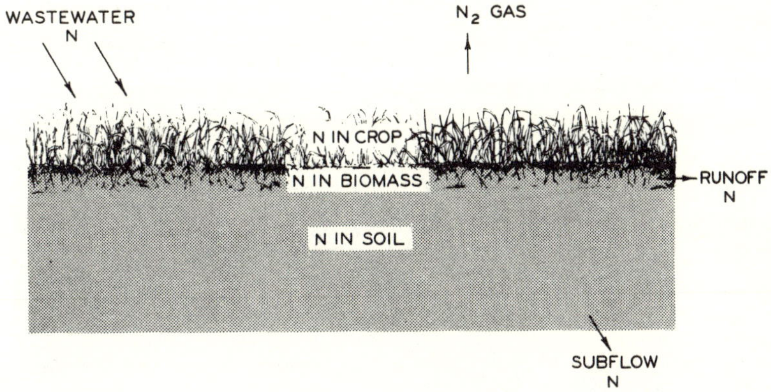

FIG. 18-1. Schematic diagram of nitrogen in overland flow treatment of wastewater.

small fraction of the applied nitrogen.[6] On the other hand, denitrification is of relatively low importance in slow infiltration systems (often referred to as crop irrigation systems), where plant intake removes considerable quantities of nitrogen.[7] Soil adsorption and incorporation of nitrogen into organic matter are also important in slow infiltration systems. Overland flow systems are intermediate in that plant uptake, denitrification, and soil and organic matter incorporation are all important.[8]

Phosphorus

Phosphorus neither appears in the varied forms nor has the removal modes of nitrogen. In particular, it does not have the microbially mediated gaseous loss. Phosphorus removal in rapid infiltration systems is primarily a soil adsorption phenomenon. The somewhat low surface areas of the coarse-textured soil for adsorption of phosphorus is compensated for by the thickness of the soils used in rapid infiltration systems. In addition, water is often forced to move considerable distances in a lateral direction to accomplish a higher degree of treatment. Slow infiltration systems are the best for phosphorus removal for in these systems there is an intimate contact of a relative small amount of wastewater with the high surface area of medium-size soil particles.[9] Overland flow systems have the poorest combination of conditions for phosphorus removal.[10] The soil contact is limited to the soil surface area and the residence time on the soil is normally less than 24 hours. However, phosphorus appears to be removed by the pronounced surface organic layer on overland flow slopes, and this layer can be turned under by plowing in order to take advantage of the fixing capacity of the heavy-textured soil.

Trace elements

Trace elements are removed in much the same manner as phosphorus in that they are absorbed by soil and organic matter. They are also removed by plants, although to a much smaller degree than is phosphorus. Yet, even small amounts can be toxic to the plants, and, in some mismanaged systems, trace elements could increase to the point that the site could no longer be used.[11] As with phosphorus, the soil thickness of rapid infiltration systems allows for more adsorption of trace elements. The slow infiltration systems, however, appear to have the greatest capacity for removing trace elements because of the high surface area contact with the wastewater. While overland flow is limited by soil contact, trace elements are readily fixed by organic material and the surface organic layer of overland flow systems has been found to be very efficient in removing trace elements. As with phosphorus, this layer can be plowed under to take advantage of the fixing capacity of the heavy-textured soils.

Oxygen-demanding substances

In most cases the oxygen-demanding substance is organic material and is removed quite well by all methods of land treatment. The material is first physically filtered as it passes through the soil or grass cover and then decomposed by the soil microflora. Part of the organic material may, however, be more resistant and contribute to an increased organic matter content in the upper layer of soil.

Overland Flow Systems

Design and Operational Description

As was stated earlier, overland flow treatment systems are located on soils of low permeability. This low permeability may result from a number of factors, but most commonly it is either a heavy-textured soil or a soil containing a barrier within the upper three feet. In addition to low permeability, a slight slope is required; slopes up to 8% are generally satisfactory. The slopes are covered with vegetation and normally are 150 to 200 ft. in length with collection channels located at the bottom of the slope.[12] A schematic of an overland flow system is shown in Fig. 18-2. The water is applied at the upper end of the slopes and flows as a sheet over the soil surface and through a surface organic mat into the collection channels. The renovated water then flows to a central channel where the water quality can easily be monitored (Fig. 18-3). The renovated water can then be used for a host of purposes varying from industrial cooling to groundwater recharge.

Operationally, overland flow systems have slow rates of application; volumes are normally

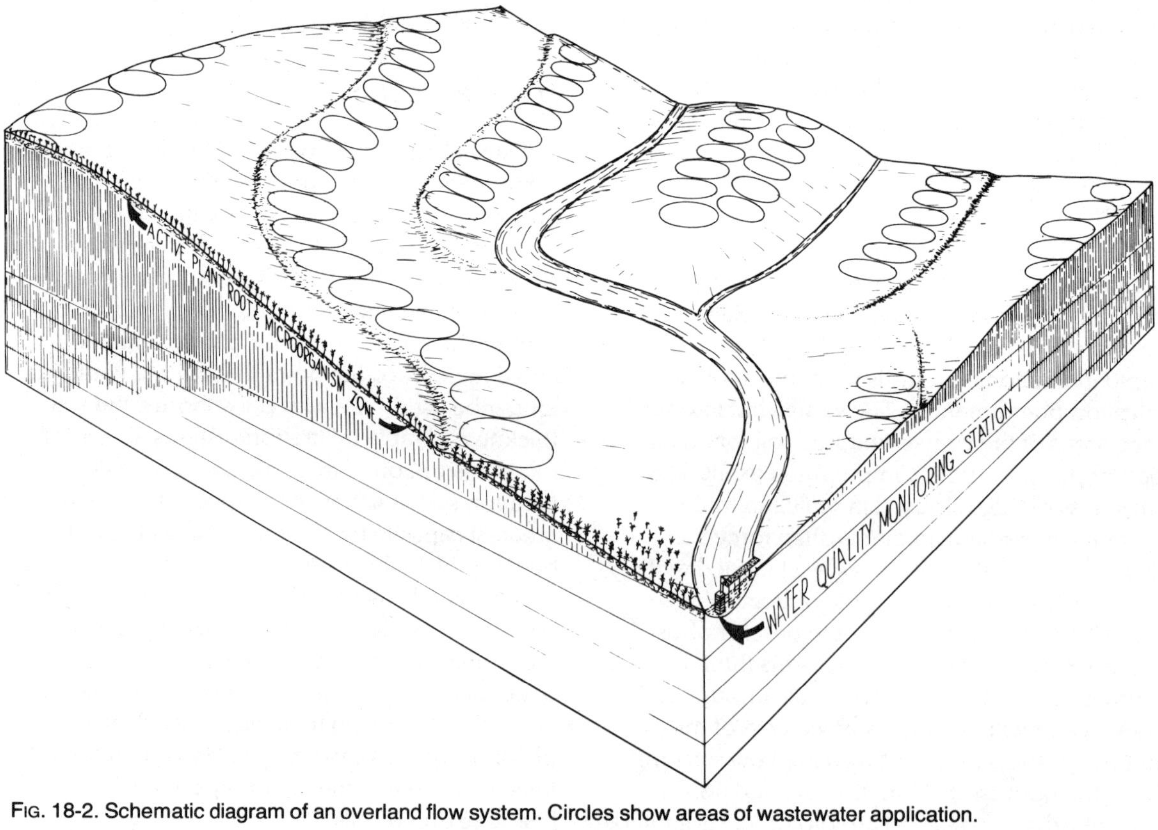

FIG. 18-2. Schematic diagram of an overland flow system. Circles show areas of wastewater application.

FIG. 18-3. Water-monitoring site for treated water leaving the overland flow treatment site.

less than ½ acre inch per day, and application rates vary from 6 to 18 hours. However, wastewater can be applied on 4 to 6 days per week giving a weekly application of 2 to 3 acre inches. These rates compare favorably with slow infiltration systems. High rates of application invariably cause poor treatment.

Mechanisms of Overland Flow Treatment

The fact that overland flow treatment of wastewater removes a very high percentage of the applied nitrogen is somewhat surprising. The wastewater does not flow into the soil where nitrogen could easily be adsorbed on the clay particle surfaces or be removed by plant uptake. In addition, the fact that the surface water is aerobic and would tend to eliminate the possibility of denitrification. However, upon further investigation other conditions make nitrogen removal seem quite likely. The organic layer on the soil surface as well as the shallow root system of the flooded cover crop usually removes from 50 to 75% of the applied nitrogen. Significant denitrification appears to be responsible for the remaining nitrogen removal.

The intricate way in which denitrification and nitrification appear to function on an overland flow system is one of the most interesting aspects of this intriguing system. The water film and underlying soil seem to form an aerobic-anaerobic double layer similar to that found in rice fields or marshes.[13] In the overlying water film and organic matter, aerobic processes can occur, and ammonium is nitrified to nitrate (Fig. 18-4). In the underlying anaerobic zone, the nitrate is converted to gaseous nitrogen via denitrification.

The redox potential at which denitrification for a soil is active has been determined with a consideration for pH. Fig. 18-5 depicts the results of redox measurements on an overland

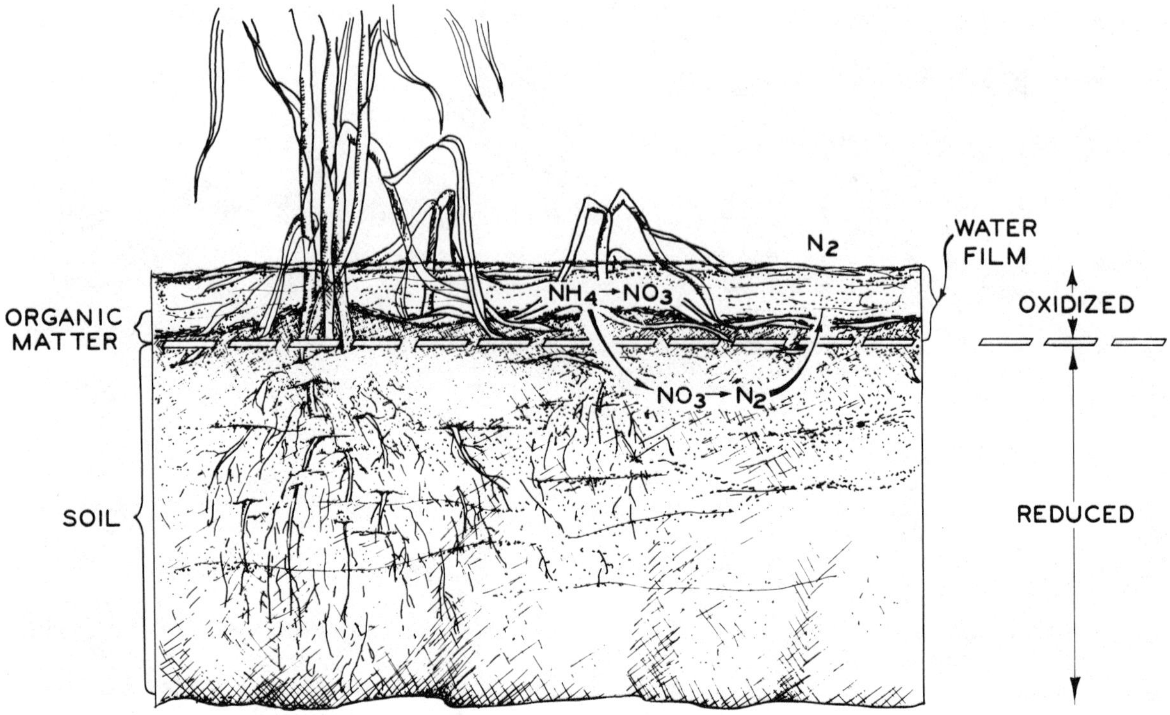

FIG. 18-4. A schematic diagram of conditions that would allow both aerobic and anaerobic processes to occur in an overland flow system.

flow model.[14] The redox potential did not reach an oxidized level during wastewater application at 5 mm depth; the presence of an overlying aerobic layer was demonstrated by oxygen in the water and rapid nitrification of ammonium to nitrate. Positive proof of substantial denitrification has not yet been obtained, but the Waterways Experiment Station is presently conduct-

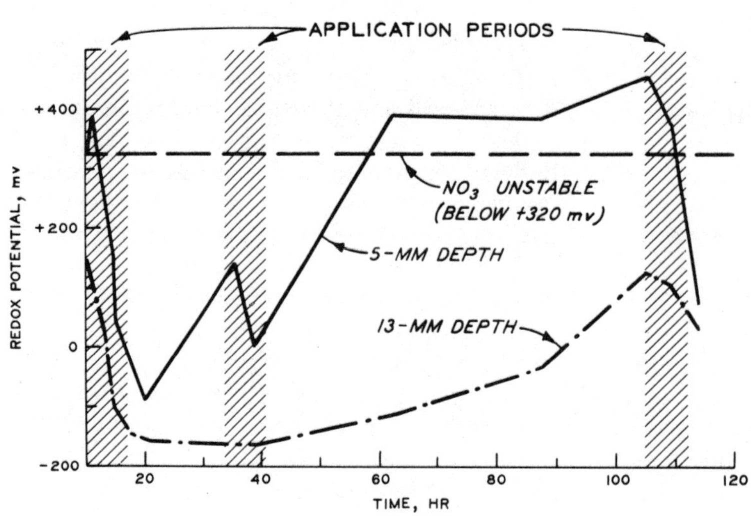

FIG. 18-5. Redox potential of a Susquehanna clay during a wastewater application-drying cycle.

FIG. 18-6. Wastewater model 43 days after start of treatment. Grass height reached 20 to 40 cm in the first 3.05 m of treatment; practically no increase in height and zero nitrogen content in surface water beyond 4.57 m.

ing studies that should yield such proof. To date the evidence is that denitrificating conditions and substantial losses of nitrogen are associated with both field and model overland flow systems.

Nitrogen removal by grass on overland flow is also an interesting phenomenon in that a grass-growth gradient is established down the slope. This gradient has been observed on slopes from 180 feet long in the field to 18 inches in the greenhouse. It appears that in both field and laboratory systems that are functioning properly, the nitrogen concentration of wastewater is reduced to 1 or 2 ppm at approximately two-thirds of the slope's length. Since grass is very responsive to nitrogen, a growth gradient associated with the decreased nitrogen content is established (Fig. 18-6).[15]

Phosphorus and Trace Element Treatment

Since phosphorus removal via land treatment is best when the wastewater is in close contact with the surface area of clay particles as it moves through the soil, overland flow would intuitively seem to be the poorest method of phosphorus removal. As was stated earlier, this is in fact true; wastewater flowing over the soil surface does not have extensive contact with the iron and aluminum compounds of the soil that normally fix massive amounts of phosphorus.

One could, however, suggest that the less-than-complete removal of phosphorus was a result of the reducing soil conditions and related very little to the soil contact phenomenon. Such a view could be supported by the fact that reduced soils often have more available phosphorus than oxidized soils.[16] While this fact is well known and the principal is used by American rice farmers, the situation is not quite that clearcut. Reduced soil seems to have a greater capacity to fix phosphorus, although the amount of soluble or available phosphorus up to 1 ppm may be higher under reduced conditions. This condition has been attributed to a difference between the fixing capacity and surface area of ferric oxide, prevalent under oxidized conditions, and a gel-like hydrated ferrous oxide or ferrous hydroxide prevalent under reduced conditions.[17] Under this hypothesis the ferric oxyhydroxide with its tight binding but low surface area characterisitcs is prevalent in an oxidized soil, and phosphorus is readily removed to very low concentrations from a diluted solution. However, if the liquid, interstitial, or overlying water concentration of phosphorus becomes greater than approximately 1 ppm, the surface area or fixing capacity of the ferric oxide is exceeded. On the other hand, under reduced conditions, they propose that neither the gel-like hydrated ferrous oxide nor ferrous hydroxide fix phosphorus so tightly, but its large surface area results in a substantial fixing capacity. Thus under reducing conditions, a soil will remove more phosphorus from a solution with greater than 1 ppm phosphorus than would be oxidized soil.

Wastewater is the kind of high phosphorus solution that would possibly exceed the fixing capacity of a thin layer of oxidized soil. If the reduced soil is behaving as reported (see note 17), the soil contact rather than fixing capacity would appear to be the reason for somewhat less-than-complete removal of phosphorus by overland flow treatment. In addition, phosphorus appears to be removed mainly by the surface organic mat in overland flow models that receive secondarily treated wastewater. This would indicate that only a fraction of the phosphorus was actually interacting with the soil and would therefore not be capable of saturating the soil's fixing capacity for phosphorus. This point is not yet resolved, but research currently being conducted should do so.

On a more practical basis, it has been shown that greater than 80% of the phosphorus in raw wastewater can be removed by overland flow if stoichiometric amounts of $Al_2(SO_4)_3$ are added before the wastewater is applied to the slopes.[18] Similar results have been found with secondarily treated wastewater by C. R. Lee of the Waterways Experiment Station, Vicksburg, Miss. If this scheme was used, the iron and aluminum phosphates could be periodically plowed under and fixed by the reduced clay soil.

Trace element removal by overland flow is very good, greater than 90% for all and greater than 98% for some heavy metals.[19] This rather high removal is attributed to the surface organic mat where most of the heavy metals in particular are bound. This is not surprising since trace elements are known to be very reactive with organic matter in agronomic soils. As with phosphorus, the surface concentration of trace elements could be reduced periodically by plowing that layer under. Although the reduced soil condition might cause certain heavy metals to be more mobile, it is the authors' opinion that this would be a minor problem in the heavy-textured soils with high organic matter content that are used for overland treatment.

Possible Modification of the Overland Flow Concept

Application to Rice Fields

Since the soil surface of an overland flow treatment system is chemically and biologically very similar to a rice field, a logical question is whether overland flow can be incorporated into rice production.[20] From a mechanistic viewpoint, it would appear that a shallowly flooded rice field would remove nitrogen, phosphorus, and trace elements very well. The diffusion distance through the water column would be five to twenty times greater than in a normal overland flow system, so the residence time should be equally increased.

The wastewater could not contain high concentrations of toxic substances, and it would appear that the wastewater would need to be secondarily treated and disinfected. Such a wastewater would eliminate rather obvious odor and public health problems.

It has also been suggested that rice so grown should be used for consumption by humans only after some processing step that would definitely eliminate any possibility of pathogen transfer.

There are also certain agronomic problems that would have to be addressed, such as the potential growth of algae associated with the wastewater and the use of pesticides on rice. However, when once considers that in rice-growing areas of the southeastern United States, water cost is roughly 20% of the production cost of rice and that in many areas the use of massive amounts of groundwater for rice production is causing ecological concern, the concept would appear worth pursuing in a vigorous manner.

Application to Marsh Areas

The marsh ecosystem also has many similarities to the overland flow mode of treating wastewater. The potential of recycling nutrients through plant and microbial populations in a marsh is great. In these systems organic waste could probably be assimiliated in a satisfactory manner. Quantities of substances such as heavy metals would have to be low to avoid accumulation to toxic levels in the ecosystems. Additionally, nutrient stimulation might increase the organic litter layer so rapidly in certain marshes that the elevation of the marsh would rise enough above the water table to cause death to the dominant plant species of the marsh. Thus application rates would have to be adjusted to avoid excessive stimulation of growth.

An interesting possibility for a marsh system would be to establish a marsh with dredged material and accelerate its establishment with nutrients from wastewater. However, this particular idea would have several obvious problems, one of which would be that a thick plant cover is required for an overland flow system. Another would be that normally artificially created marshes are small in size relative to the area needed for a moderate-sized wastewater treatment system.

Combined Forage Grass-Pulp Forest System

In such a system trees that were water tolerant and fast growing would be selected to grow with somewhat shade- and water-tolerant forages. A combination that would appear to have promise would be cottonwood trees and reed canary grass. The cottonwoods could be spaced wide enough apart to allow forage harvesting and adequate sunlight. A slight ridge could be raised for the trees or they could be planted on a normal slope.

A similar combination of forages and agronomic plants might be possible if ridges that would allow aerobic soil conditions were used for the agronomic plant. The many ramifications of such a concept, including plant disease, competitiveness, and harvesting methods, would have to be evaluated. Yet it would appear that there might be a possibility of increasing the productivity of many rather infertile soils while treating wastewater to advanced levels.

Fiber might also be produced from an overland flow system by a plant similar to kenaf. Such a concept has the appeal of avoiding any pathogenic or toxic metal problems related to human consumption. However, kenaf requires at least moderately aerated soil. The authors do not at this time have a good candidate for a thick-cover fiber crop that will tolerate reduced soils. If, however, a productive use could be found for plants such as *Phragmites communis*, a really excellent system of recycling could be designed.

Chemical and Physical Filters

Some of the attributes of an overland flow system could be used in special situations such as removing the bulk of the organic load before waste was applied to a water hyacinth lagoon for nutrient stripping. Such an approach would appear to have promise for certain cannery wastewaters. An overland flow system might also function quite well in physically and chemically removing suspended material and nutrients that may pass from land disposal areas for dredged material.

Use in Recreational and Rest Areas

Another possible modification is not with the overland flow concept, but with the type of wastewater. Recreational areas such as reservoirs, national parks, and highway rest areas have rather seasonally and daily variable wastewater loads. Overland flow as well as other modes of land treatment are ideal for treating these wastes if sufficient land is available. Often, in forest or reservoir areas there is sufficient land available to allow for such a low rate of application that there is no runoff or percolation. Such systems are referred to as land containments and offer distinct advantages in many areas.

Future Needs

Overland flow has not been used or studied in detail as have slow and rapid infiltration systems. Demonstration systems using the present concepts of overland flow need to be constructed immediately in several areas of the United States. In addition, research into concept refinement and expansion needs to be pressed forward. If both of these tasks are carried out successfully, administrators, engineers, and scientists will come to recognize the proven value of overland flow treatment of wastewater and will recommend and adopt its use where appropriate.

Notes

1. S. Reed, "Wastewater Management by Disposal on the Land," *U.S. Army Cold Regions Research and Engineering Laboratory Special Report 171* (1972):183; R. H. Sullivan, M. M. Cohn, and S. S. Baxter, *Survey of Facilities Using Land Application of Wastewater,* EPA 430/9-73-006 (Washington: U.S. Government Printing Office, 1973), p. 377.

2. Sullivan, Cohn, and Baxter, *Survey of Facilities;* W. E. Sopper and L. T. Kardos, eds., *Recycling Treated Municipal Wastewater and Sludge through Forest and Cropland* (University Park, Pa.: Pennsylvania State University Press, 1973). See also Chapter 31.

3. Sopper and Kardos, *Recycling Treated Municipal Wastewater.*

4. H. Bouwer, "Renovating Secondary Effluent by Groundwater Recharge with Infiltration Basins," *In Sopper and Kardos, Recycling Treated Municipal Wastewater,* pp. 164-75. See also Chapter 30.

5. C. A. Carlson, T. B. Delaney, Jr., and P. G. Hunt, "Overland Flow Treatment of Wastewater," *U.S. Army Engineer Waterways Experiment Station Miscellaneous Paper* Y–74–3 (1974):63; L. C. Gilde et al., "A Spray Irrigation System for Treatment of Cannery Wastes," *J. Water Poll. Control Fed.* 43 (1971):2011-25; R. E. Hoeppel, P. G. Hunt, and T. B. Delaney, Jr., "Wastewater Treatment on Soils of Low Permeability." *U.S. Army Engineer Waterways Experiment Station Miscellaneous Paper* Y–74–2 (1974):84; J. P. Law, R. E. Thomas, and L. H. Myers, "Nutrient Removal from Cannery Wastes by Spray Irrigation of Grassland," *FWPCA Report* No. 16080 (1969).

6. Bower, "Renovating Secondary Effluent"; Reed, "Wastewater Management."

7. Sopper and Kardos, *Recycling Treated Municipal Wastewater;* Bower, "Renovating Secondary Effluent."

8. Carlson, Delaney, and Hunt, "Overland Flow Treatment"; Hoeppel, Hunt, and Delaney, "Wastewater Treatment."

9. Reed, "Wastewater Management"; Sopper and Kardos, *Recycling Treated Municipal Wastewater.*

10. Carlson, Delaney, and Hunt, "Overland Flow Treatment"; Law, Thomas, and Myers, "Nutrient Removal"; R. E. Thomas, "Spray-Runoff to Treat Raw Domestic Wastewater" (Paper delivered at the International Conference on Land for Waste Management, Ottawa, Canada, 1–3 October, 1973).

11. Reed, "Wastewater Management"; Sopper and Kardos, *Recycling Treated Municipal Wastewater.*

12. Gilde, et al., "A Spray Irrigation System"; Law, Thomas, and Myers, "Nutrient Removal."

13. D. S. Mikkelsen and W. H. Patrick, Jr., "Plant Nutrient Behavior in Flooded Soil," In *Fertilizer Technology and Use* (Madison, Wis.: Soil Science Society of America, 1971), pp. 187-215; W. H. Patrick, Jr., R. D. Delaune, and R. M. Engler, "Soil Oxygen Content and Root Development of Cotton in Mississippi River Alluvial Soils," *Louisiana State University Agricultural Experiment Station Bulletin* 673 (1973); F. M. Ponnamperuma, ed., *The Mineral Nutrition of the Rice Plant* (Baltimore: The Johns Hopkins Press, 1965).

14. Carlson, Delaney, and Hunt, "Overland Flow Treatment."

15. *Ibid.*

16. Mikkelsen and Patrick, "Plant Nutrient Behavior"; Patrick, Delaune, and Engler, "Soil Oxygen Content"; Ponnamperuma, "The Mineral Nutrition"; R. A. Khalid and W. H. Patrick, Jr., "Phosphate Release and Absorption by Soils and Sediment," *Science* 186 (1974):53-55.

17. Khalid and Patrick, "Phosphate Release."

18. R. E. Thomas (EPA, Ada, Oklahoma), personal communication.

19. Carlson, Delaney, and Hunt, "Overland Flow Treatment."

20. Hoeppel, Delaney, and Hunt, "Wastewater Treatment"; Mikkelsen and Patrick, "Plant Nutrient Behavior"; Ponnamperuma, "The Mineral Nutrition."

19

Experimental Use of Emergent Vegetation for the Biological Treatment of Municipal Wastewater in Wisconsin

FREDERIC SPANGLER
WILLIAM SLOEY
C. W. FETTER

Biology Department
University of Wisconsin, Oshkosh

Introduction

Economic pressure has intensified the search for more efficient waste treatment methods.[2] Currently, biological treatment occupies the position of choice in almost all cases. Biological treatment need not be synonymous with activated sludge or trickling filter. The list of alternatives is in its infancy but is growing, and among the alternatives is the use of aquatic macrophytes as effectors of degradation of waste, or at least of uptake of excess materials. Systems using algae and higher plants, whether submergent or emergent, have the attraction of doing the job with a minimum of hardware, energy, and chemicals and thus of reducing the economic demands. For small municipalities, campgrounds, some agricultural operations, or other waste-producing units with the necessary land area available, one possible alternative for biological water quality control is the use of natural or artificial marshes.

The use of bulrush, *Scirpus lacustris* L., for treatment of domestic and other wastes has been studied in Europe with the pioneering investigations conducted by Dr. Seidel and her coworkers at the Max Planck Institute in Germany.[3]

The study reported in this chapter is an attempt to assess the effect of native Wisconsin marsh plants growing in both natural and artificial situations on the quality of water flowing through them. The goal is to demonstrate the applicability of the concept that a marsh is a living system that will oxidize organic compounds, take up nutrients, reduce the number of coliform bacteria, and at the same time function as a community with minimal management requirements.

Experimental Studies

Greenhouse Investigations

After screening a variety of emergent aquatics for regrowth and vegetative propagation capabilities, bench-scale studies were conducted in the greenhouse during the winter of 1972. *Iris versicolor* L. (blue flag), *Scirpus validus* Vahl. (softstem bulrush), and *S. acutus* Muhl. (hardstem bulrush) were planted in pea-sized gravel 7 cm deep in plastic-lined 80 x 90 cm basins. Unchlorinated primary effluent was batch-fed daily Monday through Friday. Retention time of 5, 3 and 1.5 days were studied. The plants were grown under continuous light at 200−300 ft-cdls minimum (night). Samples were analyzed two times per week for biochemical oxygen demand (BOD_5), chemical oxygen demand (COD), ortho phosphate (O-PO_4), total phosphorus (total-P), and dissolved solids according to the American Public Health Association.[4]

All beds, including the gravel control, were effective in reducing BOD_5 and COD (Table 19-1). The bulrush beds were more effective than the control or the *Iris* bed at reducing O-PO_4 and total-P, especially at the longer retention times. Total-P removal averaged over 80% and on occasions reached 98%. Dissolved solids reduction was best in the five-day retention experiments, but the presence of the plants seemed to have little effect.

Pilot Plant

During the summer of 1972, a pilot plant consisting of ten plastic-lined basins was constructed at Seymour, Outagamie County, Wisconsin (pop. 2,267). Secondary activated sludge effluent is piped from the clarifier of the overloaded municipal treatment plant to the pilot plant (about 300 ft). In the original design and during initial operation in 1973, four of the 9.29 m^2 basins contained plants growing on top of floating hardware cloth racks such that their roots penetrated into 0.3 m of free standing water in a hydroponic-type situation. A fifth basin served as a control. That arrangement was tested for one summer only and then abandoned because of difficulties encountered. Plant growth was not as vigorous or uniform as it was in the gravel beds (see below), and the plants lodged very easily. Thus, these five basins were converted into a single trench-type basin with gravel 0.70 m deep. This trench was planted with *Scirpus validus* during the summer of 1974 and will be the primary study basin during 1975 studies. The remaining five basins each contain six inches (15 cm) of pea-sized gravel and were planted with *Scirpus validus* (2 basins in 1973, 3 basins in 1974), *S. fluviatilis* (Torr.) Gray, and *Iris versicolor* (one basin each in 1973). The fifth basin is a control and contains only gravel. Constant flow rates have not been possible, but the minimum rates (maximum retention) attainable within the design constraints were attempted (about 0.3 gpm). Meters on the influent and effluent lines permitted periodic adjustments to the flow so that an average retention time of five hours was achieved.

The summer of 1973 was regarded as a period of growth and establishment of the plants and as a period for saturation of the substrate with nutrients. Most of the data reported here were collected during the operation of the five gravel basins during summer of 1974.

Samples were collected from the influent line that delivered treated wastewater to the pilot plant and from the effluent line of each of the test basins. Samples were packed in ice to keep them cool during transfer to the laboratory and

TABLE 19-1. *Greenhouse Investigations: Effects of Emergent Aquatic Plants on Unchlorinated Primary Effluent.*

	Retention Time	Iris	Control	% Reduction* Hardstem	Softstem	Softstem
	5 day	98	97	98	98	97
BOD_5	3 day	94	86	92	96	95
	1.5 day	89	86	90	88	81
	5 day	89	89	87	86	83
COD	3 day	73	75	89	85	82
	1.5 day	49	54	61	57	29
Ortho-	5 day	58	52	68	76	70
Phosphate	3 day	8	2	41	63	49
	1.5 day	–3	–1	28	57	30
Total	5 day	80	76	84	83	80
Phosphorus	3 day	34	–4	45	62	54
	1.5 day	30	32	59	60	51
Dissolved	5 day	54	60	60	51	44
Solids	3 day	21	28	25	13	13
	1.5 day	12	19	2	–9	–14

*Each value represents the mean of 6 analyses.

analyses were undertaken within 24 hours of collection. Samples were collected once each week. In addition, three intensive studies were conducted in order to account for diurnal and day-to-day variations in operating efficiency and to determine the effect of harvesting of the plant shoots on the effluent quality.

In the intensive studies, grab samples were taken at four-hour intervals for 72 hours in two studies and for 96 hours in one study. Parameters measured in each sample include: BOD_5, COD, conductivity, orthophosphate, total-P, dissolved solids, suspended solids, pH, turbidity, and temperature;[5] amonia nitrogen, Kjeldahl nitrogen, nitrate, and coliform bacteria.[6] Flow meter readings were taken at each sampling time. Data from a basin containing *S. validus* were compared with those from the control (Tables 19-2 and 19-3).

Harvesting had little impact on the quality of the effluent water (Table 19-2). Even with only a 5-hour exposure time, the pilot plant was effective at reducing BOD_5 (87 to 92%), turbidity (77 to 91%), and coliform bacteria (90 to 99.7%). Ammonia reduction (-72.5 to 40% in the control pond and 90 to 92% in the bulrush pond, NO_3 reduction (-47 to 59%), and suspended solids (-17.3 to 68%) were extremely variable. Reduction of dissolved solids and total-P were quite low (0.3 to 9.5% and 5 to 25% respectively). The most surprising result was that the control basin was just as effective in improving water quality as any of the other three bulrush ponds.

Unlike the control basin, the artificial marsh system maintained aerobic conditions in the gravel substrate at all times. Dissolved oxygen concentrations 2.0 mg/l or more were maintained. On days when the secondary clarifier was not operating efficiently, quite high concentrations of solids entered the basins; but odors, solids accumulation, or anaerobic conditions did not arise as would be expected. Insects or insect larvae did not appear in nuisance numbers unless water was allowed to accumulate above the surface of the gravel. Deep water also led to anaerobic conditions since the plants were apparently incapable of supplying oxygen in sufficient amounts to aerate the greater volume.

TABLE 19-2. *Pilot Plant Investigations*
Numbers are the average values of parameters in the influent and effluent.

Parameter	July 14–18, 1974 Pre-harvest* Infl.	Control	Bulrush	July 21–24, 1974 First Post-harvest† Infl.	Control	Bulrush	August 4–7, 1974 Second Post-harvest† Infl.	Control	Bulrush
BOD_5‡	38.5	4.7	4.5	65.6	8.8	5.3	39.3	4.2	3.2
NH_3	1.5	2.0	1.2	0.4	0.2	0.1	0.05	0.4	0.004
Total N (organic)	0.7	0.9	0.7	ND	ND	ND	ND	ND	ND
NO_3	ND	ND	ND	6.6	3.7	9.7	9.5	4.6	3.9
Turbidity (JTU)	8.6	1.8	2.0	23.2	3.1	2.2	ND	ND	ND
Conductivity (umho)	1436	1415	1440	1075	1005	950	1092	1076	1092
COD	41.6	30.8	36.0	ND	ND	ND	ND	ND	ND
Total P	22.6	18.2	21.5	21.0	18.1	15.9	22.2	18.2	17.9
Coliforms (X 10^3)	485.0	29.5	49.5	2546.0	49.9	58.3	2081.0	5.8	12.1
Dissolved Solids	902.0	930.0	910.0	758.0	704.0	696.0	762.0	740.0	749.0
pH (values)	7.44	7.85	7.45	7.42	7.67	7.37	7.50	7.66	7.52

*In most cases n = 24 samples taken at 4-hour intervals over a four-day period.
†In most cases n = 19 samples taken at 4-hour intervals over a three-day period.
‡All values not specified by parentheses are expressed as mg/l.

TABLE 19-3. *Pilot Plant Investigations*
Numbers are percentages by which concentrations were reduced.

Parameter	July 14–18, 1974 Pre-harvest* Control	Bulrush	Percent Reduction July 21–24, 1974 First Post-harvest† Control	Bulrush	August 4–7, 1974 Second Post-harvest† Control	Bulrush
BOD_5	87.8	88.2	86.5	92.0	89.4	91.3
NH_3	–35.7	20.9	40.5	83.4	–725.5	92.0
Total N (organic)	–26.3	56.8	ND	ND	ND	ND
Turbidity	79.3	77.4	86.7	90.7	ND	ND
Conductivity	1.5	–0.3	6.5	11.6	1.47	0.0
COD	25.8	13.4	ND	ND	ND	ND
Total P	19.5	4.7	13.8	24.5	17.8	19.4
Coliforms	93.9	90.0	98.0	97.7	99.7	99.4
Dissolved Solids	–3.1	0.9	7.2	9.5	2.9	1.8
pH (values)	7.85	7.45	7.67	7.37	7.66	7.52
NO_3	ND	ND	44.5	–47.1	52.2	59.2
Suspended Solids	ND	ND	28.0	68.0	9.0	–173.0

*In most cases, n = 24 samples taken at 4-hour intervals over a four-day period.
†In most cases, n = 19 samples taken at 4-hour intervals over a three-day period.

Duckweed (*Lemna minor* L.) developed rapidly in the control pond if standing was allowed.

Natural Marsh Investigations

Spring Creek in Calumet County, Wisconsin, drains a 7.4 square mile (1920 ha) agricultural basin and accepts the effluent from a foundry and the village of Brillion before entering a 1 square mile (258 ha) cattail (*Typha* spp.) marsh. During typical low-flow periods, the effluent volume from the activated sludge treatment plant may equal the stream flow of about 1 cfs (1.7 m^3/min).

Water samples were collected from the stream at points 200 m above and 200 m below the municipal outfall at the entrance to Brillion marsh. A number of samplings of the effluent were also made. The stream diffuses into the marsh and drains at the lower end into a channel about 20 m wide by 1 m deep. Water samples were also collected at this point. Periodic water quality samplings were made from June through December 1974.

The fluctuation in conductivity from 500 to 3100 μmho was attributed to the periodic discharge of electrolytes into the stream, probably from the foundry (Table 19-4). Total-P was rather high, although not unexpectedly for a stream draining agricultural lands. The BOD_5 was also slightly high for a stream in this region.

The treatment plant effluent had a mean BOD_5 of 109 mg/l and values as high as 170 mg/l were recorded. Turbidity and phosphate were also high. After mixing with the stream water for 200 m a combination of dilution and physical and biological attenuation in the stream channel reduced most paramaters to between 30 and 40% of the value of the effluent. Dilution is the principal mechanism.

The water quality improvement due to movement through the marsh was striking. The polluted streamwater entering the marsh underwent an additional 90% BOD_5 reduction. Chemical oxygen demand was not reduced as much, but was still 35% less. Both O-PO_4 and total-P underwent a 67% reduction. Coliform bacteria were reduced 97% and turbidity 78%.

Removal mechanisms are postulated to include biological oxidation, phosphate coprecipitation, nutrient uptake by marsh vegetation, and precipitation.

Studies in unpolluted Wisconsin marshes also recorded phosphorus removal during summer and fall.[7] However, the same studies revealed that during spring runoff in Wisconsin large amounts of soluble phosphate were removed from the marsh by flushing. Pilot plant studies discussed below substantiate this finding. They concluded that over an annual cycle little phosphate was permanently removed from the marsh. However, some benefits may be obtained by keeping phosphate out of lakes during the summer growing season. Also, the possibility of phosphorus removal by harvesting (see below) appears promising.[8]

The present study will be continued until June 1975, in order to study seasonal effects in winter and spring. At this point, we can conclude that there were significant water quality improvements during the summer and fall.

Phosphorus Removal by Harvesting

Theoretically, large quantities of phosphorus and other nutrients which are taken up by emergent aquatic plants could be conveniently removed from the system by harvesting of the plant shoots in a manner similar to cutting hay in agricultural practice. There have been numerous reports of massive standing crops achieved by marsh plants[9] and, on the basis of single harvest experiments, projections have been made of nutrient removal potential.[10] Most of these reports, however, were conducted in the sub-tropics or were based upon uprooting the entire plant rather than harvesting of the emergent shoots only.

Studies of natural stands have shown highest concentrations of phosphorus in young tissue

TABLE 19-4. *Natural Marsh Investigations: Effects of a Natural Wisconsin* Typha *Marsh on a Stream Receiving Secondary Effluent.*

	Spring Creek Above Outfall	Spring Creek 200 m Below Outfall	Channel Below Brillion Marsh	Treatment Plant Effluent
BOD (mg/l)				
Mean	7.7	.35.1	3.5	109.3
Maximum	14.9	100.0	5.0	170.0
Minimum	3.5	2.8	2.3	81.0
COD (mg/l)				
Mean	29.6	78.6	50.9	224.8
Maximum	61.8	137.4	76.8	309.8
Minimum	00.0	15.7	31.7	133.3
Coliforms (x 10^3)				
Mean	47	240	6	43
Maximum	226	980	23	160
Minimum	1.1	0.42	0.1	3
Turbidity (JTU)				
Mean	8.7	17.7	4.0	43.0
Maximum	21.0	37.0	5.6	57.5
Minimum	4.4	6.0	2.2	21.0
Ortho-phosphate (mg/l)				
Mean	1.16	3.75	1.21	9.42
Maximum	2.92	6.95	2.42	13.50
Minimum	0.27	0.88	0.47	7.23
Total phosphorus (mg/l)				
Mean	1.20	4.28	1.41	10.30
Maximum	3.61	8.20	2.95	15.84
Minimum	0.11	0.79	0.43	7.23
Conductivity (μmhos/cm^2)				
Mean	1690.	1150.	1170.	1045.
Maximum	3100.	1850.	1600.	1130.
Minimum	500.	500.	500.	950.
pH				
Mean	8.02	7.91	7.99	7.67
Maximum	8.81	8.75	8.50	8.01
Minimum	7.64	7.46	7.49	7.48

TABLE 19-5. *Effects of Harvesting on the Phosphorus Concentration of Various Aquatic Plants Found in Mixed Natural Stands.*

Species	Percent P in harvested shoots: Biweekly	Monthly	Control
Scirpus fluviatilis	0.623 (9)*	0.428 (6)	0.145 (2)
Typha angustifolia 1	0.476 (8)	0.355 (6)	0.140 (2)
Typha angustifolia 2	0.401 (9)	0.325 (6)	0.155 (2)
Scirpus validus	0.368 (10)	0.387 (6)	0.135 (2)
Sparganium eurycarpum	0.432 (6)	0.388 (4)	not present

*Each value represents the mean of samples collected over two growing seasons (1970, 1971) from a single representative 4 m^2 quadrat and is expressed as percent of dry weight. The number in parentheses is the number of samples analyzed.

TABLE 19-6. *Effects of Harvesting on the Removal of Dry Matter from Mixed Natural Stands of Aquatic Plants.*

Species	Biweekly	Grams dry weight/m² Monthly	Control
Site 1			
Typha angustifolia	197.28*	150.41	260.67
Scirpus validus	60.17	30.61	64.66
Sparganium eurycarpum	35.71	37.59	not present
TOTAL	293.16	218.61	325.33
Site 2			
Typha angustifolia	21.5	27.4	38.45
Scirpus fluviatilis	169.3	225.7	406.65
TOTAL	190.8	253.1	445.10

*Values represent the mean of the accumulated harvest from three (Site 1) or six (Site 2) replicate 4 m^2 quadrats during two growing seasons (1970, 1971).

and decreasing concentrations with age of tissue.[11] This suggests that the quantity of phosphorus removed could be amplified by frequent harvesting so that tissues are not permitted to mature completely.

In 1970, a series of twenty-seven quadrats, each 2 meters on a side (4 m^2), were established in two natural marsh stands a few miles from where the experimental pilot plant is now located. During the growing seasons of 1970-1971, one-third of the plots were harvested biweekly, one-third monthly, and the remainder at the end of the growing season in September (controls). The shoots were harvested by cutting with a hand sickle about 20 cm above the surface. The samples were dried at 60°C, ground in a Wiley Mill over a #20 mesh screen, and analyzed for dry weight, organic weight, and phosphorus content.[12] The phosphorus concentration in the five native species present ranged from an average of 0.37% in *Scirpus validus* to 0.62% in *S. fluviatilis* in the biweekly harvested plots (Table 19-5). At the sites sampled monthly, the phosphorus concentrations were much lower, averaging from only 0.288% in *Sparganium eurycarpum* Englm. to 0.43% for *S. fluviatilis.* The plants in the control plots contained only 0.14/ to 0.16/P.

Each successive harvest of any given plot resulted in a smaller quantity of new material. For example, the average standing crop of *Typha angustifolia* L. at one site decreased in the biweekly harvested quadrats from 99 g d.w./m^2 on 6 June 1971, to only 7.8 g d.w./m^2 at the fifth harvest on 7 August 1971.[13] Also, a smaller percentage of shoots continued to provide new growth from the intercalary meristem after each harvest, and the subsequent growth rates were slower. Similar responses have been noted in harvesting studies of *Typha* and the reed *Phragmites* in Czechoslovakia.[14]

As a result of this, along with a smaller cellulose deposition in the less mature tissue, the total amount of material removed from the multi harvested quadrats was less than that from the control quadrats in both the *Typha*-dominated and *Scirpus*-dominated communities (Table 19-6). Nonetheless, the high phosphorus concentration in the younger tissue permitted a greater yield of phosphorus via multiple harvests. In the *Typha*-dominated community, more than three times as much phosphorus was removed by biweekly harvesting than by a single, end-of-the-season harvest (Table 19-7).

In the pilot plant operation discussed above, harvests were scheduled so that the plants would

TABLE 19-7. *Phosphorus Removal from Mixed Natural Stands of Aquatic Plants by Harvesting at Various Intervals.*

Species	Grams phosphorus/m² Biweekly Harvest	Monthly Harvest	End-of-season Harvest
Site 1			
Typha angustifolia	0.79*	0.50	0.40
Sparganium eurycarpum	0.52	0.10	0.00
Scirpus validus	0.22	0.12	0.09
TOTAL	1.53	0.72	0.49
Site 2			
Typha angustifolia	0.13†	0.13	0.11
Scirpus fluviatilis	0.91	0.97	0.59
TOTAL	1.04	1.10	0.70

*Values represent the mean of three replicate 4 m² quadrats sampled over two growing seasons (1970, 1971).
†Values represent the mean of six replicate 4 m² quadrats sampled over two growing seasons (1970, 1971).

have a chance to mature to the point of developing flowering heads (on new shoots, not previously severed) between harvests. This was done in order to optimize conditions for other water quality experiments and still realize a near maximum phosphorus transport into the emergent shoots. Thus, *Scirpus validus* and *S. fluviatilis*, were harvested four times during 1974, at approximately one month intervals. Only 1.13 g P/m² was removed from a very dense stand of *S. fluviatilis*, but 3.5 to 3.8 g P/m² were removed from ponds containing *S. validus* (Table 19-8). This is equivalent to 35 g P/ha (31.25 lb/acre), a quantity sufficient to remove the phosphorus from thirty-three persons if a per capita loading of three pounds per annum is assumed over a four-month sampling period.[15] This would, however, also represent the annual removal capacity as it encompasses the full growing season. Thus, an absorption capacity of only 10 persons per acre could be assumed on a year-around basis.

Obviously, not all of the phosphorus retained in any plant-substrate system (natural or artificial) will be translocated to the harvestable shoots; some will be incorporated into new unharvestable root mass.[16] The substrate should eventually achieve equilibrium so that no additional nutrients are retained and all increases (phosphorus uptake) should be recorded in the plant biomass. In order to determine the phosphorus distribution in the pilot plant basins, randomly spaced, triplicate, 5-inch (12.8 cm) diameter cores were taken at the time of harvest-

TABLE 19-8. *Phosphorus Removal from Experimental Bulrush Ponds by Harvesting of Plant Shoots.*

Harvest Date	Grams phosphorus/m² Pond 3 *Scirpus validus*	Pond 5 *Scirpus validus*	Pond 7 *S. fluviatilis*
17 June 1974	1.53	1.86	0.58
19 July 1974	1.32	1.06	0.33
8 Aug. 1974	0.30	0.61	0.16
9 Sept. 1974	0.36	0.30	0.06
TOTAL	3.51	3.83	1.13

TABLE 19-9. *Distribution of Phosphorus in Two Experimental Ponds in 1974.*

	17 June		23 Sept.		4 Dec.	
POND 5 (Scirpus validus)						
	g P/m²	%	*g P/m²*	%	*g P/m²*	%
Shoots harvested or date indicated	1.86	14.1	0.30	1.6	0	0
Shoots harvested since last date	0	0	1.67	8.9	0	0
Unharvested shoots	0.82	6.2	1.74	9.2	0.58	7.3
Rhizomes	1.42	10.8	3.12	16.5	2.30	28.9
Roots	0.35	2.6	3.41	18.0	2.68	33.6
Total P in biomass	4.45	33.7	10.24	54.2	5.56	69.8
Gravel zone P	8.75	66.3	8.65	45.8	2.41	30.2
Total P	13.20	100	18.89	100	7.97	100
CONTROL POND (no plants)						
Gravel zone P	7.91	100	5.18	100	0.38	100

ing on 17 June, 23 September, and 4 December 1974. The unharvested portions of the plants were separated from the gravel and the gravel was washed repeatedly with small aliquots of distilled water totalling 2 liters. The eluate was then analyzed for total-P (Persulfate-$SnCl_2$ method).[17] On 17 June, 14% of the total-P present in Pond 5 was located in the harvested shoots (Table 19-9), but 66% was non-plant material associated with the gravel. Between 17 June and 23 September, 1.97 g P/m² was harvested from the stand. The amount of phosphorus in the gravel remained constant at about 8.7 g P/m², but there was an increase in phosphorus in the unharvested portions of the plants. As a result, there was an uptake of 8.22 g P/m² by the system. Between 23 September and 4 December, there was no new harvestable growth and 9.92 g P/m² was lost from the unharvested portions and gravel. It appears that prior to termination of flow on 22 November, there was a leaching from the roots and rhizomes of the plants. There was apparently also a die-off of a considerable mass of microflora and microfauna in the gravel which then washed out with the effluent. A loss of 7.53 g P/m² occurred in the control pond, and a large loss was recorded for each of the five ponds.[18] These studies of phosphorus distribution will continue through the 1975 growing season.

On the basis of these experiences, a loss of phosphorus can be expected from any emergent plant facility at temperate latitudes which continues to operate after killing frosts.

Several factors must be carefully considered if natural marshes are to be used to remove nutrients from a drainage system. Harvesting affects each species of plant in a different manner and to different degrees.[19] *Scirpus acutus* Mühl. and especially *S. validus* seemed to recover well and sustain a yield. *Sparganium eurycarpum, Typha, Iris,* and *S. fluviatilis,* however, did not respond well. In mixed stands, *Sparganium* was replaced by *Typha* and *S. validus* after harvesting. Studies in natural marshes have also shown that harvesting at very low water levels may adversly affect the plants.

The phosphorus concentrations observed in the effluents of natural or artificial marsh systems receiving sewage effluents have always been at least 1–2 mg P/l, regardless of the size of the system. This may represent a limit to the efficiency of this type of system; or it may represent the point at which some other factor, such as nitrogen or copper, becomes limiting.[20]

Conclusions

While it is much too soon to draw conclusions on the functional efficiency or economic advantages of using marsh plants for the treatment of domestic sewage, enough performance patterns in marsh plant systems have become apparent to permit some speculations.

Marsh plant (emergent aquatic macrophyte) systems are effective at rapidly reducing coliform bacteria and BOD_5, but this action appears to be associated with microflora in the plant substrate rather than with plants themselves. Considerable quantities of phosphorus (probably 10 g P/m² yr or more) are taken up by the marsh system. Only 3 to 4 g P/m² yr can be removed by harvesting of the shoots, however. The remainder is associated with the root-rhizome and microflora-substrate complexes and much is subject to loss from the system between growing seasons. It is unlikely that the residual phosphorus in the effluent of a marsh system will be below 1 mg P/l.

Aerobic conditions are maintained in the rhizosphere and undoubtedly favorably influence the microflora and microfauna. If standing water is not permitted, insects and insect larve do not appear in nuisance numbers and odors do not arise. Standing water is likely to develop blooms of algae and *Lemna* (duckweed) and to become anaerobic.

For some waste treatment needs, it appears that emergent vegetation provides a suitable alternative for secondary or tertiary treatment. Should continued research on a broad basis lead to general acceptance of the concept, the authors express their preference for artificial systems rather than an indiscriminate exploitation of natural marshes in much the same manner as has led to the degradation of our lakes and streams.

Notes

1. We wish to thank Ms. Kathleen Garfinkel for her assistance in laboratory and field work. Russell Hanseter supplied some natural marsh data. Roy Willey of the East Central Wisconsin Regional Planning Commission and Gerald Paul provided invaluable direction. Research was supported by U.S. Environmental Protection Agency Grant Number S-801042 to the East Central Wisconsin Regional Planning Commission and by the Wisconsin Department of Natural Resources.

2. Council on Environmental Quality, *Evaluation of municipal sewage treatment alternatives: A final report prepared for CEQ and US EPA,* Contract EQC 316 (Washington: U.S. Government Printing Office, 1974).

3. H. Althaus, "Biological Waste Water Treatment with Bulrushes," *Das Gas–und Wasser fach* 107 (1966):486-88; R. Kickuth, "Higher Water Plants and Water Purification: Ecochemical Effects of Higher Plants and Their Functions in Water Purification," *Schriften-reihe der Vereiningung Deutscher Gerwässerschutz VDG* 19 (1969); Käthe Seidel, "Macrophytes as Functional Elements in the Environment of Man," *Hydrobiologia* 12(1971):121-30; Käthe Seidel, "Wirkung Hohere Pflanzen auf Pathogene Keime in Gewassern," *Naturwissenschaften* 58 (1971):150-51. See Chapters 14 and 33.

4. American Public Health Association, *Standard Methods for the Examination of Water and Wastewater,* 13th ed. (Washington, D.C., 1971).

5. *Ibid.*

6. Environmental Protection Agency, *Methods for Chemical Analysis of Water and Wastes* #16020 (Washington, D.C.: U.S. Government Printing Office, 1971).

7. G. F. Lee, E. Bently, and R. Amundson, "Effect of Marshes on Water Quality," *Water Chem. Prog.* (Madison: University of Wisconsin mimeograph, 1969).

8. Lee et al. also reported significant denitrification occurring in marshes, an aspect not included in this study. *Ibid.*

9. C. E. Boyde, "Production, Mineral Accumulation and Pigment Concentrations in *Typha latifolia* and *Scirpus americanus,*" *Ecology* 51 (1971):285-90; C. E. Boyde, "Further Studies on Productivity, Nutrient, and Pigment Relationships in *Typha latifolia* Populations," *Bulletin of the Torrey Botanical Club* 98 (1971):144-50; D. Dykyjová and S. Husák, "Influence of Summer Cutting on the Regeneration of Reed," in *Ecosystem Study on Wetland Biomass in Czechoslovakia,* PT-PP/IBP Report No. 3 (Tebon: 1973), pp. 245-50; S. Husák, "Destructive Control of Stands of *Phragmites communis* and *Typha angustifolia* and Its Effects

on Shoot Production Followed for Three Seasons, in *Littoral of the Nesyt Fishpond, Ecological Study,* ed. J. Kvet, Studie CSAV 15 (Praha: 1973), pp. 89-91; W. T. Penfound, "Primary Production of Vascular Aquatic Plants," *Limnol. & Oceanog.* 1 (1956):92-101; K. K. Steward, "Nutrient Removal Potential of Various Aquatic Plants," *Hyacinth Control Journal* 8 (1970):34-35.

10. C. E. Boyde, "Vascular Aquatic Plants for Mineral Nutrient Removal from Polluted Water," *Economic Botany* 42 (1970):95-103; D. Dykyjová and J. Kvet, "Comparison of Biomass Production in Reedswamp Communities Growing in South Bohemia and South Moravia," in *Productivity of Terrestrial Ecosystems Production Processes,* PT PP/IBP Report No. 1 (Tebon, 1970) pp. 71-78; Seidel, "Wirkung Hoherer"; Steward, "Nutrient Removal"; S. Husák and S. Hejny, "Marginal Plant Communities of the Nesyt Fishpont (South Moravia)," *Polish Archives of Hydrobiology* 20 (1973):461-67.

11. C. E. Boyde, "Seasonal Changes in the Proximate Composition of Some Common Aquatic Weeds," *Hyacinty Contr. Journal* 8 (1970):42-44; J. Kvet, "Mineral Nutrients in Shoots of Reed," *Polish Archives of Hydrobiology* 20 (1973):137-47.

12. Association of Official Agricultural Chemists. *Methods of Analysis* (Washington, D.C., 1960).

13. R. Hanseter, unpublished data.

14. D. Dykyjová and S. Husák, "Influence of Summer Cutting"; S. Husák, "Destructive Control."

15. Steward, "Nutrient Removal."

16. D. Dykyjová and J. P. Ondok, "Biometer and Productive Stand Struction of Coenosis of *Sparganium erectum* L. *Preslia (Praha)* 45 (1973):19-30; D. Dykyjová and S. Husák, "Influence of Summer Cutting"; S. Husák, "Destructive Control."

17. American Public Health Association, "Standard Methods."

18. William Sloey, unpublished data.

19. D. Dykyjová and S. Husák, "Influence of Summer Cutting"; S. Husák, "Destructive Control."

20. Steward, "Nutrient Removal"; G. Gerloff, *Plant Analysis for Nutrient Assay of Natural Waters,* EPA-Ri-73-001 (Washington, D.C.: U.S. Government Printing Office, 1973).

20

The Potential Use of Freshwater Tidal Marshes in the Management of Water Quality in the Delaware River[1]

DENNIS F. WHIGHAM
ROBERT L. SIMPSON

Biology Department
Rider College

Introduction

In 1973 we began a study of the potential use of a natural ecosystem, a freshwater tidal marsh (Hamilton Marshes), in the management of water quality in a portion of the Delaware River. Our initial emphasis was on vegetation composition and primary production.[2] In 1974 we expanded our research to include other functional aspects of the marsh ecosystem, initiating experiments on soil algae distribution and biomass, soil nutrient and organic matter content, plant nutrient content, nutrient movement through the marsh by analysis of surface water, and movement of detritus into and out of the marshes. Preliminary results of this work, parts of which are reported in this paper, clearly demonstrate that the marsh is highly productive and that it is an efficient nutrient processer.

The Delaware River Ecosystem

Degradation of water in the Delaware River estuary system is a problem of local and regional concern. Compared to other east coast estuarine systems, much more biochemical oxygen demand (BOD) is discharged into the Delaware River basin.[3] A recent Delaware River census further demonstrated that areas within the estuarine system are polluted.[4] Organic enrichment, evidence of toxic pollution, and periods of low dissolved oxygen were shown for sections of the estuarine system between Trenton and Salem, N.J. Walton and Patrick suggested that degrada-

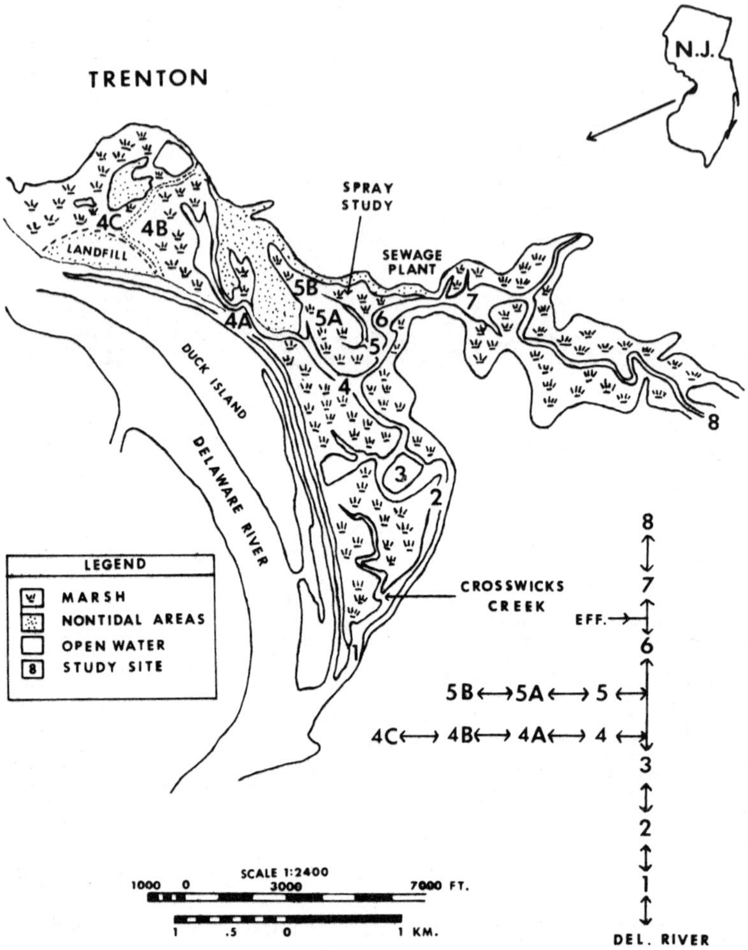

FIG. 20-1. Schematic diagram of the Hamilton Marshes. The pattern of water movement into and out of the marsh is shown at the lower right.

tion was greatest around Philadelphia. Sewage effluent is a major source of pollution in the estuarine system.

Of the approximately 158,000 acres of marshland in the Delaware River estuarine system, 5% are freshwater tidal marshes. These are located along the northern section of the tidal area from south of Philadelphia to Trenton. An analysis of the RANN study shows that the freshwater marshes vary in size from about 20 acres to over 2000 acres and that the average size is approximately 850 acres.[5] Compared to the salt marshes, these freshwater marshes show more evidence of perturbation. There was evidence of organic enrichment in all of the marshes mentioned in the RANN report with most of it coming from sewage treatment plants. In addition approximately 50% of the freshwater marshes are used as sites for landfills or for disposal of industrial wastes. The report suggested that several of the marshes are heavily impacted and in a severely degraded condition.

Of what value are the freshwater tidal marshes and is it important that they be managed? The State of New Jersey has declared that they are valuable natural areas and has included them in the New Jersey Wetlands Act of 1970. The marshes act as impoundments that store excess water during periods of high water flow.

Between Trenton and Bordentown, N.J., more than 30 million gallons per day (mgd) of secondarily treated sewage effluent are discharged either directly or indirectly into the Delaware River.[6] Except for the Ewing-Lawrence and Hamilton Township Sewage Treatment Plants, most of the effluent is discharged directly into the Delaware River. Hamilton Township discharges approximately 7 mgd into Crosswicks Creek, a tributary of the Delaware River. Crosswicks Creek is the largest stream that flows through the Hamilton Marsh ecosystem (Fig. 20-1). To account for a projected increase in utilization of their facility, officials in Hamilton Township are presently contracting for the expansion of their sewage treatment plant to handle 17.5 mgd. They must also decide whether they will continue to discharge their effluent into Crosswicks Creek or pump it directly into the Delaware River. As population increases within the Trenton metropolitan area, other municipalities will also have to expand their capacities and without tertiary treatment or other suitable means of disposal, enrichment of the upper Delaware River estuarine system will surely increase. To prevent further pollution of that system, other disposal alternatives must be examined. This paper will demonstrate the tertiary treatment capabilities of the Hamilton Marshes through analyses of several important marsh functions.

Study Site

The Hamilton Marshes occupy approximately 500 ha of tidal and nontidal land along an old meander adjacent to the Delaware River near Trenton, N.J. (Fig. 20-1). This is the northernmost tidal marsh in the Delaware River basin. Trenton has a typical temperate climate with a cold but not long winter (Fig. 20-2).[7] Precipitation averages 1020 mm annually and it is fairly evenly distributed with a maximum in July and a minimum in October. The average temperature is 12.1°C and the growing season lasts from mid-

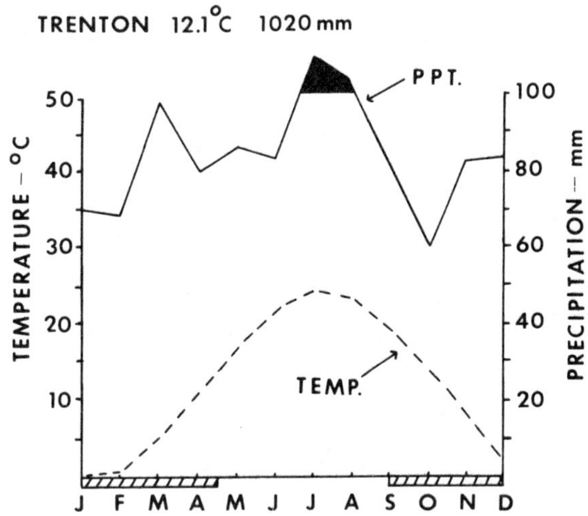

FIG. 20-2. Climatic diagram for Trenton, N.J. The mean annual temperature and precipitation are shown next to the climate station name. Because the precipitation curve is always above the temperature curve, the climate is interpreted as not having a dry season. Along the abscissa are plotted the months of the year (January–December) and a shaded area which shows the number of months during which there is a chance of temperatures that fall below 0°C. Trenton has a Type VI climate according to the system of Walter and Lieth (see note 7). Precipitation above 100 mm per month is plotted on the scale of 1:3 (10°C: 30 mm ppt.).

April until late September. Within the marsh, water supply is always adequate, and temperature, along with photoperiod, seem to be the most important factors controlling biological activities of organisms. The silt and silt-clay soils are highly organic (15–50%).

Vegetation Analysis

The marsh can be divided into 4 habitats: (1) streams and stream banks—including channels that connect streams to high marsh areas, (2) high marsh areas that are covered by shallow water (usually less than 15 cm) at high tide only, (3) areas that are pond-like during much of each tide cycle and are drained only at low tide, and (4) areas that are continuously water covered. Unlike salt marshes, vascular plant diversity is high in freshwater tidal marshes. For purposes of

TABLE 20-1. *Aerial Extent and Total Aboveground Production Estimates for Dominant Vegetation Associations of the Hamilton Marshes.*

VEGETATION TYPE	COVERAGE (ha)	ANNUAL ABOVEGROUND PRODUCTION (t/ha)	TOTAL PRODUCTION (t)
Mixed	137	9.1	1246.7
Cattail	19	13.2	250.8
Giant Ragweed	3	11.6	34.8
Arrow Arum	11	6.5	71.5
Spiked Loosestrife	10	21.0	210.0
Wild Rice	24	9.4	225.6
Yellow Water Lily	58	7.8	452.4
TOTAL	262	X = 9.5	2491.8

analysis, we have divided the marsh into seven major vegetation associations based upon the dominant species (Table 20-1). Vegetation dominated by cattail, *(Typha latifolia, T. angustifolia, T. glauca)*, giant ragweed *(Ambrosia trifida)*, and spiked loosestrife *(Lythrum salicaria)* are common on the high marsh but the mixed vegetation association is most expansive and covers approximately 137 ha. Bur marigold *(Bidens laevis)* is the dominant species in the mixed vegetation type. Other important components are arrow arum *(Peltandra virginica)*, arrowhead *(Saggitaria latifolia)*, halberd-leaved tearthumb *(Polygonum arifolium)*, and arrow-leaved tearthumb *(Polygonum sagittatum)*. Wild rice *(Zizania aquatica)* is also common in this vegetation association.

Primary Production

Measurements of primary production are frequently used as indicators of how ecosystems function. High rates of production are indicative of high rates of nutrient assimilation. Primary production of each vegetation association was examined in 1974. Aboveground biomass was harvested throughout the growing season using

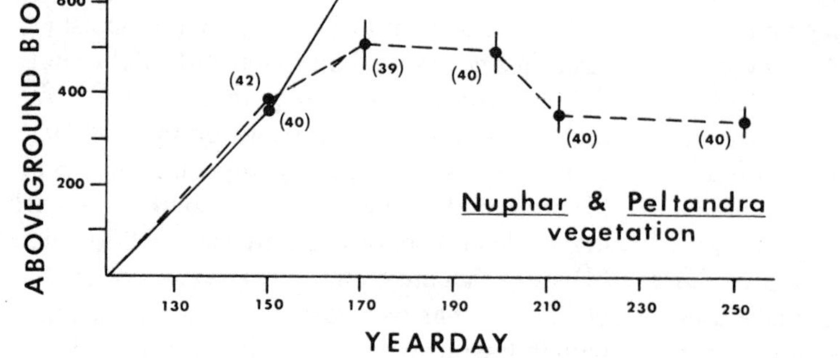

FIG. 20-3. Aboveground primary production for all study sites. As described in the text, data were divided into two categories: (1) vegetation types dominated by *Nuphar advena* and *Peltandra virginica*, (2) all other vegetation types. Numbers in parentheses represent sample size. ($g/m^2 \times 10^{-2}$ = t/ha)

Fig. 20-4. Aboveground primary production of mixed vegetation, primarily bur marigold *(Bidens laevis).* ($g/m^2 \times 10^{-2}$ = t/ha)

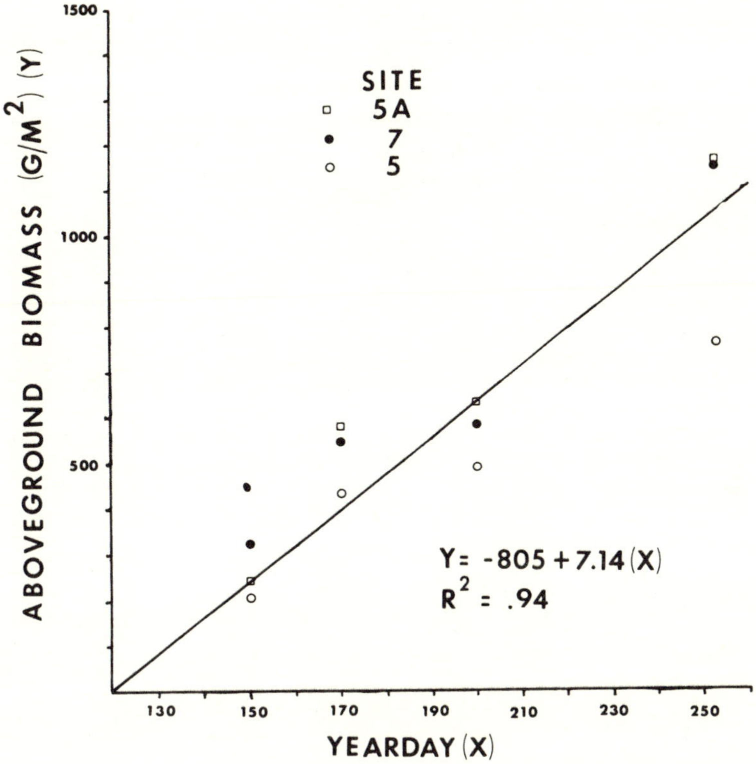

0.25 m^2 quadrats. On each sampling date, three quadrats were harvested from each study site. Fig. 20-3 is a composite of our productivity data for the entire marsh. The data have been separated into two categories: (1) sites dominated by arrow arum *(Peltandra virginica)* and/or yellow water lily *(Nuphar advena),* and (2) all other sites. This separation was necessary because of bimodal patterns of production for both arrow arum and yellow water lily. Both species assumed aspect dominance throughout the marsh during the early part of the growing season. As other species became dominant, both species began a widespread dieback as seen in Fig. 20-3. We estimate overall aboveground net primary production at 9.5 t/ha/yr (Table 20-1). Production values for the dominant community types varied from 6.5 t/ha/yr for arrow arum dominated communities to 21.0 t/ha/yr areas dominated by spiked loosestrife. Cattail and giant ragweed dominated stands were also highly productive. The mixed vegetation type was most expansive and averaged 9.1 t/ha/yr. Fig. 20-4 shows seasonal changes at one of the high marsh sites dominated by bur marigold. There was a linear increase in aboveground biomass throughout the growing season and we estimated that a total of 1246.7 tons of biomass were produced during the growing season. That value was approximately 50% of our estimated total aboveground net primary production for the entire marsh. We have completed preliminary analysis of the vegetation for total nitrogen content using a modified Kjeldahl technique.[8] Average nitrogen content for the species in the mixed vegetation type is 2.49%. The value falls well within reported data for plants from other marsh ecosystems.[9] We estimate total nitrogen uptake within that vegetation type at 32 tons.

TABLE 20-2. *Summary of Production Values for Marsh Plants. Most data is compiled from Keefe, "Marsh Production."*

COMMUNITY TYPE (DOMINANT)	ABOVEGROUND NET PRODUCTION ($g/m^2/yr$)	LOCALE	REFERENCE
	(I - *Freshwater Tidal Marshes)*		
Wild Rice *(Zizania aquatica)*	605–1547	Pa.	McCormick, "The Natural Features of Tinicum Marsh."
	659–1125	N.J.	Present study
	1390	N.J.	McCormick and Ashbaugh, "Vegetation of a Section of Oldmans Creek Marsh."
	1699	N.J.	R. A. Jervis, "Primary Production in a Freshwater Marsh Ecosystem," thesis, Rutgers University, 1964.
Giant Ragweed *(Ambrosia trifida)*	1211–1250	Pa.	McCormick, "The Natural Features of Tinicum Marsh."
	1160	N.J.	Present study
Yellow Water Lily *(Nuphar advena)*	1166–1188	Pa.	McCormick, "The Natural Features of Tinicum Marsh."
	516	N.J.	McCormick and Ashbaugh, "Vegetation of a Section of Oldmans Creek Marsh."
	775	N.J.	Present study
	245	Va.	M. L. Wass and T. D. Wright, "Coastal Wetlands of Virginia," Interim report to the Governor and General Assembly, Virginia Inst. of Marine Sci., Spec. Rept. in *Appl. Mar. Sci. and Ocean Eng.* 10 (1969).
Cattail *(Typha* sp.)	874–2063	Pa.	McCormick, "The Natural Features of Tinicum Marsh."
	987	N.J.	McCormick and Ashbaugh, "Vegetation of a Section of Oldmans Creek Marsh."
	1119–1528	N.J.	Present study
	930	Va.	Wass and Wright, *"Coastal Wetlands of Virginia."*
	1905	N.J.	Jervis, "Primary Production in a Marsh Ecosystem."
Mixed *(Bidens laevis)*	516– 897	Pa.	McCormick, "The Natural Features of Tinicum Marsh."
	756–1162	N.J.	Present study
Primrose Willow *(Jussiaea repens)*	403– 583	Pa.	McCormick, "The Natural Features of Tinicum Marsh."
Arrowhead *(Saggitaria* sp.)	628	Pa.	McCormick, "The Natural Features of Tinicum Marsh."
Arrow arum *(Peltandra virginica)*	269	Pa.	McCormick, "The Natural Features of Tinicum Marsh."
	500– 800	N.J.	Present study
Sweet Flag *(Acorus calamus)*	712– 940	N.J.	Present study
Loosestrife *(Lythrum salicaria)*	1749	Pa.	McCormick, "The Natural Features of Tinicum Marsh."
	2104	N.J.	Present study
Waterhemp *(Acnida cannabina)*	762	Pa.	McCormick, "The Natural Vegetation of Tinicum Marsh."
	(II - *Salt Marshes between New York and Va.*)		
Saltwater Cordgrass *(Spartina alterniflora)*	1332	Va.	Wass and Wright, *"Coastal Wetlands of Virginia."*
	445	Del.	M. H. Morgan, "Annual Angiosperm Production on a Salt Marsh," thesis, Univ. Delaware, 1961

TABLE 20-2. (cont.)

COMMUNITY TYPE (DOMINANT)	ABOVEGROUND NET PRODUCTION (g/m²/yr)	LOCALE	REFERENCE
	300	N.J.	R. E. Good, "Salt Marsh Vegetation, Cape May, N. J.," *Bulletin N. J. Acad. Sci.* 10 (1965): 1-11.
Salt-meadow Grass *(Spartina patens)*	805	Va.	Wass and Wright, *"Coastal Wetlands of Virginia."*
Spike Grass *(Fimbristylis* sp.*)*	360	Va.	Wass and Wright, *"Coastal Wetlands of Virginia."*
	(III - *Freshwater Ponds, Lakes, and Streams)*		
Bulrush *(Scirpus americanus)*	150	S.C.	C. E. Boyd, "Production, Mineral Nutrient Absorption, and Biochemical Assimilation by *Justicia americana and Alternanthera philoxeroides, Arch. Hydrobiol.* 66 (1969): 139–60
Cattail	684	S.C.	Boyd. "Production by *Justicia americana."*
(Typha latifolia)	1527	Okla.	W. T. Penfound, "Production of Vascular Aquatic Plants," *Limnol. & Oceanogr.* 1 (1956): 92–101.
	1358	N.Y.	R. M. Harper, "Some Dynamic Studies of Long Island Vegetation," *Plant World* 21 (1918): 33–46.
Typha glauca	416	Neb.	S. J. McNaughton, "Ecotype Function in the *Typha* Community-type," *Ecol. Monogr.* 36 (1966): 297–325.
	1360	Minn.	J. R. Bray, "Estimates of Energy Budgets for a *Typha* Marsh," Science 136 (1962): 119–20.
Typha sp.	730	Okla.	McNaughton, "Ecotype Function."
	1336	Texas	McNaughton, "Ecotype Function."
Sedges *(Carex* sp.*)*	1340	N.J.	Jervis, "Primary Production in a Marsh Ecosystem."
Rice Cutgrass *(Leersia oryzoides)*	1545	Va.	Wass and Wright, *"Coastal Wetlands of Virginia."*
Water Hyacinth	1276	La.	Penfound. "Production of Vascular Aquatic Plants".
(Eichhornia crassipes)	1478	La.	W. T. Penfound and T. T. Earle, "The Biology of the Water Hyacinths," *Ecol. Monogr.* 18 (1948): 447–72.
Water-willow *(Justicia americana)*	640	Ala.	Boyd, "Production by *Justicia americana."*
Alligator-weed *(Alternanthera philoxeroides)*	841	Ala.	Boyd, "Production by *Justicia americana."*

Table 20-2 compares data for salt marshes, freshwater tidal marshes, and other freshwater marshlands. With the exception of the freshwater tidal marsh data, most of the data has been taken from C. W. Keefe's summary of marsh production.[10] It is apparent that there is much variability in the data and that it is difficult to determine which habitat supports the highest overall primary production. In two closely related studies, biomass accumulation was observed in two freshwater tidal marshes along the Delaware River.[11] For similar vegetation types, production values are comparable. It is obvious that freshwater tidal marshes are highly productive and that they are as productive as estuarine salt marshes. Production values for salt marshes between Virginia and New York ranged from 300–1331 g/m²/yr. The reported range of aboveground production for nontidal marshes is approximately 150–2000 g/m²/yr.

Soil Algae

Studies are currently assessing the role edaphic algae play in the Hamilton Marshes. Working with the top two centimeters of marsh soil, we have estimated soil algal standing crop using chlorophyll extraction techniques outlined by H. L. Golterman and modified for our system.[12] Fig. 20-5 summarizes our findings for chlorophyll *a* and its degradation production phaeophytin. Mean chlorophyll *a* levels in the top two centimeters of soil show definite seasonal patterns and range from a high of 6.29 μg/top 2 cm³ in early summer to a low of 1.96 μg/top 2 cm³ in mid fall. These values are considerably lower than those reported for estuarine mudflats.[13] Mean phaeophytin values always exceed chlorophyll *a* reaching a maximum of 16.29 μg/top 2 cm³ in early fall.

Soil algal standing crop appears to be influenced by the dominant vascular plant communities in the marsh. Areas dominated by yellow water lily consistently have chlorophyll *a* levels greater than mean values while high marsh areas dominated by mixed vegetation (bur marigold and others) have chlorophyll *a* values below the mean. This relationship appears to be a function of differences in soils in the marsh. Silty sand soils of low organic content (about 15%) found in the yellow water lily areas provide the best substrate for algal growth, and silty clay soils of high organic content (25–50%) found in the mixed vegetation and cattail communities provide the poorest substrate. Shading by the higher plants also influences algal standing crops with the highest values occurring in the spring and early summer while the vascular plants are still relatively small. As the higher plants grow, chlorophyll *a* values decline and phaeophytin levels rise correspondingly.

Peak algal biomass for the marsh, estimated from chlorophyll *a* values using Wetzel's factor of 60 for conversion of chlorophyll *a* to organic matter in nonnutrient limiting environments, was 37.7 kg/ha, which was two to three orders of magnitude less than the peak biomass of the vascular plants.[14] Nevertheless the edaphic algae cannot be overlooked, since they are the only functioning producers in the marsh for almost eight months of the year. Furthermore, it has been found that soil algae may contribute up to 25% of the total annual production in Delaware River salt marshes with a substantial part of this production coming during the winter and spring when the vascular plants are dormant.[15]

Water Quality

Surface water quality is being examined at eleven sites (Fig. 20-1) located on Crosswicks Creek and the major side channels in the marsh with emphasis on chemical species known to reflect metabolic processes in aquatic environments. Water is collected at morning high slack water (hsw) and afternoon low slack water (lsw) biweekly in the summer and at monthly intervals otherwise. All samples are analyzed for dissolved oxygen (azide modification[16]), carbon dioxide,[17] nitrogen (reactive nitrate, reactive nitrite, and ammonia plus amino acids[18]), and reactive phosphate.[19] Based on water quality differences, the marsh may be conveniently subdivided into three regions, the main channel of Cosswicks Creek, side channels draining the high marsh, and pond-like areas. Selected sites from each of these subdivisions will be discussed separately.

The major unnatural perturbation of the surface waters in the Hamilton Marshes is the Hamilton Sewage Treatment Plant, which releases 7 million gallons of secondarily treated effluent into Crosswicks Creek daily. This impact is noticed in the main channel of Crosswicks Creek at hsw at Site 7 above the effluent release point and at Sites 1, 2, and 6 downstream from the release point at lsw. Water quality at Site 2

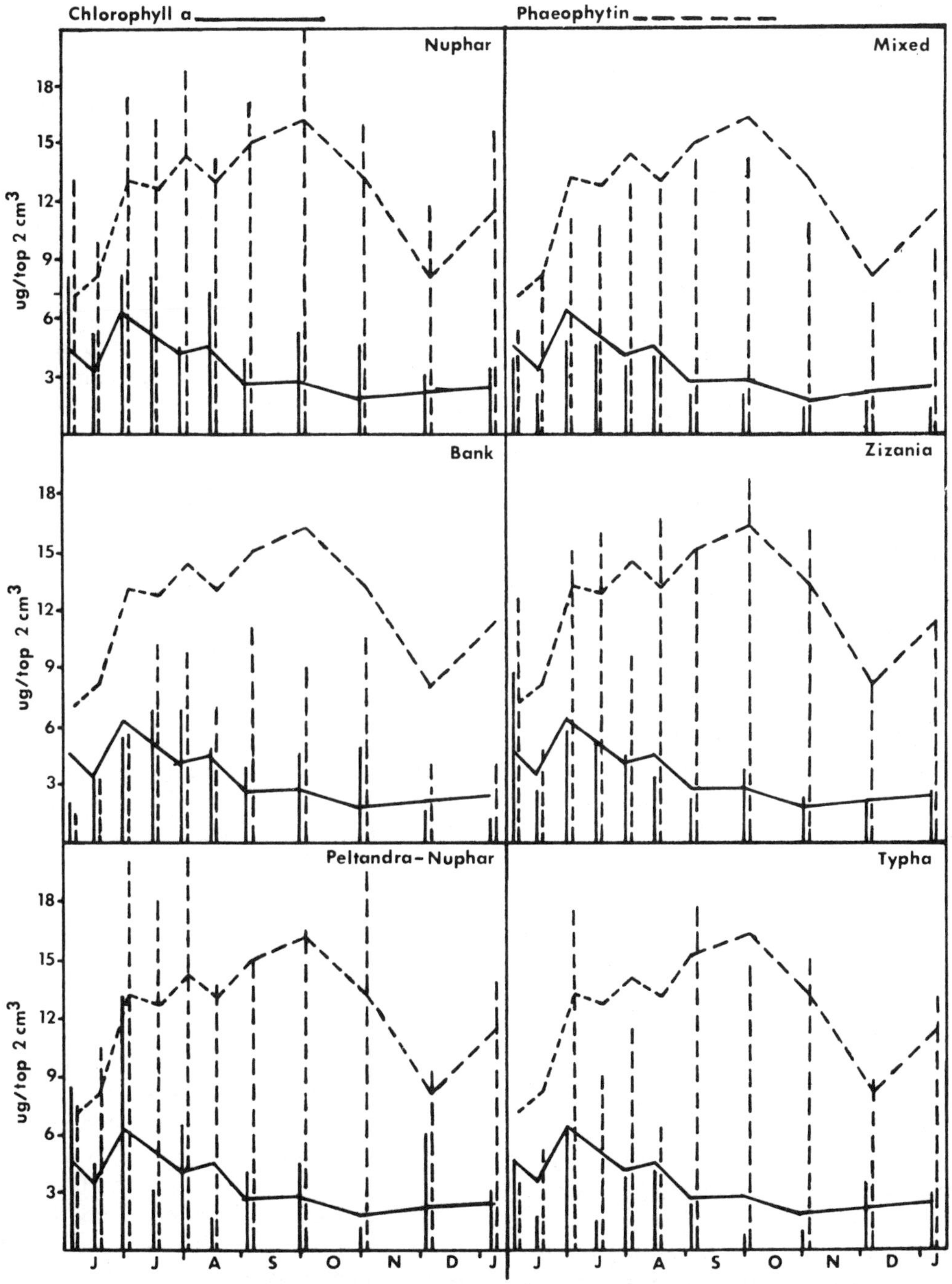

FIG. 20-5. Changes in chlorophyll *a* and phaeophytin concentrations in the top two centimeters of the marsh soil from June 1974 through January 1975. Vertical lines represent values for each dominant vegetation type sampled. Horizontal lines give mean chlorophyll *a* and phaeophytin values for the entire marsh. In each case, solid lines represent chlorophyll *a* and dashed lines represent phaeophytin.

(Fig. 20-6) is typical of these main channel sites. Dissolved oxygen levels, while following expected seasonal patterns, are always lower at lsw than at hsw. Carbon dioxide shows the reverse pattern with carbon dioxide levels being twice as high at lsw as they are at hsw. All nitrogen species and reactive phosphate are similarly elevated at lsw.

Site 8 (Fig. 20-6) on Crosswicks Creek at the upper end of the marsh is minimally influenced by tidal action and shows little difference between hsw and lsw samples except for nitrate nitrogen. Thus the quality of water at Site 2 on Crosswicks Creek may be compared with that of Site 8 to assess the impact of the sewage treatment plant on water quality. Such a comparison shows that water entering the Hamilton Marshes from the Delaware River with the flood tide is somewhat higher in phosphate, ammonia, and nitrite than is the water entering the marsh from Crosswicks Creek, but neither source is as high in these materials or in nitrate as the Hamilton Township sewage effluent.

The pond-like areas of the marsh are typified by Site 4B (Fig. 20-7), which shows a fluctuation in water level of less than 50 cm with each tide cycle. At this site, water quality parameters behave as they would in very productive freshwater ponds with summer oxygen and nutrient depletion and elevated carbon dioxide levels. During the winter, oxygen levels are markedly higher and show considerable diurnal variation due to a lush growth of *Rhizoclonium* sp. that develops as the higher vascular plants die back. Nitrate levels are consistently higher at hsw than at lsw. However, in the late fall when flood tide nitrate levels increase dramatically, lsw levels show only modest increases. Site 4C, which is also pond-like, shows similar marked increases in nitrate during the late fall, but at this site the lsw values are similar to hsw values. Since *Rhizoclonium* does not appear at Site 4C, it would appear that perhaps this algae is acting as a sink for nitrate during the winter months. Unlike nitrate, phosphate levels never exceed 5 μg/l and show no tidal influence at Site 4B.

The high marsh areas represented by Site 5A (Fig. 20-7) appear to be intermediate between the main channel sites and the pond-like sites. Dissolved oxygen and carbon dioxide show the same pattern with respect to the tide as in the main channel of Crosswicks Creek, but oxygen levels are consistently 1 – 2 mg/l lower and carbon dioxide levels consistently about 10 mg/l higher than in the main channel. The highest carbon dioxide levels occur in October and November corresponding to the rapid dieback of vascular plants in the high marsh during this period. This further verified by results of litter decomposition studies of arrow arum, bur marigold, and wild rice (Fig. 20-8). All three species lost approximately 50% of their weight between September and November. From June through September, nitrate, nitrite, and usually ammonia levels are higher at hsw than at lsw. The fall nitrogen levels show little difference between hsw and lsw. In fact, at lsw ammonia and nitrate levels at Site 5A are higher than at Site 5 located on the same side channel closer to Crosswicks Creek. Thus it appears that the high marsh areas may be acting as sinks for nitrogen during the summer months and then slowly releasing this nitrogen back into the marsh during the fall and winter. A similar mechanism may be acting for reactive phosphate which shows similar seasonal patterns.

Conclusions

Some researchers have suggested that brackish-water tidal marshes may act as nutrient sinks,[20] while others have suggested a similar role for freshwater tidal marshes.[21] Our data suggest that the high marsh areas may be acting as a nutrient sink during the summer months and that perhaps the pond-like areas of the marsh may be playing a similar role in the winter.

Salt marsh plots fertilized with sewage sludge have been shown to retain substantial amounts of the applied nutrients and thus may be poten-

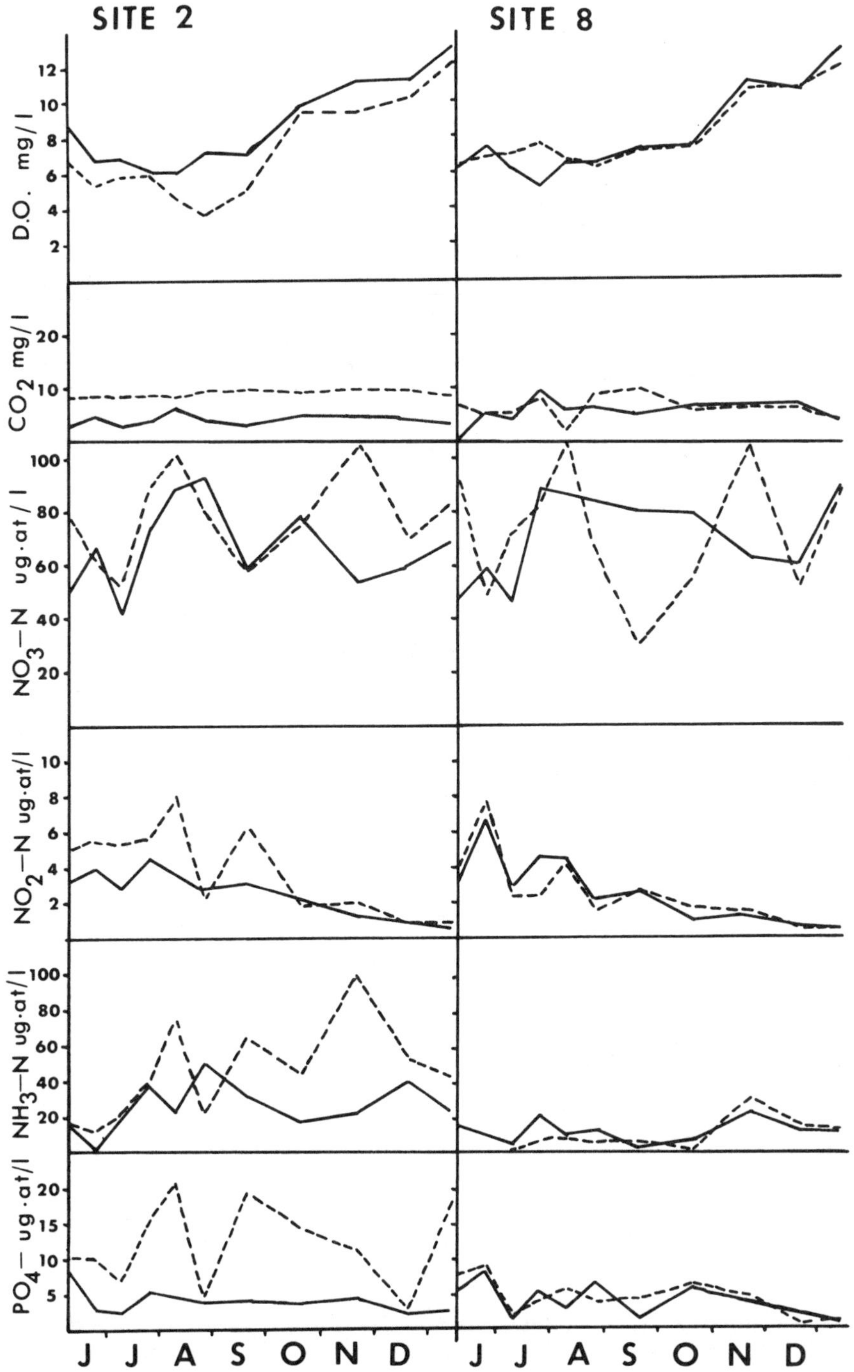

FIG. 20-6. Changes in water quality parameters at Sites 2 and 8 on Crosswicks Creek between June 1974 and January 1975. Solid lines represent high slack water (hsw) and dashed lines low slack water (lsw).

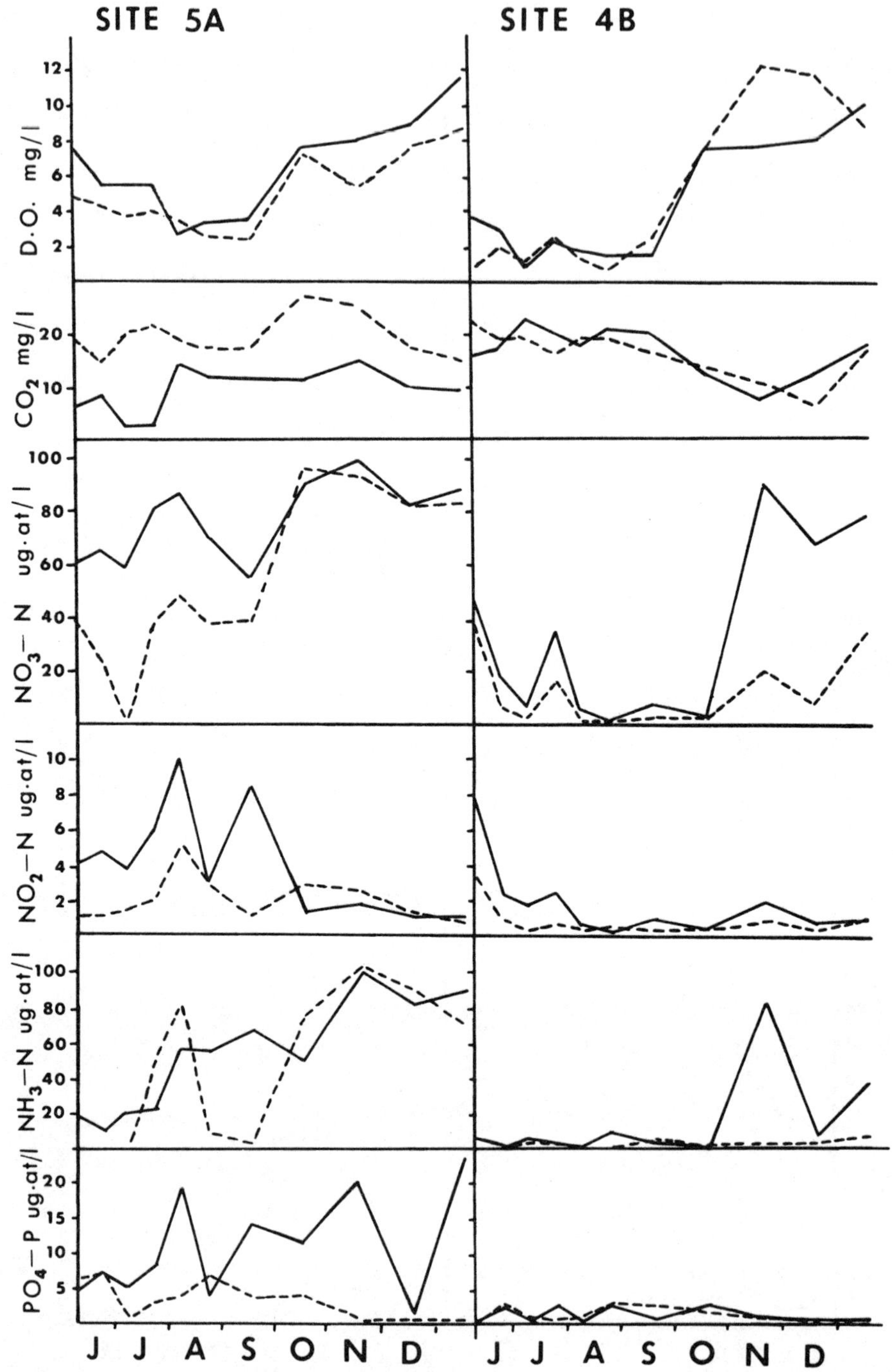

FIG. 20-7. Changes in selected water quality parameters at Sites 4B (pond-like area) and 5A (high marsh area) between June 1974 and January 1975. Solid lines represent high slack water (hsw) and dashed lines low slack water (lsw).

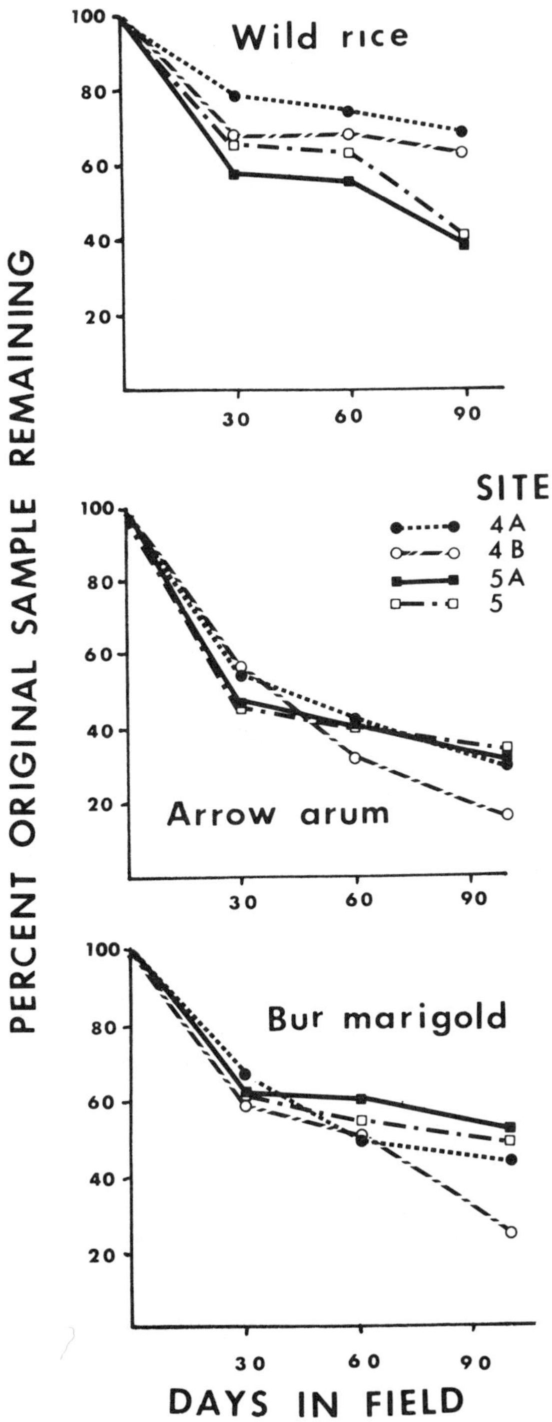

FIG. 20-8. Decomposition of wild rice *(Zizania aquatica)*, arrow arum *(Peltandra virginica)*, and bur marigold *(Bidens laevis)* litter. Site locations are shown on Fig. 20-1.

tially valuable as tertiary treatment systems.[22] It was estimated that each acre of salt marsh was worth as high as $83,000 as a tertiary treatment facility.[23] Based on our data, and that of others, it would appear that freshwater tidal marshes may also be capable of performing tertiary treatment.[24] We believe that, these tidal marshes can process greater amounts of effluent each treatment period than many biological treatment systems, perhaps as much as 2–5 inches of wastewater per day. In April 1975 we began experiments to assess the tertiary treatment capabilities of the Hamilton Township Sewage Treatment Plant. The effluent was sprayed on 10 by 20 m study plots located in a high marsh area at Site 5A dominated by mixed vegetation according to the treatment regime given in Table 20-3. The results of this experiment will determine when and how much effluent can be applied during each treatment period. If the marsh can assimilate secondarily treated effluent at our low treatment level (2 inches per day), then the high marsh areas

TABLE 20-3. *Design of the Spray Irrigation Experiment to be Used at Site 5A. Duplicate plots will be used for each experiment.*

I. Sprinklers on continuously:
   - IA. Effluent applied at 2 inches per day.
   - IB. Effluent applied at 5 inches per day.

II. Sprinklers on when tidal water is not on the high marsh.
   Two 9-hour applications daily:
   - IIA. Effluent applied at 2 inches per day.
   - IIB. Effluent applied at 5 inches per day.

III. Sprinklers on when tidal water is on the high marsh.
   Two 3-hour applications daily:
   - IIIA. Effluent applied at 2 inches per day.
   - IIIB. Effluent applied at 5 inches per day.

IV. Controls:
   - IVA. Sprinklers on continuously. Tap water applied at a rate of 5 inches per day.
   - IVB. No application of tap water or effluent.

dominated by mixed vegetation should be able to process over 18 million gallons of effluent per day, about two and a half times the current daily flow from the Hamilton plant into Crosswicks Creek.

Notes

1. Financial support for our work has been provided by the Hamilton Township Environmental Commission, the National Geographic Society, Sigma Xi, and a Rider College Grant-in-Aid. We thank Jerry Herrera, Dick Klockner, Haig Kasabach, Earl Wood, and other Hamilton Township officials and personnel for assisting us in countless ways during this study. Thanks also go to our wives and our students.

2. D. Whigham, "Preliminary Ecological Studies of the Hamilton Marshes" (mimeographed progress report, 1974).

3. D. C. Sweet, *The Economic and Social Importance of Estuaries* (Washington, D.C.: Environmental Protection Agency, Water Quality Office, 1971).

4. T. E. Walton III and R. Patrick, "Delaware River Estuarine marsh survey," in *Delaware Estuary System: Environmental Impacts and Socio-economic Effects,* a report prepared for the National Science Foundation (RANN), (Philadelphia: Academy of Natural Sciences, 1973).

5. *Ibid.*

6. *Ibid.*

7. H. Walter and H. Lieth, *Klimadiagram—Weltatlas,* VEB (Jena: Gustav Fischer Verlag, 1967).

8. American Society of Agronomy, Inc. *Methods of Soil Analysis, Part 2. Chemical and Microbiological Properties.* (Madison, Wisc: American Society of Agronomy, 1965).

9. C. W. Keefe, "Marsh Production: A Summary of the Literature," *Contributions to Marine Science* 16 (1972):163-81.

10. *Ibid.*

11. J. McCormick, "The Natural Features of Tinicum Marsh, with Particular Emphasis on the Vegetation," In *Two Studies of Tinicum Marsh, Delaware and Philadelphia Counties, Pa.,* ed. J. McCormick, R. R. Grant, Jr., and R. Patrick (Washington: The Conservation Foundation, 1970); J. McCormick and T. Ashbaugh, "Vegetation of a Section of Oldmans Creek Tidal Marsh and Related Areas in Salem and Gloucester Counties, New Jersey," *Bulletin of New Jersey Academic Sciences* 17 (1972):31-37.

12. H. L. Golterman, *Methods for Chemical Analysis of Fresh Waters,* IBP Handbook No. 8, (Oxford: Blackwell Scientific Publications, 1969).

13. J. H. Leach, "Epibenthic Algal Production in an Intertidal Mudflat," *Limnol. & Oceanogr.* 15 (1970):514-21; R. Z. Riznyk and H. K. Phinney, "The Distribution of Intertidal Phytopsammon in an Oregon Estuary," *Mar. Biol.* 13 (1972):318-24.

14. R. G. Wetzel and D. F. Westlanke, "Periphyton," in *A Manual on Methods for Measuring Primary Productivity in Aquatic Environments.* IBP Handbook 12, ed. R. A. Vollenweider (Oxford: Blackwell Scientific Publications, 1969), pp. 33-40.

15. J. L. Gallagher and F. C. Daiber, "Primary Production of Edaphic Algal Communities in a Delaware Salt Marsh," *Limnol. & Oceanogr.* 19 (1974):390–95.

16. American Public Health Association, *Standard Methods for the Examination of Water and Wastewater,* 13th ed., (New York: APHA, 1971).

17. *Ibid.*

18. J. D. H. Strickland and T. R. Parson, *A Practical Handbook of Seawater Analysis* (Ottawa: Fishing Reserve Board of Canada, 1968).

19. *Ibid.*

20. J. L. Blum, "Salt Marsh Spartinas and Associated Algae," *Ecol. Monogr.* 38 (1968):199-221; J. L. Blum, "Nutrient Changes in Water Flooding the High Salt Marsh," *Hydrobiol.* 34 (1969):95-99; D. Aurand and F. C. Daiber, "Nitrate and Nitrite in Surface Waters of Two Delaware Salt Marshes," *Chesapeake Science* 14 (1973):105-11; D. M. Axelrad, M. E. Bender, and K. A. Moore, *Function of Marshes in Reducing Eutrophication of Estuaries of the Middle Atlantic Region* (Blacksburg, Virginia: Office of Water Research, 1974).

21. R. R. Grant, Jr. and R. Patrick, "Tinicum Marsh as a Water Purifier," in *Two Studies of Tinicum Marsh, Delaware and Philadelphia Counties, Pa.,* ed. J. McCormick, R. R. Grant, Jr., and R. Patrick (Washington: The Conservation Foundation, 1970), pp. 105-23.

22. I. Valiela, J. M. Teal, and W. Sass, "Nutrient Retention in Salt Marsh Plots Experimentally Fertilized with Sewage Sludge," *Estuarine Coastal Marine Science* 1 (1973):261-69.

23. J. G. Gosselink, E. P. Odum, and R. M. Pope, *The Value of the Tidal Marsh,* LSU-SG-74-03 (Louisiana State University: Center for Wetland Resources, 1974).

24. Grant and Patrick, "Tinicum Marsh."

21

The Use of Bulrushes for Livestock Feed

B. POMOELL

Jakobstad, Finland

In June 1955, Dr. Kathe Seidel and her assistant, Miss Hilse, arrived at Kronoby, where I was druggist at the time. They came to see me in order to get more information on the location of the so-called Kronoby rush *(Scirpus lacustris)* and on how it was harvested, stored, and finally processed into rush products that were in great demand in Finland and were even exported to other countries, particularly to Sweden.

It was raining cats and dogs all afternoon and evening, and we could not, therefore, make the expedition to the rush fields down by the mouth of the stream of Kronoby and in the nearby bays. Instead we devoted our time to exhaustive discussions about Dr. Seidel's and Miss Hilse's investigations of the different possible uses of the rush. I was at once impressed with Dr. Seidel's thorough knowledge of the plant, and of all the new uses to which it could be put not only by farmers in Kronoby but also by those living near our thousands of lakes, our long coasts, the large archipelagos in the different parts of the Gulf of Finland, the Åland Islands, the Åboland Archipelago, and the different archipelagos of the Gulf of Bothnia.

Ever since my training at the Pharmacy at Rautu on the Carolian Isthmus, I had been interested in plants, especially medicinal plants. While I was studying for my Fully Qualified Chemist's exam, I also studied plant physiology at the University of Helsinki. Now thanks to the visit of Dr. Seidel, my interest in plants began to grow again and was, for obvious reasons, centered on the rush, which had an indefatigable advocate in Dr. Seidel.

At four o'clock the next morning, we arose in brilliant sunshine and cycled down to a small dock. From there we went in a simple boat to the rush bay at the mouth of Kronoby Stream. We had hardly arrived in the bay before our boat was changed into a floating scientific research laboratory containing instruments to measure many things. Dr. Seidel started a speedy investigation: testing the water, the bottom, the light circumstances, air temperature, density of vegetation, and the remaining flora, as well as taking samples from rush plants of different sizes. Rootstocks, roots, leaves, and flowers were put into special containers, and photography was accomplished efficiently with a routine obviously the result of long experience. The expedition to the rush territory took only two hours, so by seven o'clock in the morning the women scientists had set out on the journey back to the laboratory in Plön and their home in Preetz. Before they left we had agreed that I should investigate potential uses of the rush for Finland.

I soon discovered that decades ago the farmers of Kronoby had used rush as feed for cows and sheep. Through the development and mechanization of agriculture, new fields and meadows had been put to the plow and the difficult harvesting of rush had fallen into oblivion, as had so many other good customs that made use of natural products. The use of the rush experienced another renaissance after the wars, when there was a shortage of other material for practical use and decoration. In almost every home, clever fingers made practical articles from the rush: table mats, baskets, carry bags, seats and backs for chairs, covers for flower-pots, etc. An incredible variety of nice, inexpensive rush products satisfied many of the needs of farming families. However, to interest modern farmers in reviving traditional uses of the rush was quite difficult. "It is no use. We do not have enough people of convenient ages to collect and prepare the rushes," was the usual excuse people offered, and therefore I felt discouraged in my attempts. My own opportunities for doing anything outside of working at the drugstore were rather limited.

I was, however, successful in interesting one farmer in my idea. He agreed to keep ninety chickens as an experimental poultry farm for my unsophisticated feeding trials with rush and a blend of rush and Icelandic moss. The experiment went on for about one year and showed clearly that those chickens that got rush were better layers than those that got grain. The eggs were bigger, the yolks beautiful and yellow, and the shells harder. Furthermore, we found that the moulting time was shorter and the laying time longer, which of course increased production. The farmer was satisfied, and people began to praise rush as a fodder-plant. I also experimented with rush as food for rabbits, pigs, cows, and horses. They all ate rush with a good appetite, although I noticed that some of the cows and horses were reluctant at first. A little starvation cure, however, and rush was eaten heartily.

In this connection it should perhaps be mentioned that briquets of dried rush, which were more than fifteen years old, were accepted by cows and poultry and preferred to fresh grain. This is probably a demonstration of the fact that the rush then contained material that even today "tastes" better than fresh green fodder. Is this perhaps a quality trial of consequence? I still have some old briquets left and will go on testing them every year in the same way.

In other experiments I transferred rush plants from one locality to another—for example, from one lake to another or from fresh water to brackish water, and vice versa. The longest transfer was from the delta of the Rhine river in the Netherlands up to Kronoby and even up to the research station of Apukka, just above the Arctic Circle near Rovaniemi. The Dutch rush survived the winter and grew well during our short summer. The rush was used in feeding trials at Apukka by owners of reindeer herds and the results were good. The Dutch rush not only passed the winter well at Kronoby but even multiplied. An attempt at composting rush and using the compost in cultivation gave good results, especially with black currant bushes.

The most significant of my experiences with the rush resulted from a lecture that I gave at Evijärvi in the summer of 1972 at the invitation of that rural district. Amont a number of especially invited persons were employees from the sewage treatment office of the district of Gamlakarleby and agricultural engineers, among whom the most prominent was Mr. Kleimola. Under his supervision a research group has been formed to study other plants *(Scirpus lacustris, Sparganum ramosum,* and *Equisetum limqsum).* This group has developed a fodder blend using certain ratios of these. The results are promising. Analysis to determine the content of minerals, vitamins, proteins, and ash in these three plants are being carried out in state laboratories. At the same time feeding trials with pigs and beef cattle are being carried out under the supervision of professors from the universities of Helsinki and Oulu.

The research group has constructed a "sub-marine mower," by means of which the above mentioned aquatic plants are cut and then collected into a machine built for this purpose, which is fixed to the stem of a motor boat. The boat has two motors, one for propelling the boat and another for driving the "mower" and the collecting machine. The edges of the cutter can be raised or lowered according to the ground profile and the water level. The shallowest depth is some centimeters, and the deepest is two meters. After the collecting equipment is full of cut material, it is towed to the shore and brought either to a special drying apparatus, made by the research group, or to the composting place. The use of this equipment and procedure provides the following benefits:

a) The lakes are rid of undesirable vegetation thus increasing the navigability and the beauty of the water as well as its circulation and oxygen content.
b) You get a high quality fodder for livestock.
c) Different waste products from industries, households, or farms accumulated by the plants are removed from the lakes.
d) An excellent compost can be made from the harvested water plants and used to grow produce and other crops.

And perhaps people will eat a little better than before.

22

The Use of Sawgrass for Paper Product Manufacture: An Examination of Properties

LUDWIG RUDESCU

Institute of Biological Sciences
Bucharest, Romania

For many years in Romania, paper products have been manufactured from marsh plants, in particular the reed *(Phragmites communis)*. The use of the reed and other plants in the paper and cellulose industry became a necessity as available supplies of coniferous wood diminished.[1] However, the use of the reed for paper manufacture is not widespread, although many countries have a wealth of reeds. The North American continent is no exception. In 1970 the author and his colleagues reported the results of their studies on certain chemical and physical characteristics of the North American and Canadian Reed (*Cladium jamaicense* Crantz).[2] In these studies the North American marsh plants were compared to reeds in other countries for their potential in paper product manufacture. In this chapter we wish to re-examine and summarize some of those findings and illustrate the various benefits that could be obtained by harvesting sawgrass from the marshes of Florida.

In recent years nutrient enrichment of the marsh waters northwest of West Palm Beach, Fort Lauderdale, and Miami has caused heavy growths of sawgrass on thousands of acres. This excess growth hinders normal water circulation between the Atlantic Ocean and the Florida marshes, including the waters of Lake Okechobee. Thus, the natural flushing action of the tides that provides an exchange of nutrients, energy, and animal species through the North and South New River Canals, and West Palm Beach Canal, is seriously restricted. Furthermore, the restricted circulation of waters leads to even more highly polluted conditions and to the excessive growth of certain floating, submersed, and marsh plants (Table 22-1). The accumulation of previous generations of plants has

TABLE 22-1. *Predominant Macrophyte Species Growing in Florida Marsh and Canal Waters near West Palm Beach, Fort Lauderdale, and Miami.*

Floating plants	Submersed plants	Marsh plants
Eichhornia crassipes (Mart) Solms.	*Myriophyllum brasiliense* Camb.	*Cladium jamaicense* Kranz
Spirodela polyrhiza L. Scheid	*Ceratophyllum demersum* L.	*Phragmites communis* Trin.
Lemna minor L.	*Elodea canadensis* Mischsc.	*Typha* sp.
Salvinia rotundifolia Willd.	*Potamogeton* sp.	*Jussiaea* sp.
Pistia stratiotes L.		*Panicum* sp.
Limnobium spongia		
Nuphar advena Ait.		
Nymphaea sp.		

TABLE 22-2. *Stem Measurements of Sawgrass and the Reed From Selected Countries.*

Country	Sample	Growth conditions	Biometrical values								
			Length of stem (m)	Diameter of stem (mm)			Thickness of stemwall (mm)			Number of inter-nodes	Weight of stem (g.)
				bot-tom	mid-dle		bot-tom	mid-dle	top		
Canada	Reed of the marshes in Winnipeg Delta	flooded marsh	2.60	8	6	5	1	0.5	0.3	23	37
USA	Washington reed	lake	2.79	12	6.5	3	1.4	0.7	0.3	17	20
USA	Florida reed	on the bank of the canal	2.90	13	7	4	1.6	0.9	0.5	19	39
Romania	Danube Delta reed	floating reed	4.60	20	12	3	2	0.6	0.3	34	125
		marsh	4.15	17	8	3.5	1.8	0.5	0.2	34	100
		river	2.36	14	7	2	1.6	0.5	0.2	25	75
		salted soil	2.60	5.7	3.5	1.7	0.7	0.2	0.3	20	45
USSR	Dnieper Delta reed	fixed floating	3.93	15	8.2	2	1.5	1	0.4	31	110
		river reed	2.03	6.6	2.7	1.2	0.7	0.4	0.1	24	40
		bank of river	2.80	7.9	2.9	1.4	0.8	0.5	0.35	22	55
		marsh	4.25	16	8.6	2.2	1.8	1.3	0.5	28	100
Iraq	El Azair	floating reed	7.77	32	29	4	5	5	2	62	570
	M. Dana	bank of river	6.90	15	16	3	2	2	1	52	125
	El Azair reed	marsh	5.31	19	16	2	2	1	0.5	46	115
Egypt	Alexandria reed	lake reed	3.61	17	11	2	2	0.8	0.3	14	95
USA	Sawgrass *Cladium jamaicense Florida*	flooded marsh	2.20	18	14	-	2.6	1.5	-	-	32

TABLE 22-3. *Reed Fiber Dimensions in Different Countries Compared to Those of Sawgrass.*

Sample analyzed	Fiber length (microns)			Total diameter of fibers (microns)		
	maximum	minimum	average	maximum	minimum	average
Reed of the Winnipeg Delta marshes	2625	370	1200	25	9	20
Washington reed	2100	400–600	1200	15	6–9	12
Florida reed	1600	400	1250	20	9	15
Romanian floating reed	3357	600	1975	23	10	17
Romanian marsh reed	2300	400	1350	13	6	10
Romanian marsh reed in saline soil	2100	400	1250	14	8	11
USSR, Dnieper Delta reed	2500	400	1300	-	-	-
USSR, Volga Delta reed, marsh	2100	400	1150	15	6	10
Egypt	2200	284	900	28	6	15
Florida Sawgrass *Cladium jamaicense*	1500	350	975	15	6	9

produced dense marsh thickets on the bottom of the marshes, and represents a rich nutrient source for further extension and thickening of the marsh vegetation.

The treatment of these marshes with herbicides such as Dalapon, diguat, endothal, 2,4-D, or silvex to eliminate the growth brings no improvement, because only the aerial stems are killed, and these remain in place, enriching the bottom of the water with new polluting substances and accelerating the rate of eutrophication.[3] In successive years further chemical treatment is required because new plants grow from the surviving rhizomes.

The only possible remedy of this problem would be to harvest these plants. Harvesting provides an integrated response to the various components of the overall problem. It removes the nutrients trapped in the plant biomass from the water environment. This in turn reduces the available nutrients for further growth. Then, gradually, the water circulation between the marsh and the sea would improve as the accumulated nutrients were removed in subsequent harvests. The water quality would improve to reach an equilibrium determined by the biomass harvested annually, the rate of natural and cultural addition of new nutrients into the ecosystem, and the new rate of flushing of the marshes by the tide.

The harvesting of the plants would require a large expenditure. It is therefore suggested that

TABLE 22-4. *Chemical Analysis of Reed and Sawgrass With Respect to Their Use for Paper Manufacture. Figures are given in % by weight.*

Analysed samples	Extractable by alcohol - benzene	Cellulose	Ash	Pentosan	Lignin
Phragmites communis Winnipeg Delta marshes	2.93	48.24	3.39	28.41	22.94
Florida	3.57	51.65	2.01	26.16	25.21
Danube Delta	2.5–4.11	41.60	1.2–5.3	18.02–26.6	19–24
Volga Delta	0.70	48.48	3.77	14.68	26.68
Kazahstan	3.22	51.55	3.15	24.90	19.80
Iraq	7.74	42.60	3.9–7.50	25.00	23.70
Egypt	4.83	44.75	4.95	22.75	20.82
China	—	59.61	2.15	8.25	14.70
West Germany	—	34–42.5	5.75	14.7–19.2	25–36
East Germany	—	37	5.20	18	29
Sawgrass *Cladium jamaicense Florida*	6.66	37.17	2.69	25.65	32.58

the costs of harvesting be offset by converting the plant fibers into marketable paper products such as paste board, cardboard, or even base-paper. We also believe that sawgrass possesses the characteristics which make it suitable for conversion to paper products.

The anatomical, morphological, and chemical properties of sawgrass are similar to those of the reed which, as noted, is used for paper product manufacture in Romania. Various stem measurements of reeds and sawgrass are compared in Table 22-2, and it can readily be seen that the morphological measurements listed for sawgrass are similar to those shown for the reeds.

Although there are microscopic anatomical differences between sawgrass and the reed, these do not affect the paper value of sawgrass. Examination of the length and width of reed and sawgrass fibers reveals that the reed fibers are slightly longer and wider than those of sawgrass (Table 22-3). Chemical analysis of the plants studied as raw material for cellulose and paper revealed that sawgrass has lower values of cellulose than for all reeds listed, except for those from East and West Germany (Table 22-4). Sawgrass has the highest percentage of lignin of the plants examined but has similar values for pentosan and ash content. Although sawgrass appears to be slightly less suitable for paper manufacture than the reed, evaluation of the above results (Tables 22-2, 22-3, and 22-4) indicates that sawgrass could be used as a raw material for cellulose and certain paper products.

Harvesting sawgrass from the Florida swamps could be accomplished by harvesting machines similar to those now used to harvest reeds in the delta region of the Danube in Romania. (Certain modifications of the cutting mechanism of these machines may have to be made to accomodate sawgrass.) There are certainly existing techniques to bale and transport the cut sawgrass. The sale of the paper products made from sawgrass would offset the harvest and transport costs, and furthermore, the direct and indirect costs of spraying the entire region with herbicides would be eliminated.

Other potential benefits, mentioned previously, are the improvement of water circulation between the ocean and the marshes, the improvement of water quality due to the removal of nutrients trapped in plant biomass, and the reestablishment of former ecological relationships altered by the dense growths of sawgrass. Furthermore, the improved circulation in the canals will tend to discourage and thus reduce the growth of floating plants (e.g., *Eichornia*). If a major program were implemented to harvest sawgrass in this resort area of Florida, it is anticipated that the landscape of the canal region will be restored to its former state within a few years.

Notes

1. L. Rodewald-Rudescu, *Das Schilfrohr* [The Reed], Die Binnengewasser, vol. xxvii, (Stuttgart: E. Schweizerbart'sche Verlagsbuchhandlung, 1974).

2. L. Rudescu and G. Popescu, "Quelques aspects anatomorphologique et chimique-papetaires du roseau nord-american et canadien et de l'espéce *Cladium jamaicense* Crantz du Florida par comparaison aux roseaux des autres pays" [A Comparison of Some Anatomical, Morphological, and Paper-Chemical Properties of the North American and Canadian Reed and of the Species *Cladium jamaicense* Crantz with Those of Reeds from Other Countries], *Hidrobiol.* 12 (1971): 273-78. The work published in this article was first presented at a symposium on aquatic and marsh macrophytes held by IPB-PF (September 2-9, 1970) in Romania.

3. D. S. Harrison et al., "Aquatic Weed Control," *University of Florida Agricultural Extension Service Circular* 219B (1966):3-15.

23

Waste Reclamation in an Integrated Food Chain System[1]

JOEL C. GOLDMAN
JOHN H. RYTHER

Woods Hole Oceanographic Institute
Woods Hole, Massachusetts

Introduction

In recent years major effort has been directed at the development of techniques for removing both organic and inorganic contaminants present in wastewater so that water quality standards established for discharge and/or reuse can be met. Current philosophy dictates that our technology should be aimed at the efficient removal and safe disposal of these contaminants. Little, if any, concern is given to the possibility that the expenditure of energy required to accomplish these goals at a point source may in reality lead to as much or more degradation of the overall environment through ancillary air and water pollution and by spent nonreusable resources.[2] Moreover, until recently, our societal needs of food production and waste disposal have been considered mutually exclusive. Thus at one end of the spectrum we are faced with the prospect of having to produce enough food to keep pace with the needs of our population, while at the other we are determined to reach levels of waste treatment and disposal that are in keeping with our established standards of public health and aesthetic quality.

Little scientific insight is required to know that the flow of chemicals in the environment is a continuous cyclic process and that energy once expended is nonrenewable. Minerals, particularly those used in the synthesis of organic matter, are in a constant state of flux, continuously shifting back and forth from the aqueous to the gaseous to the solid states. The stresses put on the environment by modern man have caused serious imbalances in these cycles. A prime example is cultural eutrophication of natural waters. Within the past several decades there has been a tremendous acceleration of the rate at which nitrogen and phosphorous have been added to the aquatic environment. This has led

TABLE 23-1. *Estimated Quantities of Nitrogen in Fertilizer, Food and Wastewater in United States, 1970.*

Source of Nitrogen	Estimated Quantity (metric tons x 10^5)
Fertilizer Produced	70
Food Consumed	8.2*
Domestic Wastewater	8†–13‡

*Estimate based on a U.S. per capita intake of 92 g/day of protein, and assumption that typical food protein contains 12% nitrogen. J. C. Abbot, "Protein Supplies and Prospects: the Problem," in *World Protein Resources,* A. M. Altschul, ed., Advanced Chemistry Series, volume 57 (New York, 1966).

†Low estimate based on average per capita wastewater discharge of 120 gal/day, average nitrogen concentration in wastewater of 25 mg/l, and a 1970 U.S. population of 203 million.

‡High estimate based on an average per capita nitrogen load in wastewater discharged of 17.5 g/day [taken from data supplied in comprehensive studies of wastewater discharges in San Francisco Bay and Southern California], and a U.S. population of 203 million. Kaiser Engineers, *Final Report to the State of California—San Francisco Bay–Delta Water Quality Control Program* (Oakland, 1969); Southern California Coastal Water Research Project, *The Ecology of the Southern California Bight: Implications for Water Quality Management* (El Segundo, 1973).

to gross overproduction of organic matter, its eventual decay, and ultimate degradation of water quality. To combat this problem a concerted effort is being made to develop techniques for removing these two nutrients from wastewaters.[3] In this regard chemical coagulation appears most promising for phosphorous removal, and nitrification-denitrification and ammonia stripping are being considered for nitrogen removal. In all cases significant quantities of energy are required, and the removed chemicals are wasted to the environment as either solid waste (phosphate sludges) or as gas (NH_3 or N_2).

At the same time, in order to meet our own and foreign commitments for food, we must add vast quantities of nitrogen and phosphorous fertilizers to the soil to stimulate crop growth. The production of fertilizers, particularly those containing nitrogen, requires the expenditure of great quantities of energy. The current worldwide shortage of nitrogen fertilizers is in part due to an accelerating demand coupled with the recent change in the economics of fuel supply. As recently pointed out, our dependence on fossil fuels to produce nitrogen fertilizers will not be alleviated for many years to come.[4] Thus as the population increases we are faced with the prospect that worldwide food shortages will continue to grow worse unless the price of fuel decreases and either alternative techniques are developed for producing fertilizer for land use or marine resources are exploited more fully or both. Even if the price of fuel were to be returned to pre-1973–74 levels, our dependence on fossil fuels for fertilizer production is a short-sighted solution because of the limited quantities of these raw materials available. At some point in the future the use of commercially made fertilizers will, through lack of raw materials, be drastically curtailed. If U.S. agricultural technology, with its massive use of fertilizer, were to be employed to feed 10 billion people with just corn grain, the world's petroleum reserves would be depleted in less than 450 years, assuming that there were no other uses for this fossil fuel.[5]

Thus there exists a serious dichotomy in our philosophy relating on the one hand to our need to conserve both energy and natural resources, and on the other hand to our desire to con-

TABLE 23-2. *Source of Protein in Typical U.S. Diet.**

Source of Protein	Percent of Total Consumed	
Agricultural Crops		
Grains	16	
Pulses, oilseeds, nuts	5	
Vegetables	4	
Starchy roots	2	
Fruit	1	
Subtotal		28
Animal Products		
Meat and Poultry	35	
Milk and Products	26	
Eggs	7	
Subtotal		68
Fish		3
Other		1
Total		100

*Abbott, "Protein Supplies."

tinually improve our established life-style. For example, as seen in Table 23-1, approximately 70 x 10^5 metric tons of nitrogen were produced in the U.S. for use in fertilizers.[6] Of this total, less than 12% (8.2 x 10^5 metric tons) was converted to consumed food. The low conversion efficiency of nitrogen applied to the soil to nitrogen present in consumed food is due to (a) the low efficiency of crop use—40–70%[7]; (b) the fact that most of the grain that is grown is used to raise animals, resulting in an approximate 10% transfer efficiency of plant nitrogen to animal nitrogen (animal products account for over two-thirds of our consumed protein [Table 2]); (c) only a portion of the nitrogen converted to plants or animals is recovered as a final food product; the rest is wasted in the preparation; (d) part of the food grown in the U.S. is exported; and (e) approximately 15% of the fertilizer produced is used on ornamental lawns and home gardens.[8]

During this same period approximately an equal amount of nitrogen present in domestic wastewaters was released to the environment (8–13 x 10^5 metric tons). Nitrogen loads in other types of wastewater raise this figure significantly (Table 23-3).

It would appear that we are incredibly wasteful not to utilize the fertilizing potential present in waste materials for producing food, particularly when we are expending much energy and money to find ways to remove nitrogen from wastewaters for the sole purpose of protecting the quality of natural waters. The question then arises as to whether or not there are alternative techniques for removing nitrogen from wastes while at the same time taking advantage of their fertilizing potential.

TABLE 23-3. *Nitrogen Contributions from Various Sources in 1970.**

Source	Nitrogen Load Metric Tons x 10^5
Industrial waste	greater than 4.5
Rural Runoff:	
Agricultural land	6.8–68
Nonagricultural land	1.8–8.6
Farm Animal Waste	greater than 4.5
Urban Runoff	0.5–5
Rainfall†	0.01–0.3

*M. C. Goldberg, "Sources of Nitrogen in Water Supplies," in *Agricultural Practices and Water Quality,* L. Willrich and G. E. Smith, eds. (Ames, Iowa; Iowa State University Press, 1970).
†Considers rainfall contributed directly to water surface.

Alternatives to Traditional Agriculture

Human and animal solid wastes have been used as fertilizers for centuries in far eastern countries such as Japan, China, Thailand, and Malaysia for growing a variety of land crops and fish.[9] However, in the western world only recently has a concerted effort been initiated to explore the possibilities of recycling waste products, both solid and liquid, into food products for animal and human consumption.[10] The major constraints have been public health effects and consumer acceptance. The former is obviously a major drawback: food for human consumption must not, under any circumstances, pose a health hazard. The latter restraint is tightly connected to our socioeconomic development. Until recently little concern was given to the possibility that many of our resources would shortly become scarce. In the western world, particularly the U.S., it was thus possible to place great distance between the production of food and materials and the disposal of waste products. At the same time, the accumulation of wealth has bred contempt and distrust for any process that closed the gap between the biological processes of growth and degradation. Thus we have created psychological barriers that prevent us from considering food grown directly on waste products as a viable alternative to more conventional food production. Hopefully, once the reality that we will need to rely more and more on closed-loop recycling systems becomes more evident these barriers will be eliminated.

The utilization of waste products to grow food can be accomplished in a number of different ways. One way is to use the solid wastes of humans and animals as a replacement for commercial fertilizers. It is estimated that one dairy cow can produce enough nitrogen in manure to fertilize one acre of corn.[11] Research in this area is active.[12]

Another possibility is to utilize photosynthetic algae to assimilate nutrients present in wastewater. The algae in turn are harvested and dried and then used as a human and/or animal protein supplement. W. T. Oswald at the University of California at Berkeley has been engaged in the development of this process for a number of years.[13] To date, the major problem is in the recovery of the algal product and its subsequent conversion to a human and/or animal protein supplement. The process does have the inherent advantage over other nutrient removal methods in that it is assimilatory, so that nutrients are removed and directly converted to proteinaceous materials rather than wasted.

Recently, the novel idea of growing marine algae on wastewater-seawater mixtures was presented.[14] Like their freshwater counterpart, the marine algae assimilate nutrients, but they are then fed to herbivorous shellfish such as oysters and clams.

These molluscs filter out the algae, but because of their relatively low efficiency in converting ingested food to their own biomass, they excrete both soluble and particulate waste products. These waste products can then be used as food for additional links in what turns out to be a complex and integrated marine food chain. As examples, the particulate waste (feces and pseudofeces) provide the food source for scavanging small crustaceans such as polychaete worms. One species of polychaetes, *Nereis virens,* is a common bait worm used in sport fishing, and thus, like the oysters and clams, represents a commercially valuable crop. Other worms such as *Capitella capitata* are the food source for omnivorous feeders such as the flounder, lobsters, and shrimp, all highly desired marine food sources. In addition, the soluble wastes of all the marine organisms can be used as nutrients for growing a variety of marine seaweeds. Seaweed extracts such as carrageenin, agar, fucellaran, and algin are extremely important stabilizing additives used in the food industry.

Thus, in theory, it should be possible to design a controlled and balanced marine food chain comprised of a number of species representing all the trophic levels so that all of the nutrient in shortest supply, when added to the system, is eventually incorporated into the biomass of several commercially valuable marine organisms.

Controlled Eutrophication

The channelling of nutrients in a marine food chain, as described above, is, of course, the basis for life in the sea. However, J. H. Ryther has pointed out that the vast majority of the world's oceans are a "biological desert."[15] This is mainly because the first link in the marine food chain, the phytoplankton, is limited by the scarcity of nutrients, primarily nitrogen and phosphorus. Only in the few coastal and updwelling areas such as off the coasts of Peru, southern California, northwest and southwest Africa, and Saudia Arabia is primary production and fish production significant. In fact, Ryther estimates that updwelling areas, which contain about 0.1% of the world's oceans, produce approximately 50% of the world's fish supply.[16] The other half is produced in coastal and nearshore waters.

The discharge of wastewaters to coastal and estuarine waters can add significantly to the nutrient budget of these areas, but this can lead to eutrophication and degradation of water quality. The current pollution problems in Chesapeake Bay, Biscayne Bay, and San Francisco Bay are good examples of how waste discharges have caused severe imbalances in the natural biological cycles of some estuaries. Only in those coastal waters in which circulation is good have the effects of waste discharges been minimal and even beneficial in terms of providing additional nutrients to increase production throughout the food chain. A particularly good example is the coast of southern California where the wastewaters from this heavily populated area are disposed of through a series of deep ocean outfalls designed to allow rapid and complete dispersion of the wastes.[17]

Phytoplankton growth in the deep oceans is limited by a lack of all nutrients. This situation is considerably different, however, in coastal and estuarine waters. Substantial evidence is available showing that nitrogen is the key nutrient controlling phytoplankton growth in these waters.[18] This is primarily owing to the fact that the nitrogen to phosphorus ratio in coastal marine waters is similar to that typically found in wastewaters (about 5:1 by atoms). Marine phytoplankton, on the other hand, normally contain nitrogen and phosphorus in the ratio of about 10:1 to 20:1 by atoms. Thus the amount of phosphorus relative to nitrogen in wastewaters is significantly in excess of that required by phytoplankton. The similar N:P ratios in wastewaters and coastal waters suggests that wastewater nutrients contribute heavily to the nutrient budget of the coastal marine environment. Nitrogen removal may therefore be required to prevent possible eutrophication of certain poorly circulated coastal and estuarine waters receiving significant wastewater discharges.

By controlling the release of wastewater-borne nitrogen to the marine environment, not only would eutrophication be prevented, but these same nutrients could be directly channelled into the biomass of many commercially valuable marine food crops in the process described above. The resources of the sea could then be utilized to their best advantage in this process of "controlled eutrophication," leading to a new industry of mariculture, based exclusively on the recycling of waste products into human food.

Status of Mariculture In the World

The approach to mariculture, in the United States and other highly developed countries, appears to emulate the modern poultry industry or the feed-lot system of rearing livestock. Marine animals, whether finfish or crustaceans, are raised in small areas in dense culture requiring the rapid exchange of water, intensive labor and/or mechanization, and heavy and frequent feeding, usually with artificial, pelletized feeds.

Food alone may account for as much as 50% of the operating costs in these intensive aquaculture practices. And when such basic ingredients as fishmeal become scarce, as happened during the recent failure of the Peruvian anchovy fishery, the cost of food may become prohibitive. Heavy feeding also creates a host of other problems such as waste removal, oxygenation, fouling, and pollution of the receiving waters, all of which increase operating costs. The net result is that such practices, on the whole, have been economically unsuccessful.

In contrast to such intensive mariculture efforts, the long established estuarine pond culture practices of Southeast Asia, where milkfish, mullet, shrimp, and a few other species are reared in shallow impoundments usually constructed from mangrove swamps, are in most cases highly successful economically and have contributed significantly to the protein needs of the local inhabitants. Juvenile animals, usually collected from the wild, are stocked in such ponds and simply allowed to forage on natural foods. Thick matts of filimentous algae and the associated fauna of small invertebrates are encouraged to grow in the brackish-water ponds by various simple techniques prior to stocking the fish, and it is this biota that provides food for the herbivores and omnivores that are reared in such systems.

Not only is the cost of food eliminated, but the capital and operating costs for such enterprises are extremely low. Other than the periods of stocking and harvesting and such simple pond maintenance as is required, a single watchman can supervise a hundred acres or more of these fish ponds.

Yields from extensive fish-pond farming are not impressive compared to the projected (though seldom realized) yields from heavily-fed, intensive mariculture operations. But milkfish and even the highly prized penaeid shrimp can be grown to market in six months to a year and yields of one metric ton per acre are not uncommon, even with the relatively primitive techniques in use today. Such yields of high-quality animal protein are impressive and, in view of the millions of acres of undeveloped mangrove swamps in the world's tropics, represent a potential food resource that can hardly be ignored.

Furthermore, by combining the best features of modern, intensive aquaculture with those of the established and already highly-successful Southeast Asian pond culture, it should be possible to increase yields from the latter substantially. Hatcheries, capable of producing virtually unlimited numbers of juvenile animals at low cost on demand, are now a reality for many species and could eliminate the dependence on wild stocks of fry, which is not only the most costly single factor in this type of mariculture, but is also, because of limited supply, a serious constraint to the further development and expansion of the industry. Genetic improvement of stocks, also possible with hatchery production; predator and competitor control; improvements in the design and engineering of pond construction and management, harvesting, and water exchange; and so on are all present state-of-the-art realities that could find ready application while still retaining the principles of extensive fish farming with natural feeding and its other advantages.

But the latter, dependent as it is on the natural productivity of the culture system, can perhaps most readily be enhanced by increasing the primary productivity of photosynthetic green plants, the basis of all natural food chains. Through proper application of essential nutrients, it should be possible to increase photosynthetic production by as much as tenfold and, if this is done under carefully-controlled conditions, such an increase may be reflected in the ultimate yields of the cultured food species. Unfortunately, the present high cost and limited availability of commercial fertilizers, particularly nitrogen, would probably make their use as economically prohibitive as the direct feeding of the cultured organisms. However, the use of organic wastes, from domestic, agricultural, food-processing, and other sources, properly prepared and treated, have the same composition and properties and can serve the same purposes as the best blend of commercial fertilizers. The recycling of such wastes to conserve fertilizers and the energy required to make them, to produce food, and to control pollution of the aquatic environment would all seem to be objectives high enough on the scale of human priorities to make such practices imperative.

Research at Woods Hole

For the past several years research has been conducted at Woods Hole Oceanographic Institution on the design and development of small, experimental systems that combine waste recycling with marine aquaculture, and in the fall of 1973 a new "pilot-plant" facility (Environmental Systems Laboratory) was occupied in which it was possible to test and evaluate the concepts, hopefully on a scale large enough for extrapolation to commercial application.

Initially, small-scale laboratory experiments established the fact that secondary sewage effluent, mixed with seawater, is a complete and effective medium for the growth of the common marine phytoplankton that are indigenous to the Woods Hole region. Samples of raw seawater enriched with sewage effluent typically developed a mixed population of algae dominated by several species of diatoms. In continuous cultures, species diversity rather quickly decreased to the point where one dominant species persisted.[19]

During the summer of 1972, the experiments were moved outdoors and carried on under natural light and temperature conditions. The size of the cultures was increased from flasks and carboys to fiberglass tanks of 400-liter capacity, and later to 2,000-liter commercial swimming pools. Each day a portion of the culture was harvested on a batch basis into holding tanks from which the algae were fed continuously to juvenile oysters, clams, mussels, and scallops, held in trays in large, rectangular tanks. Nutrient removal, growth rate and yield of algae, and the rates of filtration (feeding), solid-waste deposition, and growth by the shellfish were all monitored.

In the early experiments, concentrations of sewage effluent in seawater varied from 10% to 25%, with batch harvest of 25% to 50% of the culture once each day. Under that regime, inorganic nitrogen removal (NH_4^+, NO_3^-) was virtually complete, while approximately 50% of the phosphorus was assimilated by the algae. The latter was expected, since it had previously been demonstrated that the amount of phosphorus relative to nitrogen in sewage is about twice that normally present in seawater and in phytoplankton, leaving a residue of half the phosphorus after the algae have assimilated all available nitrogen.[20] But bioassay experiments showed conclusively that the phytoplankton culture effluent was incapable of supporting further growth of algae, thus fulfilling the primary purpose of tertiary sewage treatment.[21]

However, it was found that the algal cultures, because they were maintained under nutrient-limited conditions or for other reasons, were unstable and tended to "crash" at intervals of one to two weeks, requiring the continual alternation between old and newly-inoculated cultures to maintain a constant yield. Later, the entire system was put on a continuous-flow basis, using 25% sewage effluent and 75% seawater with a 50% turnover rate per day (retention time of 2 days). This practice returned the same high yields and nutrient removal by the algae but markedly improved the stability of the cultures.

Although the phytoplankton were effective in removing nutrients (i.e., nitrogen), the shellfish returned some of these from their own excretion and from the decomposition of their solid wastes. It was therefore necessary to add at this stage a final polishing step consisting of the seaweed *Chondrus crispus* (Irish moss). The latter proved successful in assimilating such regenerated nutrients, providing an overall effectiveness for the system as a whole of as much as 95% to 100% nitrogen removal when all components were in proper balance.[22]

In these same experiments, it was found that five hundred 2–4 cm seed oysters *(Crassostrea virginica)* were capable of removing over 85% of the algal cells from 500 liters/day of culture consisting of *Phaeodactylum tricornutum*, *Chaetoceros simplex*, and small numbers of other diatoms, flagellates, and green algae fed to the shellfish at a concentration of approximately 10^6 cells/ml or about 13 mg organic carbon/liter.

During the winter of 1972–73, a small but complete physical model of the entire system was designed and constructed. This unit, depicted in the schematic flow chart (Fig. 23-1) and in the photograph (Fig. 23-2), consisted in part of two circular algal growth ponds (fiberglass-lined) each of 2.27 m diameter and 50 cm deep (2,020 liters). Continuous mixing and aeration were accomplished with both motor-driven rotating arms and recirculation through pumping.

Secondarily treated effluent was collected daily from the trickling filter treatment plant at Otis Air Force Base located 15 km from the Woods Hole Oceanographic Institution dock where the experiment was performed. Inorganic nitrogen concentrations in the effluent were typically 10-20 mg/l (Fig. 23-3).

The wastewater was pumped from a 1,000-liter polyethylene storage tank along with 1 μ-filtered seawater to headboxes above the ponds, and then blended into each pond at the desired flow rate and mixture. The algal ponds were elevated on steel frames so that the pond effluents flowed by gravity through the remainder of the system, discharging into Woods Hole Harbor.

The continuous effluents from both algal ponds were mixed in a common line and were considered as one flow for evaluating the performance of the remaining system (Fig. 23-1). The

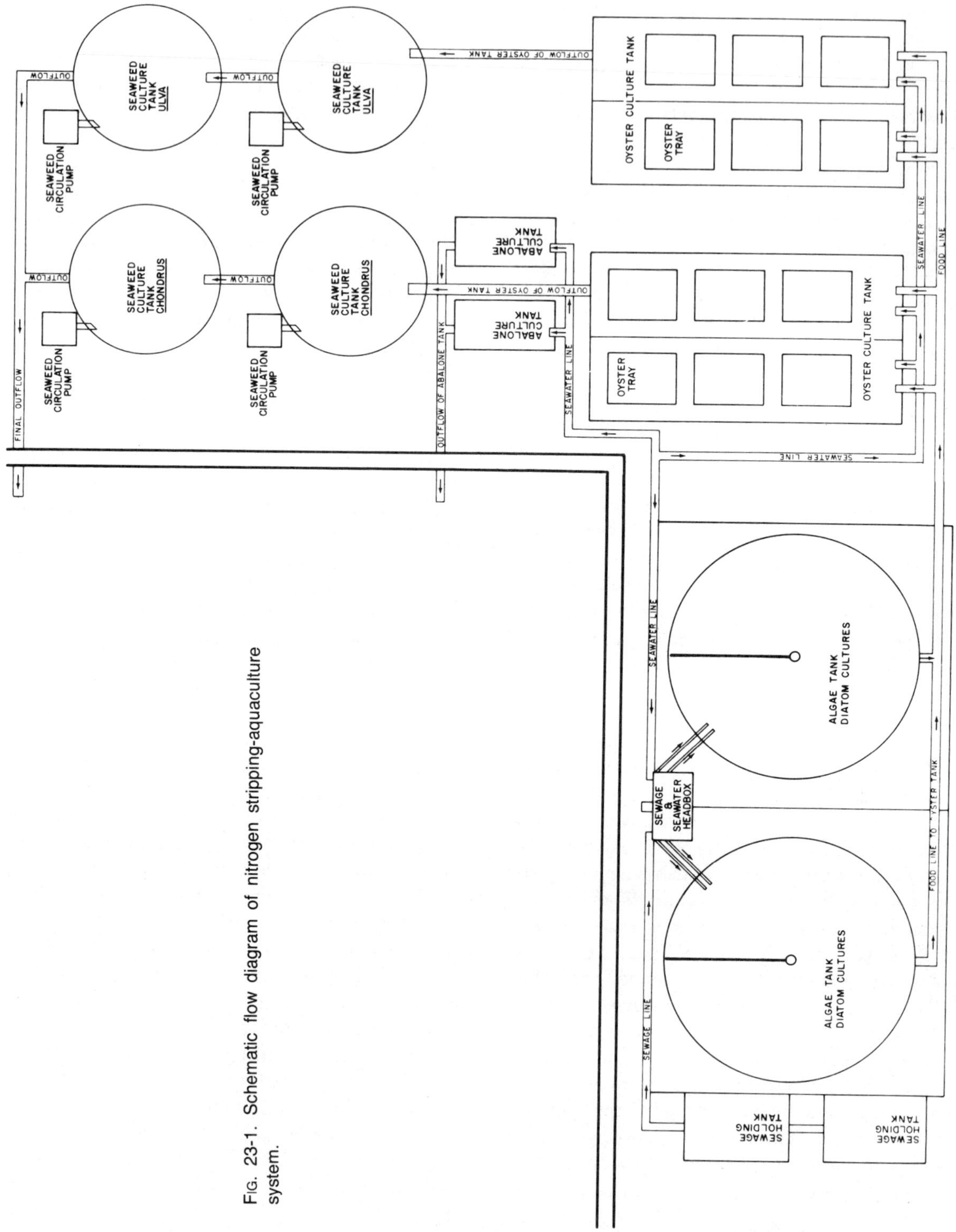

FIG. 23-1. Schematic flow diagram of nitrogen stripping-aquaculture system.

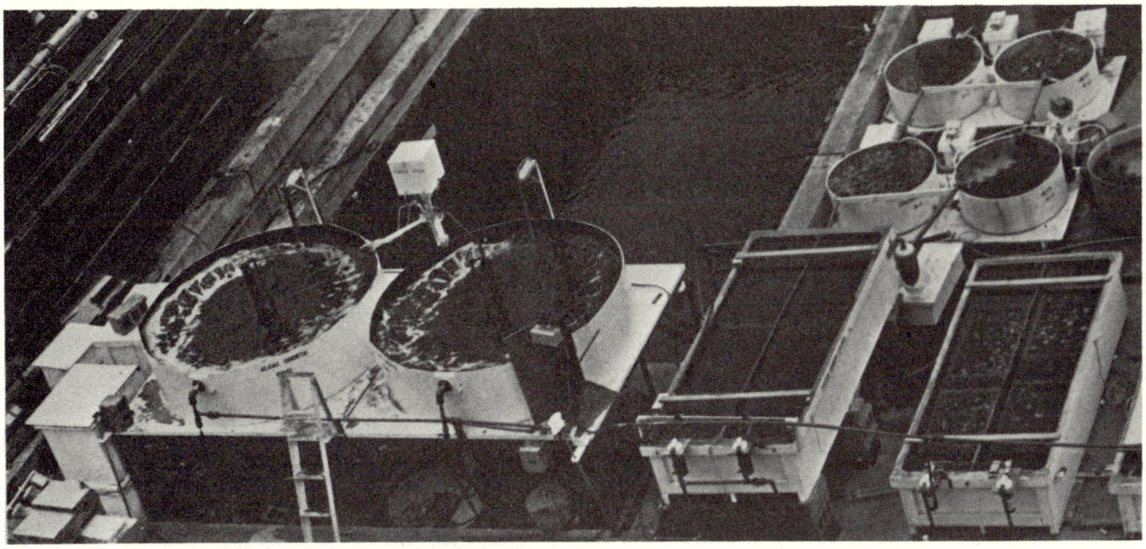

FIG. 23-2. Photograph of nitrogen stripping-aquaculture system.

combined pond effluent was then passed through a high-speed homogenizer (Tekmar Model SD-45) so that any clumped algal cells were separated intact to provide single-cell food to the oysters. Only a portion of the total effluent from the algal ponds was used as food for the oysters, with the excess culture used in concurrent but separate experiments.

The remaining portion of the process, consisting of the oysters, worms, and seaweeds was

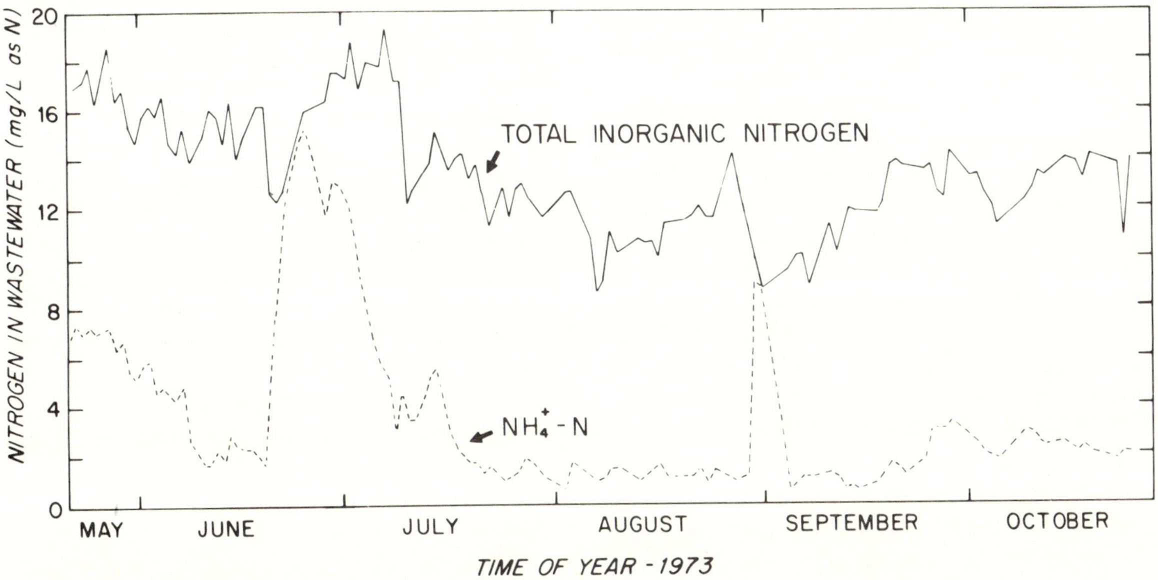

FIG. 23-3. Variations in total nitrogen and ammonia concentration in secondarily treated wastewater during study period (May - October, 1973).

divided into two parallel systems (Fig. 23-1). System A consisted of oysters and worms (*C. capitata*) combined in a growth tank similar to the one used previously. Two thousand 2–4 cm seed oysters were used in each tank, representing a significant increase in oyster biomass as compared to the five-hundred oysters of similar size used in the previous study. The algal culture was added to the oyster tank at a constant rate of about 0.35 liter/min supplemented by 100 μ-filtered seawater pumped in at rates varying from 5 to 8 liters/min. The entire overflow from this tank was then discharged to the first of two seaweed ponds in series, each containing *C. crispus* which was initially collected in local coastal waters. The seaweed growth ponds were 355 liter circular tanks (1.12 m diameter and 36 cm deep), constructed of fiberglass. Mixing was accomplished through recirculation.

A nominal standing crop of 5,000 g (wet weight) of *C. crispus* was maintained by removing weekly the entire crop, shaking free the excess water, and then weighing the biomass. Any biomass over 5,000 g was harvested.

System B was a duplicate of System A except that the polychaete worm was *N. virens* and the seaweed species was *Ulva lactuca* (sea lettuce) at a standing crop of 3,000 g (wet weight).

During the six months of continuous operation of the experimental model, primary attention was focused upon the production of unicellular algae as a function of both wastewater concentration and exchange or dilution rate (reciprocal of retention time) and upon the mass balance of nutrients, especially nitrogen, through the entire system from wastewater input to final discharge. This information is published elsewhere,[23] but is reviewed briefly below.

In one of the algal ponds the mixture of wastewater and seawater was varied from 30% to 67% sewage, representing an increase in the total nitrogen load from 5 to 10 mg/liter and a corresponding decrease in salinity from 20% to 10%. The dilution rate was held constant at 50% per day. The dilution rate in the second pond was varied at 25%, 50%, 75%, and 100% per day and the sewage effluent fraction held constant at 50% of the total pond influent. For each effluent-seawater mixture and dilution rate, steady-state conditions were established by maintaining the desired flows for a minimum of one week. From these experiments, it was concluded that maximum yields of phytoplankton were achieved at 75% dilution rate per day and a mixture of 50% effluent and 50% seawater. This combination is not, however, the best for nutrient removal (tertiary treatment), taking only 50–60% of the nitrogen from the wastewater, but the complete system including the seaweeds removed virtually all of the nitrogen originally present in the sewage effluent (Table 23-4).

No attempt was made to inoculate a particular algal species into the ponds. Rather, when the ponds were started up early in May, 1973, natural populations were allowed to develop. During the entire experiment marine pennate diatoms were the exclusive algal species, although there was a succession of species among the pennate forms.

From May until mid-June *P. tricornutum* was dominant, followed by an unidentified naviculoid diatom which prevailed until mid-August. *Amphora* sp. then prevailed until mid-September when the other naviculoid species returned as the dominant. These results are very similar to the changes in diatom species that occurred in the previous study in which *P. tricornutum* was dominant for most of the experiment except during the period from late August to mid-September when *Chaetoceros* sp., a small centric diatom, prevailed. The switch from larger pennates to smaller diatoms appears strongly correlated with temperature, as the highest pond temperatures (25° -28°C) were observed when the smaller species were dominant, a phenomenon that typically occurs in natural marine waters.

Relatively little effort was devoted to the feeding, growth, and trophic dynamics of the oysters and the two species of polychaete worms during

TABLE 23-4. *Inorganic Nitrogen Transformations in an Integrated Food Chain System. Average total inorganic nitrogen loads-g/day.*

System	Algal* ponds input	*Oyster tanks input* Algal* ponds output	Seawater fraction	Total input	*Seaweed ponds input* I	II	*Final effluent* Total	Less seawater fraction
A	3.06	1.43	0.55	1.98	2.46	1.61	0.85	0.30
B	3.09	1.34	0.53	1.87	2.56	1.25	0.56	0.03

*Based only on fraction of wastewater-seawater load treated through complete system.

the 1973 experiments, primary attention having been focused on monitoring the algal growth and nutrient mass balance characteristics under varying operating conditions. Growth of the oysters was disappointedly small, due primarily, it is believed, to the food source. *Phaeodactylum tricornutum* is not a particularly good food for bivalve molluscs, as numerous shellfish culturists have discovered. The pennate diatoms (*Amphora* sp. and others) that succeeded *Phaeodactylum* tended to clump into large particles, 1 mm or more in size. Although the algal harvest was passed through a commercial homogenizer, as mentioned above, this treatment did not eliminate clumping completely and many of the particles were probably still too large for assimilation by the shellfish. Thus the need for controlling species in the algal growth system and of selecting for those forms that are the best food for shellfish is apparent.

In the fall of 1973, the Woods Hole Oceanographic Institution's Environmental Systems Laboratory was completed and occupied. This laboratory was designed and constructed to serve as a pilot-plant for the WHOI Waste Recycling-Aquaculture Project as well as to provide space and facilities, including a relatively large-flow treated seawater system, for related studies in aquaculture and in pollution assessment, control, and management (Fig. 23-4).

Because of temporary logistic and financial problems in obtaining daily shipments of the required large volumes of treated wastewater effluent from the nearest treatment plant (Otis Air Force Base or Wareham, Massachusetts) to the tank storage system, the algal farm was initially operated with "artificial wastewater effluent," a mixture of monobasic sodium phosphate and ammonium chloride. Together with filtered seawater, these salts, made up and delivered in concentrations comparable to secondary wastewater effluent were added continuously to the 130,000 liter algae ponds beginning in December 1973. Two of the algal ponds are capable of being heated and circulated by heat exchangers in the mechanical space of the laboratory, and they were operated at 15°C throughout the winter. Following the initial heating and enrichment, a population of mixed diatoms rapidly developed in the ponds, evolving after a few weeks to a virtually pure culture of *P. tricornutum*. This culture in each of the two ponds has been maintained for 8 months at a concentration of approximately 10^6 cells/ml with a turnover rate of 30% of the pond volume per day and an input of seawater and nutrients comparable to 50% treated wastewater effluent. Little or no change in algal production has occurred in response to variations in solar radiation associated with storms, cloudy weather, or the like. Beginning in April 1974, the algal ponds were operated at ambient water temperature and all six ponds put into operation, with algal production comparable to that obtained earlier in the season with the heated ponds. In July 1974, daily deliveries of 30,000 liters of secondary effluent from a nearby treatment plant were begun. The two heated ponds were then switched to a 1:1 mixture of secondary effluent and seawater and have been operating in this mode throughout the year.

The output from each algal pond (ca. 45,000 l/day), is fed by gravity into one of the 12 x 1.2 x 1.5 m deep cement raceways which contain 150,000 seed oysters *(Crassostrea virginica)* approximately 2.5 cm long or 150,000 seed hard clams *(Mercenaria mercenaria)* approximately 1.25 cm in length (longest dimensions). The shellfish are held in stacked wooden trays lined with plastic (Vexar) mesh, at approximately 3,000 animals per tray. A flow of seawater variously filtered and/or heated and at different flow rates is added to each of the shellfish raceways to dilute the food and provide additional flow for oxygen, waste removal, etc. The raceways containing the shellfish are covered with plywood sheets to prevent fouling of the animals with filamentous algae and to reduce heat loss.

The small polychaete worm *C. capitata* was inoculated in the bottom of one of the raceways containing the oysters, where it feeds upon the feces and pseudofeces produced by the molluscs. These organisms, inoculated in January 1974, were observed to be multiplying but the population has not been assessed since it was stocked. Five hundred (ca. 3–6 cm) winter flounder *(Pseudopleuronectes americanus)* were subsequently stocked in this raceway to feed upon the capitellid worms.

The raceway containing the hard clams was stocked with 1,400 juvenile "bait worms" *(N. virens)* approximately 2 cm in length. It is expected that these worms, which may reach lengths of 25 cm at maturity, will grow on the clam biodeposits, as occurred in the small-scale experiments during the summer of 1973. To provide the worms with shelter and reduce cannibalism, the bottom of this raceway was lined with beach stones to a depth of 5–10 cm. (Fine sand tends to become anoxic and unsuitable as a substrate.)

The discharge from the raceways containing the molluscs flows into and through adjacent raceways stocked with seaweed *(C. crispus)*, which is kept in suspension by aeration along one edge of the bottom of the raceway (providing a circular rotation of the water). The purpose of the seaweed, as in the earlier experiments, is to provide the final polishing step to remove nutrients not initially assimilated by the diatoms in the algae ponds together with soluble nutrients added to the system by the shellfish, worms, and fish.

In the year since the Environmental Systems Laboratory was first occupied, the expanded "pilot-plant" model of the waste recycling–marine polyculture systen has gradually been put into full operation, the various components stocked with organisms, and the complete unit empirically balanced to achieve the dual objectives of advanced waste treatment and aquaculture. Performance data are now being taken and, after sufficient experience, will be reported separately.

An interesting new departure recently initiated at the Environmental Systems Laboratory has been the growth, in one of the 15 m diameter, 130,000 liter ponds, of a mass continuous culture of brine shrimp *(Artemia salina)*. Started from nauplii hatched from eggs in the laboratory, the pond now contains a dense culture of adult *Artemia* that are actively producing living young. The daily harvest from an algal pond (45,000 l/day of diatom *P. tricornutum* at approximately 10^6 cells/ml) is fed into the *Artemia* pond from which the brine shrimp completely remove the algae. The corresponding harvest of 45,000 l/day from the *Artemia* pond contains both adult and larval brine shrimp, which are fed into a raceway containing trout and other plankton-eating fin fishes.

Although currently operated on an "artificial" wastewater food chain in which full seawater (30%) is enriched with highly concentrated nutrients as described above, preliminary laboratory experiments indicate that the *Artemia* will continue to grow and reproduce at a salinity of 15%, consistent with a 50% wastewater effluent enrichment of the algal ponds.

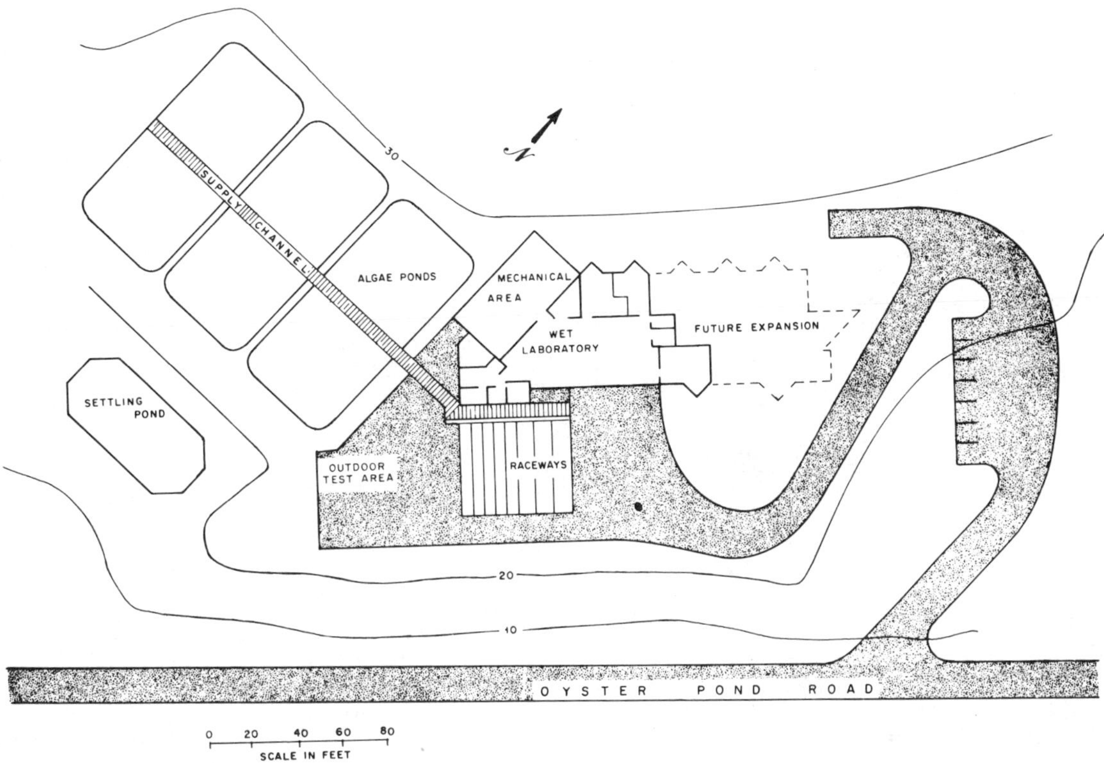

FIG. 23-4. Environmental Systems Laboratory: Plan view.

The wastewater–algae–brine shrimp–trout food chain, or modifications thereof, may represent an attractive alternative to shellfish production in cases where the concentration of pathogens from wastewater by the molluscs cannot be satisfactorily corrected or resolved. It is recognized, however, that introduction of an additional link in the food chain will lead to reduction in the production of the final production (fin fish) by 80–90% from that which could be expected from the herbivores.

Research in Florida

The systems described above appear to operate effectively and with a high degree of efficiency for nutrient removal, but only insofar as the inherent biological processes are working at or reasonably near their optimal conditions. In temperate climates during the winter months, biological processes for many, if not most, species come to a virtual standstill as water temperatures fall below 10°C and, in many locations such as New England, approach 0°C.

The heating of seawater and wastewater effluent to permit biological activity to continue and the system to operate throughout the winter would be prohibitively expensive. A possible solution would be to use the heated effluent from the once-through cooling water system of a coastal power plant, but this would require side-by-side locations of the utility and waste-treatment plants and other constraints that would impose severe limitations on the application of the concept. The rationale can also be used that advanced wastewater treatment (nutrient removal) is most critically needed in summer, particularly in coastal resort communities where both population and coastal water use are at their peaks. But all things considered, the system developed

FIG. 23-5. Harbor Branch Laboratory aquaculture facilities.

and described above is clearly most practical for tropical and semitropical climates where the biological processes will operate continually throughout the year at ambient water temperatures. It is also obvious that both the physical-chemical environment and the biota, including the organisms that could be used in the system, are, at least to some extent, different in the tropics from those that occur in temperate latitudes and that experience obtained in the latter (i. e., Woods Hole) is not necessarily easily translated to a more southern climate.

For these reasons, it was considered advisable to initiate experiments in Florida, and it was possible to transport the entire system used in the 1973 Woods Hole experiment and described here to the Harbor Branch Foundation Laboratory in Fort Pierce, Florida, under a grant from the Atlantic Foundation. The system was installed in January 1974 and is currently in full operation (Fig. 23-5). Treated wastewater effluent from a small extended aeration plant serv-

TABLE 23-5. *Urban Population Areas in the U.S., 1970.*

Area	Coastal Zone	Population $x10^6$
New York, N.Y.–N.J.	Yes	9.97
Los Angeles–Long Beach, Calif.	Yes	7.03
Chicago, Ill.	No	6.98
Philadelphia, Pa.–N.J.	Yes	4.82
Detroit, Mich.	No	4.43
San Francisco–Oakland, Calif.	Yes	3.11
Washington, D.C.–Md.–Va.	Yes	2.91
Boston, Mass.	Yes	2.90
Nassau–Suffolk, N.Y.*	Yes	2.55
St. Louis, Mo.–Ill.	No	2.41
Total		47.11

*Only coastal urban area without major sewerage and sewage treatment systems.

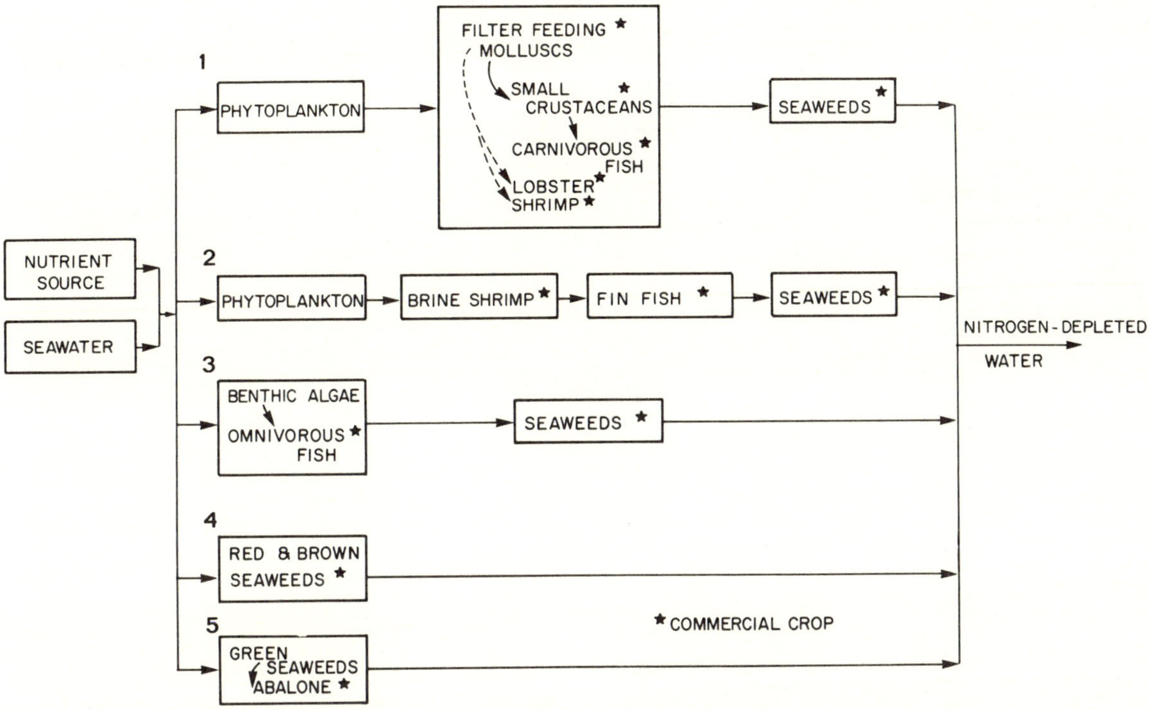

FIG. 23-6. Flow schemes of candidate integrated food chains for waste recycling-aquaculture systems.

ing the Harbor Branch Laboratory community and located adjacent to the aquaculture facilities is used in the various experiments. Not only does the location provide excellent year-round climatic conditions, but it is possible to test the feasibility of utilizing a variety of native species in the system. For example, native seaweeds such as *Hypnea musciformis, Eucheuma isoforme, Gracilaria foliifera,* and *Agardhiella tenera* (all commercially-valuable algae) are being tested in the seaweed growth systems and juvenile white shrimp, *Penaeus setiferons,* have been stocked in the oyster system as a detrital feeder. Ancillary experiments dealing with the feasibility of operating a benthic marine algae – grey mullet *(Mugil cephalus)* – seaweed food chain as an alternative to the diatom – mollusc – seaweed system are under way. In addition, a one-step wastewater-fed seaweed system is being examined.

Process Applicability

Application of a waste recycling-mariculture system is obviously restricted to coastal areas. This is not a particularly serious drawback, at least in the U.S., for although coastal lands contribute only about 15% to the total U.S. land area, over one-third of the population lives there.[24] About four-fifths of this population live in urban areas. Of the ten largest urban areas in the U.S., seven are located on the coast, and only one of these—Nassau-Suffolk, N.Y.—does not contain a major sewerage and sewage treatment system (Table 23-5). Areas south of Washington, D.C., on the east coast, south of San Francisco on the west coast, and all along the Gulf Coast have the climate to support mariculture on either a year-round basis or for a good part of the year. These areas represent a significant population and waste load.

Research on waste recycling in mariculture is

TABLE 23-6. *Some Marine Organisms That Have Commercial Value and May be Cultured in Waste Recycling-Aquaculture Systems.*

Food Source	Species Grown	Commercial Value
Dissolved nutrients: domestic wastes, food processing wastes, agricultural wastes, fertilizers	Phytoplankton Benthic Algae Seaweeds	Agar, food stabilizers fertilizers, human food
Phytoplankton	Filter-feeding molluscs: oysters, clams, mussels, scallops	Human food
	Brine shrimp	Tropical fish food
Seaweeds	Abalone Lobsters	Human food
Benthic algae	Omnivorous fish: grey mullet Shrimp	Human food
Brine shrimp	Fin fish: rainbow trout, flounders, puffers, juvenile lobsters	Human food
	Tropical fish	Recreation
Detritus, feces, pseudofeces	Small crustaceans: polychaete worms	Bait in sport fishing
Small crustaceans, worms, detritus	Carnivore fish: flounder Lobster Shrimp	Human food

at an early stage and there are a number of still unresolved problems. Yet, the process has great versatility as evidenced by the various food chain schemes that may be used (Fig. 23-6), and the potential uses for the grown and harvested products (Table 23-6). Some of the attractive advantages to this type of waste recycling include the following:

a) Potentially, large quantities of fossil fuels that will normally be required for the production of nitrogen fertilizers could be saved for other uses. Energy requirements for waste recycling-mariculture systems are as yet unknown, but because the systems must be connected to already existing sewerage and sewage treatment systems, in most cases additional pumping requirements would be minimal. In addition, it may be possible through engineering design to take great advantage of tidal power for salt water supply and circulation.

b) By coupling these systems to thermal power plants it may be possible to utilize the dissipated heat energy for heating the biological systems so that they can be used in northern climates on a year-round basis. Not only would this improve the efficiency of the mariculture process, but thermal pollution effects would be reduced.

c) If waste recycling-mariculture were to be employed on a large scale, it would have a strong impact on conventional agricultural practice, which has traditionally been to grow single-species crops over large tracts of land. This practice, although productive in the short run, can possibly lead to a catastrophic result in the long term.[25] By eliminating diversity, which is nature's regulator, modern agriculture is constantly in a state of instability: the food chain is broken, fertilizer must be applied in massive quantities, and pesticides must likewise be used in large amounts to control predators. In the western world man has virtually ignored the detrital component of nature as a food source. Not only does this component serve to recycle nutrients, but it is more stable than the primary producer component because it is higher up the food chain. Marine species such as filter-feeding molluscs, lobsters, shrimp, fish, and the like in integrated food chains would comprise an ecologically stable community that would be less susceptable to destruction than, say, a field of wheat or corn.

d) By exploiting the marine resources, pesticide application would be reduced. Pesticides not only require the expenditure of energy during production, but also have longterm effects on the environment that, although relatively unknown, are suspected of being devastating.

On the other side of the coin, the most serious obstacle to use of closed-loop systems for converting waste products to human food is, of course, the problem of public health. Pathogenic bacteria and viruses, trace metals, and refractory compounds all are present in wastewater in varying concentrations and could accumulate in components of the marine food chain, particularly in the filter-feeders, so that the organisms would be unfit for human consumption. Solutions to this major problem override the application of any waste conversion process. In this regard, it will either be necessary to add additional treatment steps between the biological wastewater treatment and aquaculture components or the organisms will have to be cleansed in contaminant-free water before they are safe for human consumption.

Current trends are to establish standards that are designed to eliminate all forms of pollution in order to meet stringent water quality criteria (Public Law 92-500). This approach has come under attack because of the need to apply advanced technology (a pollution generator and energy consumer) to eliminate pollution at a point source.[26] However, by coupling mariculture with complete advanced wastewater treatment, not only would it be possible to produce a high quality water that met established water quality standards, but also a food crop that was safe for human consumption could be realized. In terms of total pollution reduction and energy conservation, such a combined process might be much more effective and economical than advanced waste treatment alone and conventional agriculture equivalent to the mariculture component.

Although it is difficult at the present time to compare these alternatives on an economic basis, the concept of combining complete waste treatment with aquaculture appears attractive enough to warrant continued support for research, particularly in regard to definition of the public health problem and its possible solutions.

Notes

1. This research was supported by NOAA Sea Grants 04-4-158-5 and 04-5-158-8, National Science Foundation (RANN) Grants GI-32140 and GI-43884, and the Atlantic Foundation.

2. F. D. Schaunberg, "Nature—an Important Factor in Management of the Total Environment" (Paper delivered at Seventh Conference of the International Association on

Water Pollution Research, Paris, 9–13 September, 1974).

3. M. R. Henzen, G. J. Stander, L. R. J. Van Vuuren, "The Current Status of Technological Developments in Water Reclamation," in *Progress in Water Technology, Vol. 3—Water Quality Management and Pollution Control Problems,* S. H. Jenkins, ed. (Oxford: Pergaman Press, 1973), pp. 307–18; T. G. Reeves, "Nitrogen Removal: a Literature Review," *J. Water Poll. Control Fed.* 44 (1972): 1895–1908.

4. D. R. Safrany, "Nitrogen Fixation," *Scientific American* 231 (1974): 64-80.

5. D. Pimentel, L. E. Hurd, A. C. Belloti, M. J. Forster, I. N. Oka, O. D. Sholes, R. J. Whitman, "Food Production and the Energy Crisis," *Science* 182 (1973): 443–49.

6. R. C. Loehr, "Agricultural Runoff—Characteristics and Control," *J. Environ, Eng. Div. Am. Soc. Civ. Engrs.* 100 (1974): 363-68.

7. *Ibid.*

8. E. H. Kone, "New Kinds of Fertilizer: Can They End the Shortage," *New York Times,* November 24, 1974, Section 2, pp. 41.

9. J. E. Bardach, J. H. Ryther, W. O. McLarney, *Aquaculture—The Farming and Husbandry of Freshwater and Marine Organisms* (New York: Wiley Interscience, 1972).

10. U.S. Environmental Protection Agency, *Wastewater Use in the Production of Food and Fiber-Proceedings,* Series EPA 660/2-74-041 (Washington, D.C.: U.S. Government Printing Office, 1974).

11. Pimental et al., "Food Production."

12. J. L. Mahloch, and E. C. McGriff, "Agricultural Wastes—a Literature Review," *J. Water Poll. Control Fed.* 46 (1974): 1280-83.

13. W. T. Oswald, "Current Status of Microalgae from Wastes," *Chem. Eng. Prog. Symp. Ser.* 65 (1969): 87-92.

14. J. H. Ryther, W. M. Dunstan, K. R. Tenore, and J. E. Huguenin, "Controlled Eutrophication—Increasing Food Production from the Sea by Recycling Human Wastes," *Biosci.* 22 (1972): 144-52.

15. J. H. Ryther, "Photosynthesis and Fish Production in the Sea," *Science* 166 (1969): 72-76.

16. J. H. Ryther, and W. H. Dunstan, "Nitrogen, Phosphorus, and Eutrophication in the Coastal Marine Environment," *Science* 171 (1971): 1008-13.

17. Southern California Coastal Water Research Project, *The Ecology of the Southern California Bight: Implications for Water Quality Management* (El Segundo, California: 1973).

18. Ryther and Dunstan, "Nitrogen, Phosphorus, and Eutrophication"; J. C. Goldman, K. R. Tenore, J. H. Ryther, and N. Corwin, "Inorganic Nitrogen Removal in a Combined Tertiary Treatment-Marine Aquaculture System—I. Removal Efficiencies," *Water Research* 8 (1974): 45-54; R. W. Eppley, "Eutrophication in Coastal Waters: Nitrogen as a Controlling Factor," *Wat. Poll. Control Res. Ser.,* 13030 ELY 12/69, U.S. Environmental Protection Agency (Washington, D.C.: U.S. Government Printing Office, 1971); W. H. Thomas, D. L. R. Siebert, and A. N. Dodson, "Phytoplankton Enrichments. Experiments and Bioassays in Natural Coastal Sea Water in Sewage Outfall Receiving Waters Off Southern California," *Estuar. Coast. Mar. Sci.,* 2 (1974): 191.

19. W. M. Dunstan and D. W. Menzel, "Continuous Cultures of Natural Populations of Phytoplankton in Dilute Sewage Effluent," *Limnol. & Oceanogr.* 16 (1971): 623-32.

20. Ryther and Dunstan, "Nitrogen, Phosphorus, and Eutrophication."

21. J. C. Goldman, K. R. Tenore, and H. I. Stanley, "Inorganic Nitrogen Removal in a Combined Tertiary Treatment-Marine Aquaculture System—II. Algal Bioassays," *Water Research* 8 (1974): 55-59.

22. Goldman, Tenore, Ryther, and Corwin, "Inorganic Nitrogen Removal."

23. J. C. Goldman and J. H. Ryther, "Nutrient Transformations in Mass Cultures of Marine Algae," *J. Environ. Eng. Div. Am. Soc. Civ. Engrs.* 101 (1975): 351-64.

24. T. A. Wastler and L. C. Wastler, "Estuarine and Coastal Pollution in the United States," In *Marine Pollution and Sea Life,* M. Ruivo, ed. (Surrey, England: Fishing News, 1972), pp. 40-59.

25. E. P. Odum, "The Strategy of Ecosystem Development," *Science* 164 (1969): 262-70.

26. Schaunberg, "Nature."

24

Aquaculture as an Alternative Wastewater Treatment System

R. LEROY CARPENTER
MARK S. COLEMAN
RON JARMAN

Oklahoma State Department of Health

Introduction

Man has cultured fish aided by the use of human wastes for several centuries. Asian and European countries have long had a history of marketing "organically grown" aquatic organisms. However, this practice had been almost nonexistent in this country primarily due to an abundance of cultivated and natural foods coupled with concern over disease problems and instilled aversion to association with human waste products. In recent years a combination of events has stirred interest in the advisability and practicability of aquaculture in wastewaters. Protein of various types has become less abundant and the associated increase in price has created a need for additional protein production. A tone of urgency was added in 1972 with the passage of amendments to the Federal Water Pollution Control Act (Public Law 92-500) part of which established optimistic deadlines and effluent standards for all discharges into streams.

Effluent Standards

The act requires all publicly owned treatment works to achieve effluent limitations based on secondary treatment as defined by the Administrator of the Environmental Protection Agency. Discharges to most stream segments must meet the defined secondary treatment requirements whereas discharges to certain other segments must generally meet more stringent treatment requirements.[1] Subsequent to publication of the effluent limitations associated with the definition of secondary treatment, the Environmental Protection Agency determined that waste stabilization ponds (lagoons) as presently designed would meet the five-day biological oxygen demand (BOD_5) limitation of 30 mg/l, but would not ordinarily meet the suspended solids limitation of 30 mg/l or the fecal coliform limit of 200/100 ml without improved pond design and operation and in some cases additional treatment steps for algae removal and disinfection.[2]

Additionally, BOD_5 standards for designated stream segments in many states are such that effluent discharges to those streams are significantly less than 30 mg/l.[3] As several thousand inexpensive and simply operated waste treatment lagoons exist, particularly in smaller communities in the less populated areas of the country, it is highly desirable that a means be developed whereby these facilities will meet the various upgraded requirements.

Lagoon Treatment

Lagoon treatment is actually a biological system that degrades and stabilizes organic compounds in wastewater. Prior to lagoon treatment, sewage may receive no treatment or mechanical treatment ranging from primary to a high degree of secondary treatment. Numerous types of lagoon designs exist and a wide range are in current use.[4]

The commonly acknowledged major steps in a lagoon treatment system are: (1) anaerobic and aerobic bacterial decomposition of organic wastes resulting in plant nutrients (carbon, nitrogen, and phosphorous) being released into the water; and (2) algal utilization of these nutrients.[5] The algal phase of the process constitutes the basic mechanism for separation of the nutrients from the water.

Proposed mechanisms to upgrade effluent quality from lagoons tend to be physical-chemical in scope and include settling, floculation, and/or filtration aimed toward the removal of the algal cells. Removal of the algae results in lower concentrations of suspended solids and removal of the nutrients comprising their organic mass.[6] Proper design and maintenance of a physical-chemical system provides a high degree of waste treatment. However, these types of treatment mechanisms have some undesirable attributes that are of concern, principally greatly increased construction, operational, and maintenance costs. In addition, waste sludges are produced imposing further disposal problems.[7]

A significant removal of pathogenic bacteria and viruses during the process of the lagoon method of sewage treatment is well documented, though not entirely understood.[8] Stabilization ponds with a 30-day retention time can achieve a reduction of up to 99%.[9] D. A. Okun reported 99.99% reduction of coliforms;[10] R. J. Drew found 99.6% reduction in summer and 96.8% in winter.[11] However, this level of treatment, in many cases, will not continually assure attaining the fecal coliform limits of 200/100 ml.

Chlorination of the lagoon effluent has been proposed to achieve these standards. However, several drawbacks exist with the use of chlorine to reduce bacterial levels. Initially the chemical is expensive and shortages have occurred in some areas of the country. An additional major use could create a critical condition in supply and demand. And, while the chemical will kill bacteria, it will kill other plants and animals as well. One result of the use of chlorine could be the creation of a biological desert in the receiving stream. Chlorination has also been implicated recently as a producer of long-lived, biologically deleterious compounds, some of which are carcinogenic.[12]

Existing waste treatment systems are biological to the algae removal point. The physical-chemical treatment phase for algae removal is replaced in the system undergoing study by employing a further biological segment based on herbivorous or filter-feeding fishes. Although either an advanced biological treatment system or an advanced physical-chemical treatment system will produce an effluent of high quality, a significant point of the biological alternative is that the removal of algae through the ecological food chain will produce a useful product in the form of fish as opposed to a product requiring further disposal. Many nonconsumptive alternatives for the use of fish are available ranging from reduction for animal food to sale of live fish for bait, restocking, or forage.[13]

Methods

Personnel of the Oklahoma State Department of Health have been conducting preliminary re-

FIG. 24-1. Aerial view of study area—Quail Creek sewage lagoons, Oklahoma City, Oklahoma.

search since 1970 in an attempt to attack and ameliorate the problems of algae removal, bacteria reduction, and upgrading the quality of lagoon effluents. The basic assumption of the project is that a biological means of removal exists for the biological phenomenon of algae production.

To determine the feasibility of nutrient removal by various fish species, a study was conducted in a large lagoon system serving a small portion of a major Oklahoma municipality. The study area consisted of a six-cell lagoon system that averaged 2.6 hectares (6.5 acres) and totaled 15.8 hectares (39 acres) (Fig. 24-1).

System Design and Operation

Facility Description

During the project period, about 1 million gallons per day (mgd) of raw domestic waste received conventional aerated treatment in the first two-cell set of the three-set system. For experimental purposes, it was decided to utilize the four nonaerated and unused ponds in conjunction with the existing aerated system to form a six-cell, serially operated system. Accordingly, in May 1973, 25,000 two-to-four-inch channel catfish were stocked in the third cell and other 25,000 into the fourth cell. The fifth and sixth cells received 85 pounds (about 1500) golden shiner adults equally divided between the two cells. An additional stocking of approximately 5 pounds of fathead minnows were introduced into the third cell in July 1973. Also in July, 175 three-inch *Tilapia nilotica* were placed in the third cell.

Problems arose in the system when black bullhead, green sunfish, and mosquito fish already present in the aerated cells contaminated the remainder of the system. However, the resulting polycultural situation lent itself to gleaning data on a wider food web than originally intended.

The study was repeated without fish present in the summer and fall of 1974 to measure the effectiveness of the lagoon only.

Analytical Methods

Water sampling consisted of removing 2 liters of the raw waste and effluent from each cell on a weekly basis. Standard laboratory analyses were

FIG. 24-2. Typical seining operation—Quail Creek sewage lagoons.

conducted by the Oklahoma State Department of Health and the City of Oklahoma City.[14] Bacteriological indicator sampling consisted of removing 100 ml of water from the raw waste and effluent of each cell on alternate weeks. Fish population analyses consisted of a preliminary food habits investigation and biomass estimates. Estimates of biomass were made from seine hauls encompassing a closed known area of each cell (Fig. 24-2). Mean weights were taken from samples of 100 fish and applied to estimated numbers. Food habit samples were taken by excising stomachs of various species at the isthmus and duodenal sphincter. Identification of stomach contents was made in the laboratory utilizing standard methodology.

For analysis to determine levels of human pathenogenic microorganisms, 179 fish on twelve different dates, and 77 water samples on eleven different dates, were collected over a seven-month period beginning July 1, 1973 and ending January 28, 1974. While there was some variation, most water collections consisted of 4-liter catch-samples of the untreated sewage and similar samples of the effluent from each of the six cells. A total of 34 samples of five channel catfish each were collected from cells #1 through #5 during twelve fish-sampling visits. *Tilapia,* green sunfish and bluegill were collected one time each from cells #3, #4, and #5 respectively. Water samples were processed according to the American Public Health Association (APHA) *Standard Methods of Examination of Water and Wastewater.*[15]

Water samples were examined for total coliform, fecal coliform, fecal streptococci, and pathogenic enteric bacteria using the membrane filter methodology recommended by the APHA.[16] For isolation of pathogenic enteric bacteria, 100 ml portions of sample were passed through 0.45 micron pore size membrane filters. The membranes were then placed in Hajna's gram negative (GN) enrichment broth and incubated overnight.

For virus studies, 50 ml aliquots of lagoon water were centrifuged at 2,000 rpm. The supernatant was decanted, treated with antibiotics, and 0.1 ml portions inoculated into rhesus monkey kidney, Wi38 and, when available, HEp2 tissue culture tubes. Cultures showing cellular destruction were examined by conventional methods for viral identification.[17] Cultures not showing cellular destruction were passed once to the same cell line before being discarded as negative.

It was recognized that the catch sample method has a comparatively low degree of sensitivity, but the elaborate concentration techniques employed by Wallis, Melnick, and Fields[18] and by Metcalf, Vaughn, and Stiles[19] were beyond the financial capabilities of this study. Control studies conducted by seeding of the lagoon water with low levels of Coxsackie A9 enterovirus followed by the above procedure showed no apparent interference to the recovery of virus by naturally occurring constituents in the lagoon water. Samples of fish were carefully dissected and the skin, upper intestines, cloaca, and muscle were cultured separately. However, after the first two collection dates, skin was combined with muscle to constitute one specimen, and all viscera were pooled to constitute a separate specimen. After homogenizing, an aliquot of each pool was used for both virus and bacterial study.

Results and Discussion

Analysis of Water

A summary of water analyses are presented in Table 24-1 and are the average of weekly samples from June 6, 1973 to October 3, 1973. The secondary treatment standard for BOD_5 of 30 mg/l was met in the effluent from the second cell and fell significantly throughout the remainder of the system. The remaining secondary standards for suspended solids of 30 mg/l and fecal coliform of 200/100 ml were met in the effluent of the fifth cell.

It should also be noted that excellent removal of both nitrogen and phosphorous was accom-

TABLE 24-1. *Mean Values of Weekly Analyses of Water Samples from the Quail Creek Lagoon System. Samples were collected June 6, 1973 through October 3, 1973.*

Parameters	Raw	1	2	3	4	5	6
Biochemical Oxygen Demand (5 day)	184	47	24	17	14	9	6
Suspended Solids	197	79	71	52	54	26	12
Total Nitrogen (as N)	18.94	10.50	7.04	6.65	3.97	3.13	2.74
Total Phosphorus (as P)	9.01	9.87	7.97	5.80	3.66	3.01	2.11
Fecal Coliform/100 ml	3.05×10^6	10880	1380	322	15	15	20
Turbidity	55	15	23	25	42	17	9

*Data reported in mg/l except where noted.

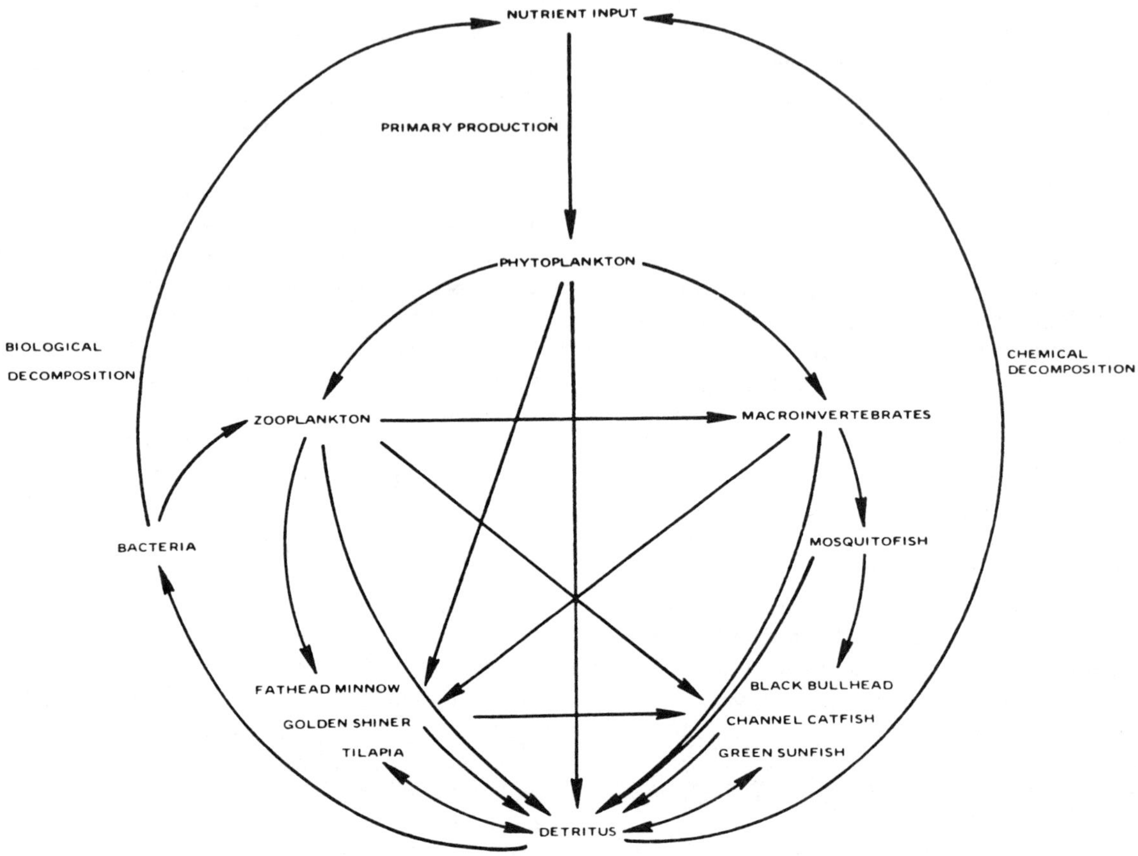

FIG. 24-3. General schematic food web of organisms inhabiting cell 3 in the Quail Creek lagoon system, 1973.

plished as evidenced by the overall 86% reduction of nitrogen and 77% reduction of phosphorous.

The overview of these data indicates a high grade effluent. These preliminary efforts are not seen however, as maximum efficiency; further research efforts should attain greater removals. Other studies have also indicated that sewage treatment through production of commercially valuable biological products is effective in producing an upgraded effluent.[20]

Food Habits

The preliminary data on food habits allowed general classification of the fishes according to position in the food web. A diagramatic representation is presented in Fig. 24-3.

Tilapia nilotica, fathead minnows, and golden shiners were found to consume phytoplankton, microcrustacea, and insect larvae. Channel catfish were found to utilize microcrustacea and insect larvae as their principal foods.

Microbiological Results

Results of the bacteriological study are shown in Fig. 24-4. The log count of fecal coliforms per 100 ml is shown in the bars and sampling dates are shown along the horizontal axis. The top shaded area shows the log of the number of coliforms in the raw influent to the system and the lower portion of the bars, shown by the dotted area, represents organisms in the treated effluent as it passed out of the last of the six cells into the receiving stream.

Inspection of this figure readily shows that there is a remarkable reduction of fecal coliform bacteria to less than detectable limits in a standard 50 ml portion in three separate samples.

The pathogens isolated and shown by the boxes in the lower portion of Fig. 24-4 were:

a) *Edwardsiella tarda* (August 13),
b) ECHO 1 enterovirus (December 17), and
c) *Salmonella newport* (January 14).

Fig. 24-5 shows the average log of the count of indicator organisms per 100 ml isolated on the eleven water-sampling dates. As the wastewater flows through the system from raw influent on the left to effluent from the last lagoon on the right, one can see a progressive reduction in the indicator organisms. By the time the effluent passes through cell #3, it is found to contain less than ten fecal coliform organisms/100 ml, and remains at this low level as the effluent passes through the last three cells. It is interesting from a public health standpoint that of the three pathogens isolated in water during the study, two were found in raw influent and one was in cell #2. No pathogens were found in wastewater beyond the first two conventional cells; nor in any of the cells containing the test fish; nor in any of the 179 fish sampled.

Biomass

Since no supplemental feeding of fishes occurred during the experiment, it follows that fish

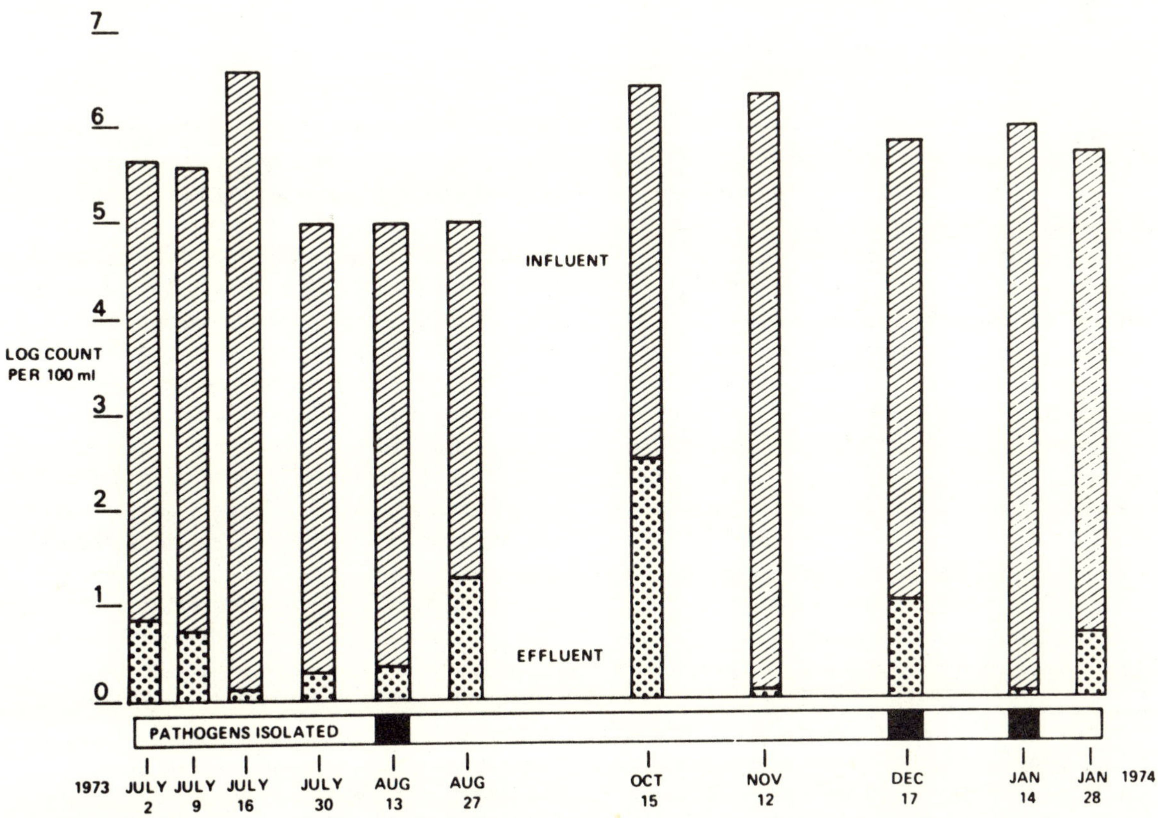

FIG. 24-4. Pathogen presence and influent/effluent fecal coliform count in the Quail Creek lagoon system (July 1973–January 1974).

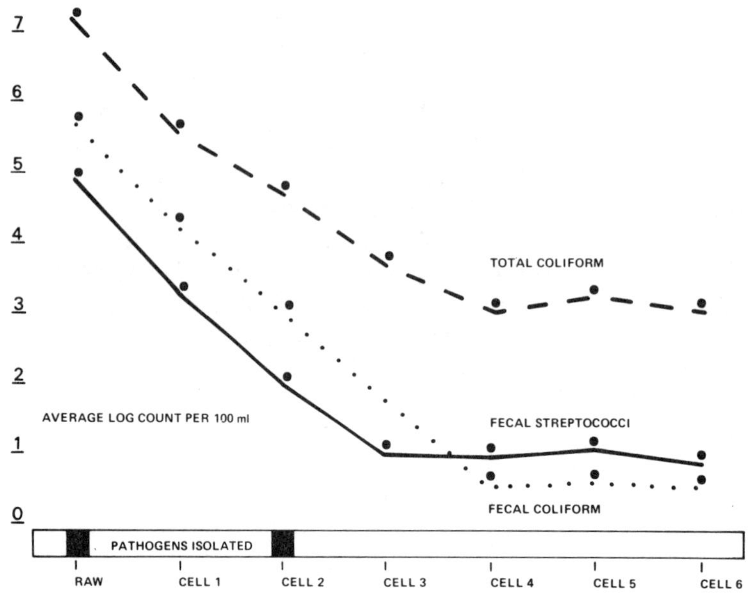

FIG. 24-5. Presence of indicator bacteria and pathogens in the Quail Creek lagoon system.

biomass must owe increases to organic constituents present in the system. It is then reasonable to assume that all biomass gained during the observation period represents entrapment of elemental constitutents that would possibly have been discharged to receiving waters if the fish had not been present.

Tilapia nilotica biomass increased from the initial 4 pounds, consisting of 175 individuals (averaging 3 inches, stocked in July 1973) to 163 pounds in October 1973. The harvest consisted of 2,339 fish with some individuals attaining lengths of 10 inches. It was felt that their planktivorous food habits, the presence of abundant algae, and their rapid reproduction rate were the major reasons for this increase. The total biomass of *Tilapia* was not determined due to procedural difficulties coupled with a sudden temperature drop that resulted in a large mortality. The biomass reported was determined from *Tilapia* removed. It is likely that a significantly greater amount of *Tilapia* were produced and not recovered.

The biomass of channel catfish increased from the initial 600 pounds to an estimated 4,400 pounds. However, the channel catfish were observed to gain the majority of this biomass in a six-week period from late May 1973 to mid-July 1973. Slow growth followed this initial rapid gain. Limited information suggests this was due to stunting because of limited food available to the larger fish. After the virtual stoppage of growth, the system was not operating at maximum efficiency for either water quality improvement or production of fish biomass.

From the initial stocking of 85 pounds of golden shiner minnows, an estimated 535 pounds were recovered. This smaller than expected amount was probably due to bullhead catfish predation. It should be noted that only the production of golden shiners in cells #5 and #6 was estimated. However, this species was found in moderate numbers in other cells.

It can be said that the microbiological portion of this study has confirmed the previous observations of others that indicator coliform organisms are efficiently removed in a lagoon-method wastewater treatment system.[21] This study has shown that human pathogens are rare in those wastewaters tested and in fish grown in those wastewaters beyond the raw or first two conventionally-operated cells.

Results with No Fish Present

Operation of the same system without fish through the summer and fall of 1974 has demonstrated that a significant reduction of BOD_5, COD, nutrients, and suspended solids results as a direct effect of the extended lagooning through multiple cells. Preliminary analysis of the data indicates that nutrient reduction is most efficient in the summer whereas suspended solids are at the lowest concentration in the winter. Average effluent data obtained to date are reasonable: fecal coliform of less than 200/100 ml; BOD_5 of 13; COD of 69; suspended solids of 39; total-N of 8.2; and phosphorous of 7.6 mg/l. It should be noted that a significantly higher quality effluent resulted with fish present in the system and that suspended solids effluent criteria were not met without them although BOD_5 and fecal coliform criteria were met.

Cost Evaluation

The replacement of the physical-chemical technology with a biological system producing a marketable product can help defray the cost of treating sewage. An estimated total cost for the system employed is $0.15 per thousand gallons. It is appropriate to note that this comparatively low figure is for an effluent of high quality, surpassing secondary standards usually attainable only through mechanical separation at a cost approaching $0.25–0.30 per thousand gallons.[22] If no monetary recovery at all were made from the fish in the system, this would remain an excellent effluent at a comparatively low price.

Future Studies

From research completed to date, several urgent areas of study have been revealed. Several of these are included in planned research by the Oklahoma State Department of Health.

Studies so far have yielded information regarding the food niche of various fish species in nutrified water. Future studies in this area should be oriented toward and include detailed niche identification and utilization to maximize production and resultant upgraded effluent quality by varied species at varied stocking rates.

To date cost analysis gleaned from studies has been sparse. Since this factor is of prime importance in the future adoption of aquaculture for wastewater treatment, more information will be required. Future studies will examine cost-effectiveness ratios and provide information on the overall usefulness of this treatment technique.

We should investigate the potential dangers to the handlers, harvesters, and processors of fish and shellfish grown in wastewaters and not limit our attention to the consumer only. We must sharpen our techniques for isolating pathogenic viruses and bacteria from wastewaters, use them more frequently in our monitoring programs, and not rely solely on coliform indicators as we have in the past. Parasites such as worms and amoebae should not be overlooked. And last, where possible, we should use sound epidemiological methods to correlate the presence or absence of disease in a community with those organisms found in that community's wastewaters.

Additional studies currently under investigation by the Oklahoma State Department of Health are examining more specifically the potential effects on public health. These studies include sampling for aerosol-transported contaminants. Water samples of up to 10,000 liters are concentrated in a viral concentrator for subsequent examination for enteroviruses. Additional samples of both air and water and biological products are examined for enteropathogens including bacteria and viruses.

Other studies are attempting to more closely delineate suitable filter-feeding species and polycultural optimizations.

Notes

1. U.S. Environmental Protection Agency "Water Programs, Secondary Treatment Information," *Federal Register* 38 (August 1973): 22298-99.

2. H. P. Cahill, Jr., "Waste Stabilization Ponds," memorandum defining secondary treatment standards, U.S. Environmental Protection Agency (September 1973).

3. Oklahoma Department of Pollution Control, *Fiscal Year—74 Water Pollution Control Program for State of Oklahoma,* mimeographed, 1974; Oklahoma Department of Pollution Control, *Permit Management Plan for Planning Basin One—State of Oklahoma,* mimeographed, 1974.

4. Oklahoma State Department of Health, Bacterial Reduction Section, *Phosphate and Nitrogen Variations and Algae Prevalence in Oklahoma Waste Oxidation Ponds* (forthcoming); U.S. Environmental Protection Agency, "Upgrading Lagoons," Technical Transaction Series (Washington, D.C.: U.S. Government Printing Office, 1973), p. 43.

5. W. Van Heuvelen and J. H. Svore, "Sewage Treatment by the Lagoon Method," *Transactions Fourth Annual Conference Sanitary Engineering,* Bulletin #30 (1954): 5-8; G. J. Hopkins and J. K. Neel, "Sewage Lagoons in the Midwest," *Missouri Department of Health, Education and Welfare* (1956), p. 23; D. E. French, "Sewage Lagoons in the Midwest," *Water & Sewage Works* (December 1955): 537-40; D. F. Kincannon, "The Role of Oxidation Ponds in Waste Treatment," *Journal of the Missouri Water Sewerage Conference* (1966): 6-11.

6. U.S. Environmental Protection Agency, "Upgrading Lagoons."

7. D. S. Parker, J. B. Tyler, and T. J. Dosh, "Algae Removal," *Water Works and Wastes Eng.* 10 (1973): 26-29; V. Kothandaraman and R. L. Evans, "Removal of Algae from Waste Stabilization Pond Effluents—State of the Art," *Illinois Water Survey Circular* 108 (1972):9.

8. Gerald Berg, "Virus Transmission by the Water Vehicle. II. Virus Removal by Sewage Treatment Procedures," *Health Laboratory Science* 2 (1966): 90; J. W. Kock, "Survival of Coliform Bacteria in Wastewater Treatment Lagoons," *J. Water Poll. Control Fed.* 43 (1971):2071-83; WHO Meeting of Experts on the Reuse of Effluents, "The Reuse of Wastewater: Methods of Wastewater Treatment and Health Safeguards, 1973 Report" (Geneva, World Health Organization Technical Report Series, No. 517) *WHO Chronicle* 27 (November 1973): 492-95; J. A. Little, B. J. Carroll, and R. E. Gentry, "Bacterial Removal in Oxidation Ponds," *Second International Symposium for Waste Treatment Lagoons,* June 1970, Kansas City, Missouri (Lawrence, Kansas: 1970), pp. 141-51.

9. WHO Meeting, "The Reuse of Wastewater."

10. D. A. Okun, "Experience with Stabilization Ponds in the U.S.A.," *Bull WHO* 26 (1962): 550.

11. R. J. Drew, "Field Studies of Large Scale Maturation Ponds with Respect to their Purification Efficiency," *Journal and Proceedings of the Institute of Sewage Purification* 3 (1966): 1-16.

12. See chapters 9 and 10.

13. J. Trimberger, "Production of Fathead Minnows *(Pimepheles promelas)* in a Municipal Waste Water Stabilization System," *Mich. Dept. Nat. Res.* (1972): 4.

14. American Public Health Association, *Standard Methods for the Analysis of Water and Wastewater,* 13th edition (New York, 1971).

15. *Ibid.,* pp. 657-60.

16. *Ibid.,* pp. 692-704.

17. A. A. Hajna, "A New Specimen Preservative for Gram Negative Organisms of the Intestinal Group," *Public Health Laboratory* 13 (1955):59-62, 83.

18. Joseph L. Melnick and Craig Wallis, "Mechanism of Enhancement of Virus Plaques by Cationic Polymers," *Jour. Virology* 2 (1968): 267-74; C. Wallis, J. L. Melnick, and J. E. Fields, "Detection of Viruses in Large Volumes of Natural Waters by Concentration on Insoluble Polyelectrolytes," *Great Britain Water Research* 4 (1970): 787-96.

19. T. G. Metcalf, J. M. Vaughn, and W. C. Stiles, "The Occurrence of Human Viruses and Coliphage in Marine Waters and Shellfish," *FAO Technical Conference on Marine Pollution and its Effects on Living Resources and Fishing,* December 1970 (Rome, 1970).

20. J. Songer et al., "Controlled Eutrophication: Sewage Treatment and Food Production," in *Wastewater Use in the Production of Food and Fiber–Proceedings.* Environmental Protection Technology, Series EPA—660/2-74-041 (Washington, D.C.: U.S. Government Printing Office, 1974):529 (See also chapter 26); J. H. Ryther et al., "Controlled Eutrophication-Involving Food Production From the Sea by Recycling Human Wastes," *Biosci.* 22 (1972):144. See also chapter 23.

21. Berg, "Virus Transmission"; Klock, "Survival of Coliform Bacteria"; WHO, "The Reuse of Wastewater"; Little, "Bacterial Removal."

22. EPA, "Upgrading Lagoons."

25

A Proposed Integrated Biological Wastewater Treatment System[1]

RAY DINGES

Division of Wastewater Technology and Surveillance
Texas State Department of Health

Introduction

Construction of an experimental facility at an area wastewater treatment plant is underway to evaluate potential employment of various animals and plants in an integrated, sequenced system for improvement of water quality. The wastewater treatment plant consists of an aeration basin, clarifier, and three stabilization ponds operated in series. Input to the experimental unit will be a portion of the effluent from the terminal stabilization pond. Stabilization pond effluent in which original organic material has been reduced by bacterial action and converted into algae cells would be of high quality except for algae present and high bacteria and nutrient levels. The study scheduled to commence in the near future will incorporate information gained from past studies and include several new concepts. It is intended to correct the deficiency of stabilization pond effluent quality.

Past Studies

Zooplankton

Extensive field studies of Texas wastewater stabilization ponds were conducted from 1970 to 1973 to identify environmental factors affecting zooplankton populations and, specifically, populations of *Daphnia.*[2] Results of studies indicated severe restriction of *Daphnia* production in most pond systems due to elevated pH levels attributable to algal growth. Ammonia dissociation precluded *Daphnia* survival at the high pH levels encountered. Microstratification and resultant sulphide occurrence in lower layers of pond waters was also found not to be conducive to *Daphnia* production.

A pilot pond, which was provided by the City of Giddings, Texas, and supplied with a portion of effluent from a stabilization pond, was used to verify field study findings by maintaining a *Daphnia* culture under controlled conditions.

TABLE 25-1. *Performance of the* Daphnia *Culture Unit during a Controlled Operational Period. The organic loading was 43.6 lbs of* BOD_5*/acre/day. The detention time was approximately 11 days.*

	Influent			Effluent		
	BOD_5	COD	VSS	BOD_5	COD	VSS
Mean concentration in mg/l	57.5	158.5	78.4	11.8	70	13
Percent reduction of influent concentration				79	55	83

Mineral acid was added to the influent to control pH within a range of 7.0-7.5 and sulphide evolution was restricted by mixing-aeration of pond contents. Sustained culture of *Daphnia* at a high population level was found to be feasible. Effluent from the culture unit was macroscopically clear and contained few coliform bacteria. Mean reductions in biochemical oxygen demand (BOD_5), chemical oxygen demand (COD) and volatile suspended solids (VSS) are presented in Table 25-1.

Phosphate content of the effluent was comparable to that of the influent and nitrate and organic bound iron accumulated in the unit. *Daphnia* harvest and potential usage as a protein additive to animal foods have been given consideration.[3] Zooplankton culture unit design and operating features were formulated from experiences gained and observations made during field surveys and pilot pond operation.[4]

Macrophytes

In 1973, a unique opportunity was afforded to observe the effect of the presence of water hyacinths, *Eichhornia crassipes,* on water quality in a drainage canal receiving effluent from the Gregory, Texas, municipal wastewater treatment plant.[5] In November 1972, hyacinths in the canal were treated on two occasions by aerial application of 2-4 D, a broadleaf herbicide. This action reduced the number of water hyacinth plants to a small fraction of the former population. Thus it was possible during the next growing season to measure the effects of a growing culture of hyacinths on wastewater quality as it flowed through the canal. The section of the canal studied is about 25 feet in width and 3 feet in depth. Sampling Station 1 was established 70 feet below the plant outfall and Station 2 at a point 0.4 miles downstream. Surface area of the study segment is about 1.2 acres. With an influent input of 140,000 gallons per day, a theoretical detention time of approximately 8 days is provided.

Water samples for mineral and organic analyses were collected weekly from July through October, 1973, when hyacinth population was expanding. Observed mean reduction in total dissolved solids content was 59.3%. Mean reductions in individual mineral parameters are presented in Table 2. Mean reduction of BOD_5 was noted to be 35.4% and volatile suspended solids were increased by some 15%.

Two heavy rains occurred during the period of study and the effect of residual dilution was evident in subsequent sample results. Greatest reductions in mineral content, however were observed during the driest portion of the study period. No groundwater intrusion into the canal was detected and it is apparent that shallow groundwaters would be quite saline as the area is near the seacoast.

Water hyacinths are known to cause significant water losses by evapo-transpiration, which contributes to mineral concentration in surface waters. Results obtained in this investigation are even more remarkable when this factor is introduced. Data indicated greater mineral reduction occurring during the latter phase of study when canal surface was covered completely with mature plants. If, indeed, hyacinths are capable of demineralization of water to the extent indicated, then a mechanism other than simple accumulation and storage of minerals within growing portions of plants would be necessary to explain why tissue saturation does not occur. Hyacinths have dense fibrous root systems and there is a continual sloughing of tissue from roots. It appears possible that mineral accumulation might occur within tissues shed from roots and be incorporated into benthic debris layer. This could serve to explain continuing mineral uptake by mature plants precluding tissue saturation and resultant toxicity.

The promising results of the preliminary studies with *Daphnia* and water hyacinths led to the idea that several separate biological wastewater treatment steps, each imposing its own distinct improvement on water quality, could be designed to operate in series on municipal effluent. Below, a proposed study is described in which a five-step biological waste treatment system is designed to accumulate water-polluting nutrients in animal and plant biomass, which are then harvested.

TABLE 25-2. *Mean Reductions in Individual Mineral Parameters.*

mg/l	Station 1	Station 2	% Reduction
Calcium	79.5	45.5	42.8
Magnesium	15.25	5.38	64.8
Sodium	173.6	59.8	65.6
Potassium	10.75	7.71	37.2
Bicarbonate	273.13	134.38	50.8
Sulphate	43.0	16.3	62.1
Chloride	291.0	97.0	66.6
Nitrate	7.24	4.06	43.9
Phosphate	16.4	4.3	73.6
Alkalinity	236.0	110.0	53.8
Hardness	262.0	136.0	48.2

Proposed Study

Study Objectives

a) Reduction of influent pH to permit zooplankton growth by use of a novel biological filter.

b) Culture of aquatic plants, invertebrates, and fish to improve organic, mineral, and bacteriological quality of water.

The components of the waste treatment system designed to accomplish the above objectives are described in the following paragraphs. A diagram of the system illustrating the sequence of biological components and the purposes of each is shown in Fig. 25-1.

pH Reduction Filter

Influent to the pH reduction filter will be by gravity flow through pipe equipped with gate valve. The filter unit is of unique design to permit retention of a high level of biological solids at a rapid rate of flow through. Aerobic conditions will be maintained in the unit by appropriate recirculation of water from culture ponds. Carbon dioxide resulting from respiration of the microorganism concentration in the filter should reduce pH level of the influent to a satisfactory level for mass development of zooplankton and permit initiation of floating macrophyte growth.

Pond Culture Units

An excavation 280 feet in length and 30 feet in width is to be divided into four segments (see Fig. 25-1). Culture units will be separated one from another by stacked rock barriers that will be about 8 feet in width at the base and 5 feet wide at the top. Length of the first unit will be 100 feet. The second culture pond will be 120 feet long and remaining sections will be 30 feet each in length.

Function of Units

Filter

Expected Results: a) Reduction of influent pH.
b) Limited nitrification.
c) Minimal suspended solids removal.
d) Minimal organic reduction.

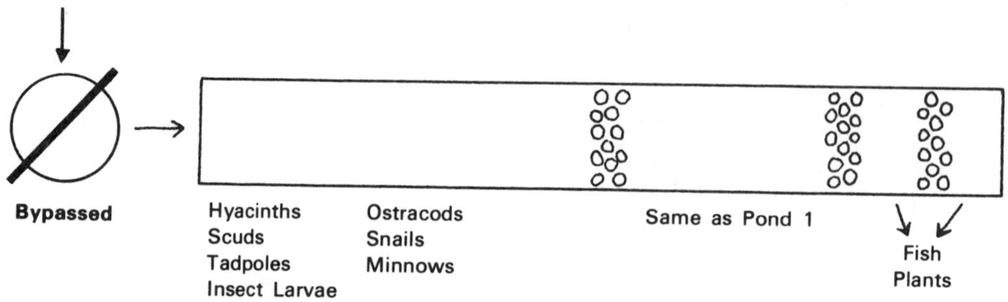

FIG. 25-1. Diagram of proposed 5-step biological treatment system.

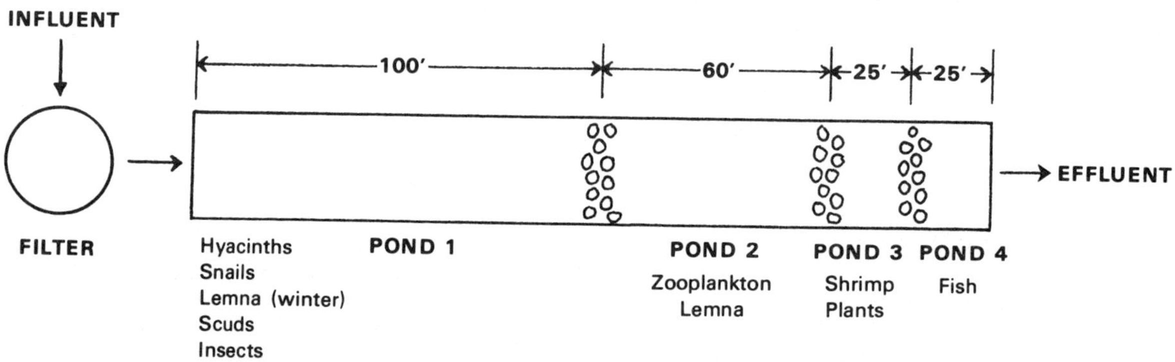

FIG. 25-2. Actual installation of 5-step biological treatment system. (See Fig. 25-1).

First Pond – 2 Feet Deep

Water hyacinths (*Eichhornia crassipes*) and duckweed (*Lemna* spp.) will be grown. Hyacinths will dominate during most of the year and duckweed will grow during winter season when hyacinths are dormant. Scuds (*Hyalella azteca*), commonly associated with hyacinths, will be abundant. Snails (*Physa* spp.) and midge fly larvae (*Chironomus* spp.) will also be present in high numbers. The three rock barriers will likely be infested with scuds, snails and midge larvae.

Expected Results: a) Mineral nutrient uptake.
b) Reduction of hardness and total dissolved solids.
c) Denitrification in bottom debris layer.
d) Suppression of algae growth.
e) Minimal organic reduction.

Second Pond

This unit will be devoted to culture of zooplankton. A 900 square foot area in the midsection of the pond will be 8 feet in depth and the remainder of the pond will be 2 feet deep. The deep area is provided to retard pH fluctuations and will be mixed using an aerator (airlift pump) supplied with 2 ft^3/min of air. Duckweed will cover most of pond surface to restrict algae growth. Effluent from pond will be recirculated to filter.

Expected Results: a) Clarification.
b) Organic reduction.
c) Coliform bacteria reduction.
d) Mineralization and nutrient mineral uptake.

Third Pond – 2 Feet Deep

Glass shrimp (*Palaemonetes* spp.), which are known to reproduce in a pond environment and have been observed to be adept predators on small invertebrates, will be cultured in this unit. Limited investigation to determine cultural methodology for the asiatic clam (*Corbicula manillensis*), which is abundant in area streams, will be conducted.[6] Asiatic clams are able to remove suspended particles rapidly and efficiently from water by filtration and deposition of pseudofeces. Bench tests have revealed that asiatic clams cannot be successfully cultured in stagnant enriched waters because deposited pseudofeces decay and deplete oxygen and carbon dioxide from clam respiration accumulates. To combat this problem, a flume structure will be constructed in the pond. Clams will be supported on a metal mesh above the bottom of the flume, and water will be pumped through the unit with flow directed near the bottom to carry away carbon dioxide and remove pseudo feces.

Post larvae of giant fresh water shrimp (*Macrobrachium rosenbergi*) will be introduced to pond to evaluate culture potential. A *Macrobrachium* from the Amazon Basin that readily reproduces in fresh water and is of a desirable size for fish bait may also be placed in pond.

Water primrose (*Ludwigia repens*) and perhaps other suitable species of attached plants will be provided to serve as cover for shrimp and to absorb residual nutrient salts.

Expected Results: a) Biomass conversion to larger life form.
b) Nutrient uptake.
c) Settling of suspended solids.
d) Minimal organic reduction.

Fourth Pond – 2 Feet Deep

Golden shiner minnows (*Notemigonus crysoleucas*), fathead minnows (*Pimephales promelas*), and goldfish (*Carassius auratus*) are used extensively in the region as bait fish. All are tolerant to somewhat extreme environmental conditions and are effective predators upon small invertebrates. Goldfish will feed upon particles of plant material that may enter the unit.

Expected Results: a) Biomass conversion to larger and higher life form.
b) Facilitation of biomass harvest.
c) Settling of suspended solids.
d) Minimal organic reduction.

TABLE 25-3. *Performance of the 5-Step Biological Treatment System, from June to November 1975.*
The theoretical detention time is 5.3 days.

Wastewater Constituent	Influent mg/l	Effluent mg/l	Percent reduction of influent concentration
BOD_3	15	3.5	77
BOD_{20}	90	18	76
COD	70	40	43
Suspended solids (total)	35	7	80
Total organic nitrogen	4.8	1.2	75
Ammonia	2.1	0.1	95
Chlorophyll a	0.23	0.019	92
Fecal coliform/100 ml	1400	10	99

Discussion

It has been domonstrated that a *Daphnia*-dominated zooplankton culture with environmental control can be maintained to effectively remove planktonic algae from water. Mineralization is also enhanced by zooplankton culture. The proposed investigation includes plants for uptake of mineral nutrients and extends the process to production of vertebrates that can be easily harvested and readily marketed. The study, if successful, will demonstrate the possibility for by-product production to offset treatment costs with water being available for irrigation or industrial use. Results of the brief field study on the effect of hyacinths on water quality should certainly be viewed as being tentative. The mineral reduction capability of hyacinth culture should receive further careful study as an economical biological method for removal of soluble salts from solution.

Addendum

During the spring, summer, and fall of 1975, the proposed 5-step biological treatment system was constructed and used to treat effluent from a stabilization pond. The proposed distribution of plants and animals (see Fig. 25-1) was altered somewhat in the actual installation (Fig. 25-2). Various water quality parameters were determined for wastewater entering and leaving the system between June and November. The results, shown in Table 25-3, are encouraging and indicate that the biological treatment system is functioning as expected.

Notes

1. I wish to express my appreciation to Henry L. Dabney, P. E. Division Director, for his past and continuing support, and to my fellow employees and the many other persons who have contributed to the studies reported upon.

2. R. Dinges, *Ecology of Daphnia in Stabilization Ponds* (Austin: Texas State Department of Health, 1973).

3. R. Dinges, "The Availability of Daphnia for Water Quality Improvement and as an Animal Food Source," *Proceedings of Conference on Wastewater Use in the Production of Food and Fiber,* Series EPA 660/2-74-041 (Washington, D.C.: U.S. Government Printing Office, 1974), pp. 142-61.

4. R. Dinges, "Biological Considerations in Stabilization Pond Design" (Paper presented to 56th Texas Water Utilities Short School, Texas A&M University, College Station, March 1974); R. Dinges, " 'Wee Beasties' May Improve Your Effluent," *Water and Wastes Engr.* 12 (1975): 35-37.

5. R. Dinges, "Biological Demineralization of Water" (Unpublished report, Austin: Texas State Department of Health, 1973).

6. R. Dinges, "Asiatics Invade Texas," *Water—Southwest Water Works J.* (June 1973), pp. 7-8.

26

Sewage Treatment by Controlled Eutrophication Using Algae and *Artemia*[1]

NORMAN M. TRIEFF AND REBECCA HINTON

Department of Preventative Medicine and Community Health

University of Texas Medical Branch

GLEN J. STANTON

Department of Microbiology

University of Texas Medical Branch

J. GLEN SONGER

Department of Veterinary Microbiology and Preventative Medicine

Iowa State University

DOV GRAJCER

Aquaculture Consultant

Menlo Park, California

The inadequacy of waste treatment and of disposal of improperly treated waste causes public health problems and endangers the environment through the presence of bacterial or viral pathogens or chemical toxins. Most conventional secondary treatment facilities are inadequate in terms of removal of phosphate, nitrate, and organics. The presence of high levels of phosphate, nitrogen, and other nutrients in both the untreated wastewater, and even in conventionally treated effluent from a biological secondary treatment plant, leads to eutrophication in lakes, rivers, and bays. Furthermore, the residual bacterial sludge from an activated sludge plant presents an expensive solids-handling problem that few communities have solved properly because of expense and inadequate technology. Finally, recent studies by the Environmental Protection Agency on drinking water in both New Orleans and Miami suggests that chlorination of sewage effluent containing organics may produce chlorinated hydrocarbons such chloroform, which may be carcinogenic to humans or, at the very least, deleterious to their health.

Recently in our laboratory a controlled eutrophication system has been developed on a laboratory scale and given a preliminary evaluation.[2] It is similar to many aquaculture, lagooning, or oxidation pond-type systems. It involved the use of the salt water alga *Tetraselmis chui* and brine shrimp *(Artemia salina)* for controlled eutrophication of raw sewage. In this system, algae fix the organic and mineral components in a photosynthetic process using an artificial light source, and the brine shrimp graze on the algae thus keeping them at an optimal level of log-phase growth that does not exceed the nutrient level present in the sewage. The beneficial products of the system are (1) a purified effluent and (2) brine shrimp that could be used as food for fish or shrimp in a mariculture operation.

The changes in physicochemical parameters in a static system[3] and the passage of bacterial pathogens[4] have been studied. Presently, the passage of viral pathogens is being investigated and some preliminary results have been obtained.[5] It is the purpose of this paper to summarize the studies already completed, describe the work currently underway and unreported and indicated the implications and potentialities of this method.

Materials And Methods

Materials

Instant Ocean (IO) (Aquarium Systems, Inc., Eastlake, Ohio 44094) was used instead of natural sea water and mixed to a salinity of 33 parts per thousand (ppt). It was made as needed and poured into a 20 liter pyrex carboy for use.

Tetraselmis algae. The marine alga *Tetraselmis chui* was obtained from the National Maine Fisheries Service, NOAA, Dept. of Commerce, Galveston, Texas, courtesy of Ms. Loretta Ross. The algal cells (approximately 75,000) were cultured in about 60 ml of NH-15 medium of Wilson. When the log phase of growth occurred, the algae were transferred to a liter of IO, supplemented nutritionally according to Ross (unpublished data). Following sufficiently high algal growth, the algae were diluted to 25,000/ml in a total volume of 9 liters. A rubber stopper containing 40 cm length of glass tubing was inserted into a 2.5 cm hole in the bottom of the carboy. The tube provided for an automatic overflow and rose to a point where 9 liters were retained. The algae were continuously aerated and at the appropriate time placed in the continuous flow system (Fig. 26-1).

Wastewater. The raw and treated wastewaters were obtained from the main sewage treatment plant of Galveston, Texas.

Brine shrimp *(Artemia salina)*. Dry eggs of *Artemia salina* (California Brine Shrimp, Inc., Menlo Park, Ca. 94025) were hatched in a funnel using a method described by Sorgeloos.[6] The funnel was filled with 1 liter of IO and the brine shrimp eggs were added. Air, continuously bubbled through the bottom of the funnel, provided the agitation to keep the cysts in motion and prevented them from settling out. This procedure appears to result in a higher hatching rate.[7] The *Artemia* were hatched in the light on the basis of Sorgeloos' discovery of improved hatching efficiency by light.[8] While hatching, the nauplii (freshly hatched eggs) were provided with 50 ml of *Tetraselmis* algae culture. The positive phototactic nauplii were separated from the unhatched cysts by directing a light beam on the hatching device[9]; they swam towards it and were easily separated from the cysts, and were then placed in a large aquarium tank (33.5 cm [width] x 77 cm [length] x 31.4 cm [height]) filled with IO and *Tetraselmis* algae and covered with a plexiglass sheet. Air was continuously bubbled into the tank.

Lighting. The tank was illuminated by three fluorescent lights in parallel, placed directly overhead, on a 12-hour light-dark cycle. The cycling had two functions: (1) it more closely simulated actual outdoor lighting than a continuous lighting system, (2) it aided brine shrimp growth since it has been found that too much light retards growth of *Artemia*.[10]

Analytical Methods

The pH was measured with a Beckman pH meter.

Total suspended solids (TSS) were determined by the aluminum dish method.

All other analytic methods were performed according to "Standard Methods."[11]

Static system. Treatment schedules developed during earlier studies[12] were observed using breakers and are schematically represented as follows:

$$\text{92\% raw waste} \xrightarrow[\text{growth-24 hr.}]{\textit{Tetraselmis} \text{ algae}} \quad (1)$$

$$\text{effluent (1)} \xrightarrow[\text{48 hours + decantation}]{\textit{A. Salina}} \quad (2)$$

$$\text{effluent (2)} \xrightarrow[\text{growth-24 hrs.}]{\textit{Tetraselmis} \text{ algae}} \text{final effluent} \quad (3)$$

Flow system. The continuous flow system depicted in Fig. 26-1 was used to study bacterial pathogen flow.[13] The pump indicated is a Buchler Polystaltic pump with 1/16'' (internal diameter) tygon tubing. The flow rate was set between 4 and 9 liters per day.

Bacterial Sampling. In this study, four sets of samples were taken at three stages—raw settled sewage, algal culture, and effluent—over a period of two months and subjected to the various tests. For bacterial analysis of brine shrimp, 100 were collected, macerated, and the tissue homogenate diluted with sterile distilled water to the required volume. All data were analyzed by analyses of variance and the multiple-range test. Except where otherwise stated, incubation was at 37°C for 48 ± 3hr and all bacteria growth media were from Baltimore Biological Laboratories, Cockeysville, Md.

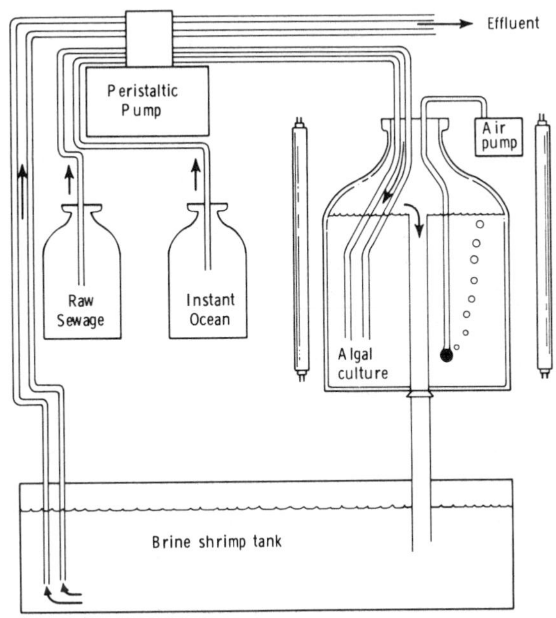

FIG. 26-1. Schematic diagram of Continuous-Flow System for Controlled Eutrophication of Sewage (Smith, Songer, and Trieff, "Sewage Treatment"). Raw sewage and IO are pumped into the algal culture, displacing algae into the brine shrimp tank. An equal volume of the system's effluent is pumped out of the brine shrimp tank.

Bacteriological methods. A coliform most probable number (MPN)/100 ml was determined by the multiple-tube fermentation technique using lauryl sulfate broth. Transfers were made from tube cultures showing bacterial growth to Enteroccosel (EC) broth. These were incubated at 44.5°C in a water bath. Bacterial growth and gas production in 24 hr indicated the presence of fecal coliforms and confirmed the MPN.

Azide dextrose broth served as a primary medium for enterocci yielding an MPN/100 ml. Bacteria from positive cultures were transferred to ethyl violet azide broth. Turbidity and the appearance of a purple button constituted a positive test; streaks were made to Columbia CNA blood agar (composed of the Columbia CNA agar base containing colistin and nalidixic acid with 5% sheep blood added). Typical colonies were picked and subcultured in EC broth, where blackening of the medium in 24 ± 2 hr indicated a completed test and confirmation of the MPN.

Growth of bacteria in Selenite-F broth was considered a presumptive test for *Salmonella* or *Shigella;* streaks were made to MacConkey's agar. Typical colonies were picked to Kligler's iron agar slants and incubated for 18 hours. Cultures giving reactions typical of *Salmonella* or *Shigella* were subjected to biochemical testing in the API-20 profile recognition system (Analytab Products, Inc., New York, N.Y.). A correct profile confirmed the presence of *Salmonella* or *Shigella.*

Enrichment cultures for *Vibrio parahaemolyticus* and *V. alginolyticus* (alkaline peptone broth from Difco Laboratories, Detroit, Mich.) showing bacterial growth were streaked to TCBS agar (BBL). Appearance of typical colonies in 24 hours indicated the presence of *Vibrios.* Suspect colonies were subjected to biochemical tests, again using the API-20 profile recognition system. A solution of NaCl (3%) was used as diluent. A correct profile confirmed the presence of *V. parahaemolyticus* (#4346106) or *V. alginolyticus* (#4146124).

Viral Methods

Viral Growth and Preparation. The virus used in this study was attenuated poliovirus MEF-1, grown in U-cells (human amnion) for 10 hours; the culture medium was then poured off and centrifuged in a Sorvall RC-2 centrifuge (head #GSA-3745) at 5000 rpm for 30 min. The supernate was decanted and dispensed into 17 x 100 mm tubes (Falcon #2057, Falcon Plastics, Oxnard, Ca., 93031), which were frozen until needed.

Plaque Assay for Polio Virus. The growth medium for the cells was MEM-Hanks (GIBCO, Grand Island, N.Y.) supplemented with 10% normal calf serum and 10 ml of 5.6% sterile bicarbonate. The maintenance medium was MEM-Earle (GIBCO, Grand Island, N.Y.) supplemented with 2% normal calf serum and 20 ml of 5.6% sterile bicarbonate.

Cells were grown in 25 cm^2 tissue culture flasks (Falcon #3013). The same flasks were used for growing cells and for the plaque assays.

Plaque assays were run according to the methods of Hsiung[14] with modification of Stanton[15] and Pagano.[16]

Virus Inactivation Studies. The inactivation of virus was determined in isotonic (0.85%) saline, IO, raw sewage, and algae. In future work inactivation studies will be performed in a tank of algae plus *Artemia* as well as on the flow system in operation. In each case 100 ml of the sample was innoculated with 1 ml of virus. Two ml samples were removed at 0, 1, 2, 4, 8, and 16 hours and frozen in 12 x 75 mm tubes at -20°C until assay. Plaques were counted after 48 hr.

Results

Table 26-1 shows the effect of brine shrimp on water quality parameters of waste seeded with *Tetraselmis.*[17] Recall that these results pertain to a static system. It may be noted from Table 26-1 that *A. Salina* (1) caused a substantial drop in total suspended solids (TSS), turbidity, and NH_3, (2) led to a slight drop or no change in PO_4^{-3}, NO_3^-, NO_2 and pH (in some cases a rise was noted), (3) led to an increase in biochemical oxygen demand (BOD) in all cases.

To reduce the BOD further an additional treatment with algae after brine shrimp treatment was considered necessary. This is described schematically under "Static System" in "Materials and Methods."

The results of such a three-stage process are summarized in Table 26-2.[18] The results of the bacterial pathogen transport studies are summarized in Table 26-3. And the results of polio viral deactivation by components of the system appear in Table 26-4.

Discussion

Physicochemical parameters

As noted in the results obtained from static experiments (Table 26-1), the system: sewage → algae → effluent is partly successful in that the parameters such as TSS and turbidity drop. However, others such as PO_4^{-3}, NO_3^-, NO_2^-, and NH_3 drop only slightly, while the major drawback is in the rise of BOD_5. When the third stage of additional algae treatment was added, as shown in Table 26-2, a significant drop in BOD_5, PO_4^{-3}, NO_3^-, turbidity, and odor also occurred, resulting in a fairly pure effluent. It must be emphasized that this is a static system and these experiments should be repeated in a flow system.

Bacterial pathogens

Previous results using the flow system of Fig. 26-1 demonstrated a significant reduction in number of fecal coliforms on the passage of raw sewage through the system (raw sewage - effluent difference was significant at $p < 0.01$).[19] Similar patterns were evident for the enterococci (raw sewage - effluent difference was significant at $p < 0.01$). Neither *Salmonella, Shigella,* nor *Vibrio parahaemolyticus* was detected at any stage. The detection of *V. alginolyticus* in three of four samples of effluent was unexpected, but was explainable on the basis of the proximity of the sewage treatment plant to Galveston Bay. The presence of *V. alginolyticus* may be due to occasional infiltration of seawater into the raw sewage lines. It is doubtful that significant numbers of this organism would be present in raw sewage (or the algal culture) due to human inte-

TABLE 26-1. *Effect of Brine Shrimp (BS) on Water Quality Parameters of Waste Seeded with* Tetraselmis.*

	Algae only (A)				Algae + Raw Wastewater (RW)				Algae + Treated Wastewater (TW)				Algae + Filtered Wastewater (FW)			
	67% H_2O + 33% A		92% H_2O + 8% A		67% RW + 33% A		92% RW + 8% A		67% TW + 33% A		92% TW + 8% A		67% FW + 33% A		92% FW + 8% A	
Parameters (mg/l)	Before BS	After BS	Before BS	After BS	Before BS	After BS	Before BS	After BS	Before BS	After BS	Before BS	After BS	Before BS	After BS	Before BS	After BS
TSS	180	50	90	30	295	80	300	100	200	20	175	15	250	40	250	45
BOD_5†	—	20	—	—	—	36	4.0	24	—	36	2.0	40	—	40	2.0	50
PO_4^{3-}	7.5	6.0	7.1	7.0	10.0	7.5	8.5	5.0	27.5	12	20	10	12.5	7.5	7.5	5.0
NO_3^-	5.0	5.0	3.0	4.0	10.0	8.0	18.0	15.0	7.5	7.0	8.8	7.8	15	13	16.5	14
NO_2^-	0.001	0.002	0.001	0	0.018	0.012	0.02	0.01	0.004	0.00	0.002	0	0.012	0.008	0.014	0.006
NH_3	0.04	0.02	0.04	0.03	0.06	0.03	0.08	0.04	0.04	0.06	0.06	0.007	0.10	0.061	0.09	0.05
Turbidity	20	6.0	15	13	30	10	37	15	20	2.0	25	6.0	20	1.0	24	2.0
pH	8.8	8.8	8.4	8.4	8.2	8.0	8.2	8.4	8.0	7.8	8.0	8.2	8.4	8.0	8.4	8.6
Percentage survival of BS	—	80	—	80	—	80	—	70	—	80	—	80	—	80	—	90

*Brine shrimp were in contact with algae for 72 hours; *Tetraselmis* was allowed to grow for 24 hr after seeding with illumination (McShan, Triff, and Grajcer, "Biological Treatment.")

†Samples indicated with dash were not run.

TABLE 26-2. *Controlled Eutrophication of Wastewater with* Tetraselmis *and* A. salina *at 30°C.**

Parameters (mg/1)	Raw	Stage 1†	Stage 2‡	Stage 3§
TSS	120.0	45.0	20.0	3.5
BOD_5	318	5.5	23	4.5
PO_4^{3-}	20.0	16.5	10	4.8
NO_3^-	30.0	14.5	7.0	1.0
NO_2^-	0.02	0.008	0	0
NH_3	0.30	0.10	0.75	0.02
Turbidity	32.0	20.0	10	1.0
pH	7.02	8.75	8.73	8.74
Odor	Rank+	Fishy	Fishy–	None
Percentage survival of brine shrimp after 48 hr	20	50	70	90

*250 ml containers employed
†Stage 1 = Effluent from wastewater and *Tetraselmis*
‡Stage 2 = Effluent from Stage 1 + *A. Salina*
§Stage 3 = Effluent from Stage 2 + Tetraselmis

stinal discharge because of its low salinity. Upon mixing with IO (or seawater) the halophilic nature of the *Vibrio* causes it to multiply and thus be detected in both the effluent and the brine shrimp. The IO and brine shrimp eggs were both assayed for halophilic organisms and found to contain no *V. alginolyticus* (or *V. parahaemolyticus*).

The results in testing of *Artemia* and effluent for sewage-related organisms give further promise for the use of this system as a combined sewage treatment-mariculture facility. While a few coliforms, enteroccoci, and vibrios were found in the *Artemia,* they were most likely associated with the algae in their guts. Further study must be done to determine if vibrionic disease is produced in shrimp by feeding them *Artemia* from this system.

Viral Inactivation Studies

Our preliminary data on viral inactivation in Table 26-4 indicates that isotonic saline (0.85% NaCl) has essentially no effect in 16 hr on plaque-forming units of virus. The available data of IO shows a 99.7% inactivation of virus in 8 hr. For two samples of raw sewage the inactivation averaged 91.9% at 16 hr. For two samples of algae the inactivation averaged 62% also at 16 hr. No experiments have been done yet on inactivation of polio virus by brine shrimp. Clearly, there is significant variance, which hopefully can

TABLE 26-3. *Summary of Coliform and Enterococcus Data.**

Material	Coliforms		Enterococci	
	$\overline{X}$†	SE‡	$\overline{X}$	SE
Raw sewage§	7.138	0.627	4.251	0.081
Algae culture§	3.080	0.280	1.953	0.257
Effluent§	1.570	0.581	0.558	0.364
Brine shrimp[11]	0.571	0.344	0	0

*Smith, Songer, and Trieff, "Sewage Treatment."
†$\overline{X}$, Mean of log x values.
‡SE, Standard error of the mean.
§Per 100 ml.
[11]Groups of 100.

TABLE 26-4. *Deactivation Studies on Polio Virus* MEF-1 on Various Components.*

Time (in hours)	Isotonic Saline Blank	Plaque-Forming Units/ml x 10^{-5} Instant Ocean	Sewage 1	Sewage 2	Algae 1	Algae 2
0	- -	>10^3	5.4	5.7	2.8	1.7
1	- -	315	2.1	0.80	5.0	0.70
2	2	21	0.5	0.30	1.8	1.8
4	1.35	- -	0.25	0.58	0.50	0.60
8	2.6	3.25	0.37	0.55	1.7	0.30
16	1.85	- -	0.40	0.50	0.80	0.80

*Hinto, "Controlled Eutrophication."

be minimized in the future as more becomes known about the system as well as about the effects of relatively low dilutions of virus on plaque assays.

Nevertheless, if it is hypothesized that the total inactivation is obtained in a sequential way by multiplying the inactivating effects of each component we may say:

$Inact_{tot} = Inact_{IO} \times Inact_{sewage} \times Inact_{alg} = 0.00325 \times 0.081 \times 0.38 = 0.00050$

This represents an inactivation of 99.95% without considering any inactivation due to *Artemia.* It must be emphasized that no viral studies have yet been done on the flow system. If our hypothesis holds true, then we can expect almost complete inactivation by the controlled eutrophication system. With a second algal step, inactivation should certainly be 100%.

Chlorine, of course, has been used as a viricidal agent and can remove polio virus activity providing that pH, temperature, and length of exposure are optimized. Thus, for example, at a temperature of 0°C, pH = 8.5, and chlorine residual of 1.95 mg/l, there was found 99.6% destruction in 7.5 min.[20] C. G. Robeck found up to 99.9% removal of virus with the effectiveness of chlorine varying with dose and contact time.[21] In another study, ozone was found to be an even more effective viricidal agent with 0.05 – 0.45 mg/l of O_3 in 2 min destroying the same percentage of polio virus as 0.5 – 1.0 mg/l of residual chlorine in 1.5 – 3 hr.[22]

Conventional sewage treatment has also been studied by various workers as a means of removing polio virus.[23] Thus, primary sewage treatment (settling) reduced polio virus content from 29–63%, while activated sludge treatment reduced virus activity from 74–94%.[24] Other workers found reductions of either 90%[22] or 76–90%.[26] Aeration is necessary for virus removal in that it permits the activated sludge process to occur, although it is not directly involved in the mechanism for virus removal.[27] Robeck found that flocculation and settling removes 95–99% of polio virus with the completeness of removal varying with dose and time.[28] Stabilization ponds are erratic in virus removal as noted by H. I. Shuval: removal rates range from 0–96%.[29]

While preliminary results suggest that the removal of polio virus by our reported system is expected to be high, much further study needs to be done. The percentage removal by the sewage is probably mainly due to settling and is in line with results reported by other workers. Again, the removal of activity by the algae falls into the range given by Shuval.[30] The main inactivation appears to occur in the highly saline IO. No previous data have been found by us in the literature relating to inactivation of virus by either seawater or IO. No doubt, part of the mechanism of this inactivation must be the osmotic effect produced by immersion of the virus in the hypertonic solution, IO. A more detailed investigation of this mechanism is required.

Comparison to other controlled eutrophication systems

The uniqueness of the present system is not the sequence of sewage → algae → animal, but two factors: (1) the high salinity and salt water alga, *Tetraselmis,* and (2) the brine shrimp, *Artemia salina.* Other systems have been used: J. H. Ryther, a pioneer in the field, has used oysters as the animal.[31] Other animals used were catfish by E. E. Prather and H. S. Swingle,[32] carp by K. Kuronuma,[23] and goldfish by C. L. Chu and G. N. Breene.[34] One of the problems with oysters is their notorious ability to concentrate sewage-borne contaminants ranging from microorganisms to heavy metals.

Further study of the brine shrimp is needed with regard to virus transmission, parasites, heavy metals, and toxic organic chemicals before their use as food for fish and shrimp can be considered positively safe. In addition, it is necessary to know whether the effluent requires further purification such as chlorination before it can be safely discharged to coastal waters.

Notes

1. It is a pleasure to acknowledge the support of the James W. McLaughlin Fund for fellowships for J. Glenn Songer and Rebecca Hinton. We must gratefully acknowledge the work of Dr. Michael McShan for his preliminary physicochemical studies and to Dr. Rodney F. Smith and Ms. Sandra Dayton for their great help in the bacteriological studies.

2. M. McShan, N. M. Trieff, and D. Grajcer, "Biological Treatment of Wastewater using Algae and *Artemia*," *J. Water Poll. Control Fed.* 46 (1974):1742; R. F. Smith, J. G. Songer, and N. M. Trieff, "Sewage Treatment by Controlled Eutrophication: Bacterial Study," *Appl. Microbiol.* 28 (1974):359; J. G. Songer et al., "Controlled Eutrophication: Sewage Treatment and Food Production," In *Wastewater Use in the Production of Food and Fiber – Proceedings,* Series EPA 660/2-74-041 (Washington, D.C.: U.S. Government Printing Office, 1974); J. G. Songer, "Controlled Eutrophication with Brine Shrimp as the Herbivore" (Diss., Graduate School of Biomedical Sciences, University of Texas Medical Branch, Galveston, 1974).

3. McShan, Trieff, and Grajcer, "Biological Treatment."

4. Smith, Songer, and Trieff, "Sewage Treatment"; Songer et al., "Controlled Eutrophication"; Songer, "Controlled Eutrophication with Brine Shrimp."

5. R. A. Hinto, "Controlled Eutrophication of Sewage Using Algae and *Artemia salina*: Polio Virus Transmission" (Thesis in progress, Graduate School of Biomedical Science, University of Texas Medical Branch, Galveston, 1975).

6. P. Sorgeloos, "The Influence of Algal Food Preparation on Its Nutritional Efficiency for *Artemia salina L.* Larvae," In *Proceedings of the Conference on Marine Invertebrate Larvae, Rovini,* (September 1973).

7. P. Sorgeloos, *Technological Improvements for the Cultivation of Invertebrates as Food for Fishes & Crustaceans, 2. Hatching and Culturing of the Brine Shrimp,* Artemia salina L. (Charleston, S.C.: Mariculture Society, January 1974).

8. P. Sorgeloos, "First Report on the Triggering Effect of Light on the Hatching Mechanism of *Artemia salina* Dry Cysts," *Mar. Biol.* 122 (1973):75.

9. Sorgeloos, *Technological Improvements.*

10. P. Sorgeloos, "The Influence of Light and Dissolved Oxygen on the Reproduction of the Brine Shrimp *Artemia salina" (Paper delivered at the Eighth European Symposium on Marine Biology, Sorrento,* Naples, Italy,October 1973).

11. Water Pollution Control Federation, *Standard Methods for the Examination of Water and Wastewater,* 12th ed. (New York: American Public Health Association, 1965).

12. McShan, Trieff, and Grajcer, "Biological Treatment."

13. Smith, Songer, and Trieff, "Sewage Treatment"; Songer et al., "Controlled Eutrophication"; Songer, "Controlled Eutrophication with Brine Shrimp."

14. G. D. Hsiung, *Diagnostic Virology* (Connecticut: Yale University Press, 1964).

15. D. C. May, G. J. Stanton, and S. C. deSt. Jeor, "Enhanced Detection of Western Equine Encephalitis Virus Plaque Variants in Crowded Cultures and Plaque Progeny and Potential Use in Genetic Studies," *Appl. Microbiol.* 28 (1974):145.

16. J. S. Pagano and A. Veheri, "Enhancement of Infectivity of Poliovirus RNA with Diethylaminoethyl Dextran (DEAE-D)" *Archiv. fur Virus Forshung* 17 (1973):456.

17. Details are given in McShan, Trieff, and Grajcer, "Biological Treatment."

18. Abbreviated from *ibid.* where details are given.

19. Smith, Songer, and Trieff, "Sewage Treatment by Controlled Eutrophication."

20. P. W. Karber, "Viricidal Effects of Disinfectants in Water," *Publ. Health Rpts.* 76 (1961):565.

21. C. G. Robeck, "Effectiveness of Water Treatment Processes in Virus Removal," *J. Am. W. Works Assoc.* 54 (1962):1275.

22. J. F. Kessel, "Comparison of Chlorine and Ozone as Viricidal Agents of Poliomyelitis Virus," *Proc. Soc. Exp. Biol. & Med.* 53 (1943):71.

23. N. A. Clarke, "Removal of Enteric Viruses from Sewage by Activated Sludge Treatment," *Am. J. Public Health* 51 (1961):1118; W. M. Mack et al., "Enterovirus removal and heat failure by activated sludge treatment," *J. Water Poll. Control Fed.* 34 (1962):1133; B. England, et al., "Virologic Assessment of Sewage Treatment at Santee, California," in *Transmission of Viruses by the Water Route,* ed. G. Berg (New York: Wiley, 1967), p. 401; S. Kelly, "Removal of Enteroviruses from Sewage by Activated Sludge," *J. Water Poll. Control Fed.* 33 (1961):1056.

24. Clarke, "Removal of Enteric Viruses."

25. Mack, "Enterovirus Removal."

26. England, "Transmission of Viruses."

27. Kelly, "Removal of Enteroviruses."

28. Robeck, "Effectiveness of Water Treatment."

29. H. I. Shuval, *Developments in Water Quality Research* (Michigan: Humphrey Science Publishers, 1970), p. 47.

30. *Ibid.*

31. J. H. Ryther, "Nitrogen, Phosphorus, and Eutrophication in the Coastal Marine Environment," *Science* 171 (1971):1008; J. H. Ryther et al., "Controlled Eutrophication—Increasing Food Production from the Sea by Recycling Human Wastes," *Biosci.* 22 (1972):144. See also Chapter 23.

32. E. E. Prather and H. S. Swingle, "Preliminary Results on the Production and Spawning of White Catfish in Ponds," *Proc. 14th Am. Conf. S. E. Assoc. Game & Fish Commission* 14 (1960):143.

33. K. Kuronuma, "Carp Culture in Recirculating Tanks," *Proc. World Symp. Warm Water Pond Fish Culture* FAO Fish. Rept. 44 (1966):123.

34. G. N. Breene and C. L. Chu, "Experiments on the Use of a Biofilter to Remove Wastes from Fish Tanks," *Proc. 20th Am. Conf. S. E. Assoc. Game & Fish Commission* 20 (1967):446.

27

Interference by Blue-Green Algae with Nutrient Recovery in Water Quality Control Schema: Management Implications[1]

K. IRWIN KEATING

Department of Environmental Science
Cook College, Rutgers

Our capacity to control the food chain dynamics in nutrient-rich fresh waters becomes increasingly significant as the emphasis in sewage treatment shifts from waste disposal to resource recovery. Nutrient-rich natural waters are usually dominated by blue-green algae. From the viewpoint of a food chain culminating in harvestable fish the blue-greens are among the least desirable primary producers in that most, or all, are considered unsatisfactory food sources for zooplankton (i.e., for the trophic level just preceding the fishes). They become even less suitable for intentional cultivation when certain of their other characteristics are considered—e.g., many produce unpleasant odors and tastes in waters; some act as allergens, or toxins—for animals ranging from invertebrates to humans;[2] some have recently been implicated as possible vectors for viruses, although the capacity to harbor viruses may provide a means to control blue-green algae.[3] It is therefore essential to develop the management information and techniques necessary either to control, or to eliminate, the blue-green algae.

A recently concluded five-year project encompassing both laboratory and on-site study has provided strong evidence that the competitive advantage of the blue-green algae is based at least in part on an allelopathic capacity (i.e., the blue-greens manufacture metabolic products that interfere with the optimum development of other, more desirable, algae). The bloom sequence of a highly eutrophied (excess N, P, C) pond in Connecticut was monitored. Dominant organisms (consistently blue-green algae) were isolated from the pond. Bacteria-free, and/or unialgal, cultures of each were established. Biologically active substances that were readily destroyed by heat were found in the cell-free filtrates of cultures of each of the blue-green bloom dominants. The biological activity of each of these filtrates was tested against each of the other bloom dominants and against a series of diatom cultures that had also been established with organisms isolated from the same pond. Diatoms were chosen as assay organisms because a) they are a desirable element in the food chain of fresh water ponds (b) they are the most common bloom producing organisms in less nutrient rich environs, and (c) the basis for the change from diatom to blue-green dominance in blooms as nutrient levels rise was unknown.

A pattern of growth effects wholly consistent with the bloom sequence of the pond was determined. That is, organisms preceding a given bloom produced fertilizing (or neutral) effects on the bloom organism, and organisms following a given bloom produced growth inhibiting (or neutral) effects on bloom organisms.[4] However, of greater significance to the management of biological water reclamation schema is the observation that the filtrates of all of the nine blue-green bloom organisms tested exerted a generalized inhibitiory effect on diatoms. This effect was most pronounced in tests of bacteria-free blue-greens against bacteria-free diatoms, but filtrates of bacterized blue-green cultures, tested against bacteria-free or bacterized diatoms, evidenced similar, though less pronounced, inhibitory effects. This diminishing of inhibition strength is to be expected since bacteria, given sufficient time, are notoriously able to break down metabolic compounds.

Population densities of sets of filtrate producing blue-green cultures were monitored. These generally ranged from less than one to several times those of natural blooms. This point is critical to the legitimacy of transfer of information obtained from laboratory studies to the understanding of a natural phenomenon, for it assures that there were no problems of dilution to interfere with interpretation. An additional, frequently cited, problem with such laboratory work is the possibility that pH differences, rather than allelopathic effects, might cause the critical growth differences. The pH levels in test and control cultures were identical. Preliminary chemical studies of the substance responsible for diatom inhibition suggest a heat-labile, dialyzable substance that retains activity after ether extraction and concentration.

These laboratory results involving blue-green inhibition of diatom growth correlate with events in the pond. The yearly appearance of spring diatoms (this is the most common time of appearance for diatom blooms in moderately eutrophied fresh water ponds) was found to vary inversely with the density of the blue-green population throughout the immediately preceding winter. In those years when, after a heavy blue-green winter bloom, no diatom bloom appeared in the spring, blue-green blooms continued to dominate the pond through spring and summer. In those years when no heavy winter blue-green blooms occurred, the commonly anticipated spring diatom bloom appeared. After these spring diatom blooms ended, blue-green bloom populations once again developed in the warm, nutrient-rich, summer waters of the pond. Since diatoms are usually considered to be competitively favored in early spring by both nutrient and physical conditions, it is reasonable to expect that the flourishing diatom blooms would out-compete any embryonic blue-green bloom population until such time as the nutrient or physical conditions in the pond weakened the dominant diatoms; however, the possibility of heretofore unknown allelopathic effects being involved here, as elsewhere, must not be summarily dismissed. As the previously noted results of filtrate studies[5] provide a basis for the involvement of allelopathic effects in the patterning of natural blooms, this generalized inhibition of diatoms by blue-green, cell-free filtrates similarly provides a basis for allelopathic involvement in the change from diatom to blue-green dominance in general as the nutrient levels in a pond rise.

The spring diatom blooms ended at times when silica available to the blooming diatoms was depleted. Bioassay of pond waters collected before and after diatom blooms indicated that silica addition alone would significantly increase diatom growth. This suggests the possibility of increasing the period of diatom dominance while forestalling any blue-green bloom occurrence by providing additional silica (an element commonly included in detergents and of exceptional fertilizer value only to diatoms and to a few other desirable forms). If this treatment were combined with an effort to diminish blue-green winter blooms, diatom dominance with its concomitant strengthening of the food chain might be significantly enchanced.

The possibility of controlling the growth of undesirable organisms in sewage lagoons or water reclamation ponds increases with each piece of additional information concerning their capacity to dominate such nutrient-rich waters. Rational design of optimal biological wastewater treatment nutrient recovery systems can only be achieved with a thorough understanding of the essential bases for a strong, pre-planned, food chain.

Notes

1. Work done at Haskins Laboratories, Inc., Yale University; supported by EPA Research Grant RA801387.
2. D. Schwimmer and M. Schwimmer, "Algae and Medicine," in *Algae and Man,* ed. D. Jackson (New York: Plenum, 1964) pp. 368-412.
3. R. Safferman and E. Morris, "Control of Algae with Virus," *J. Am. W. Works Assoc.* 56 (1964):1217-24; M. Shilo, "The Ecology of Cyanophages," *Proceedings: International Association for Theoretical and Applied Limnology* 18 (1973):1882-85.
4. K. I. Keating, *The Participation of Algal Extracellular Metabolites in the Determination of Bloom Sequence in Linsley Pond, North Branford, Connecticut* (New Haven: Yale University, 1975).
5. *Ibid.*

28

The Use of Bacteria to Reduce Clogging of Sewer Lines by Grease in Municipal Sewage

NAZIR BAÏG
EDWARD M. GRENNING

The Maryland-National Capital Park and Planning Commission
Silver Spring, Maryland

Introduction

The flow in a sewer system generally consists of domestic, commercial, and industrial wastewater, and water from infiltration and inflow. In an urbanizing area largely residential in character, domestic and commercial effluents would be expected to dominate the wastewater component of flows in the sewer system. Prior to 1940, residential sewage basically consisted of human waste, cellulose, paper, and water. The development of garbage disposals and the greatly increased use of cosmetics, various petroleum products, and detergents has caused existing wastewater to contain large amounts of grease materials. For example, the installation of a garbage disposal unit in a residence increases the grease content of its waste water by about 35%. Growth in the food franchising industry, with the resultant increased number of restaurants, is estimated as also having substantially increased the input of grease to the sewer system. These greasy materials build up on the inner surfaces of the sewers a coating of grease with embedded mineral particles and eggshell fragments, thus reducing the effective capacity of the sewers to carry their design flow. This effect is believed to be a major contributor to sewer backup, overflow, foul odors, and related pollution problems, particularly at locations in the system where flow velocity is low.

Based on annual averages, infiltration and inflow is a second major cause of raw sewage overflows and related environmental problems in the Montgomery and St. George's Counties, Maryland (Bi-County), sewage collection

TABLE 28-1. *Annual Maintenance and Cleaning Cost for Sewers.*

Location	Total Milage in System	Year of Data	Maintenance Cost Cents/Ft.
Phoenix, Arizona	1747	1966–1967	5.25
Richmond, Indiana	156	1967	4.58
Knoxville, Tenna.	191*	1967	9.93*
Walla Walla, Washington	72	1967	3.39
South Bend, Indiana	275*	1967	7.7*
Columbus, Ohio	1678*	1967	11.72*
Levittown, Pa.	224*	1967	21.78
Allegheny Co. Sanitary Authority	69	1960	36
East Bay	21	1967–1968	2.16
Philadelphia, Pa.	2463*		45.07*
Decatur, Illinois	48	1967	2.96
Estimated Washington D.C. †			25.00*

*Comparable to Washington Metropolitan Area in labor cost.
†Based on averaging maintenance cost of Philadelphia, Levittown, South Bend, Knoxville, and Columbus and adjusted to 1973 prices.

system. Infiltration and inflow consists of water that enters the collection system either through damaged pipes and pipe connections or directly via downspouts and other storm water conveyances. This component of the flow is strongly influenced by a number of external variables, such as weather conditions, local soil permeability, and the like. Experience with similar municipal sewer systems indicated that 50 to 80% of the average total flow is probably due to infiltration and inflow.[1]

The potential for raw sewage overflows is particularly great when both grease accumulation and damaged pipes exist at the same location in the collection system.

Sewer System Improvement

Minimizing the two collection system inefficiencies described in the previous section would very likely eliminate existing raw sewage overflows and prevent the occurrence of new overflows. Section 201–(3) of the Federal Water Pollution Control Act Amendments of 1972 (Public Law 92-500), specifically identifies the reduction of infiltration and inflow on a system-wide basis as being a necessary prerequisite for the approval of grants for the construction of new treatment works.[2] Section 201–(4) of this Act provides Federal financial assistance to specifically deal with infiltration/inflow problems in existing sewer systems. The reduction of infiltration and inflow and the cleanup of accumulated grease in existing sewer systems are operationally related and should be pursued simultaneously. Clean, grease-free sewers are highly desirable for efficient detection and location of infiltration and inflow, using modern, closed-circuit television observation techniques.

Traditional Technique

Traditionally, cleaning of sewers is accomplished mechanically, using rodding and bucket machines, high velocity waterjets, and hydraulically propelled tools and devices.[3] While these techniques are effective, they are generally unpleasant, expensive, and, in some cases, hazardous to sewer workers. Table 28-1 presents annual sewer maintenance and cleaning costs based on a recent EPA study of the sewerage systems of twelve metropolitan areas.

TABLE 28-2. *Cost Comparison of Clean Up Techniques.*

Data Available	
1. Length of sewage system served by Washington Suburban Sanitary Commission	3,000 miles
2. Total amount of sewage generated	150 mgd
3. Cost of bacteria cultures	$6.00/pound
Assumptions	
1. Rate of Dosage—Manufacturers suggested rate of dosage. These are conservative figures, and in practice, lesser amounts have been used.	
a. Cleaning	
First two days—7 lbs/day per 100,000 gallons	
Next five days—2 lbs/day per 100,000 gallons	
b. Maintenance	
0.75 lbs., 3 times each week per 100,000 gallons	
2. Total number of manholes in the system	50,000 approximately
3. Rate of common labor	$4.00 per hour
4. Rate of pickup truck	$2.00 per hour
5. Manhole dosage time	15 minutes/manhole
6. Manhole fumigation	$1.00 per manhole
7. Travel time per day	2 hours
8. Number of days to innoculate ½ of all manholes $\frac{(15 \times 25,000)}{(6 \times 60)}$	1000 days
9. Estimated cost for cleaning sewers by traditional means. (See Table 1)	$0.25 per foot
Alternate A	
Annual Cost of Cleaning Sewers by traditional means	
i) Cleaning cost at 25 cents/ft. 0.25 x 3000 x 5680	= $3,960,000
ii) Manhole Fumigation to control roaches	$50,000
iii) Water for Periodic flushing quantity not available but very substantial	—
Total Cost:	$4,010,000
Alternate B	
Annual Cost of Cleaning Sewers by bacteria culture	
a) Materials	
i) 1st 2 days—7lbs/day/100,000 gallons $\frac{150,000,000 \times 7 \times 2 \times 6}{100,000}$	= $126,000
ii) Next 5 days—2 lbs/day/100,000 gallons $\frac{150,000,000 \times 2 \times 5 \times 6}{100,000}$	$90,000
iii) Thereafter (51 weeks)-3/4 lbs. 3 times/week/100,000 gallons $\frac{150,000,000 \times 0.75 \times 3 \times 6 \times 51}{100,000}$	$1,032,750
b) Labor Cost: 1000 x 8 (4+2)	$48,000
Total Cost Material and Labor	$1,296,750

The above estimate is quite conservative since:

a) Dosage of bacteria culture is on the conservative side.

b) Computations are based on the assumption that the total flow of 150 mgd will be flowing in each segment of the system.

TABLE 28-3. *Survey of Bacteria Users.*

Name & Address of Company	Population Served	Area	Total Length of Sewerage System	Type of System	Capacity	Comments
Ralph Shook Yorba Linda Water Dist. Yorba Linda, Calif. 92686 (714) 528–6231	2,500	N/A	N/A	Collect & Transport to Regional System	N/A	Bacteria cultures were successfully used to remove localized grease build-up.
Clearance Cardener Water & Sewer Super. 80001 Ralston Road Arvada, Colo. 80002 (303) 424–6441	70,000	Arvada, Colo.	280 miles	Secondary with chlorination. Trickling filter with 2-stage digester.	1 mgd	Symptoms indicating inefficient digester operation were: a) Inadequate production of methane; b) 4–6 ft. blanket in the digester; c) Sludge had excessive amounts of organic & volatile contents. Within 30–40 days gas production doubled, tripled in 60. Also there was a tremendous improvement in the sludge.
Wayne Coloney, P.E. P.O. Drawer 3966 Tallahassee, Fla. 32303 (904) 385–8171	10,000	City of Perry, Florida	20–30 miles	Primary	.75 mgd dry day 1.25 mgd wet day (Badly infiltrated system)	This system operated at 28–30% BOD_5 & suspended solids (SS) removal. In the first year of bacteria use BOD & SS removal increased to 50% to 70%, the second year to 75%. A polishing pond increased this to 95%. Conventional treatment would have cost $300,000 while the pond cost $22,000 and $200/month for bacteria.
Morris C. Allen Tahoe City Public Utility Tahoe City, California 95730 (916) 583–3796	25,000 Average 75,000 Peak	Tahoe City	City	Primary	3.5 mgd	This plant uses 1.5 lbs. of bacteria culture per day, costing about 2.5 cents per million gallons. Tremendous decrease in odor, grease, solids, and BOD was observed.
Russ Bond District Supt. Laguna City Sanitation P.O. Box 1068 Santa Maria, Calif. 93454 (805) 925–1475	19,000	Laguna County	N/A	Secondary, filter	1.25 mgd	One pound of bacteria culture per day was used in the digester to remove grease and scum.

TABLE 28-3 (cont.).

Name & Address of Company	Population Served	Area	Total Length of Sewerage System	Type of System	Capa-city	Comments
Paul Cygan, Chief Bureau of Sewers Dept. of Public Works City of Erie Erie, Penn. 16512 (814) 456–8561	180,000	Erie & outlying community	N/A	Activated Sludge	65 mgd	Bacteria cultures have been used very successfully for several years in clean up of lift stations, scum wells and lines, skimmer lines, primary and aeration tanks. Lift station energy requirements reduced by factor of 3 and offensive odors eliminated.
Charles Caswell President Environ. Audit Corp. 5100 Center Avenue Pittsburgh, Penn. 15232 (412) 682–1031	N/A	Chicago	N/A		.108 mgd	Bacteria cultures were used primarily for removal of grease & industrial waste from oil tankers. Treatment was 1 lb. every other day, gradually reduced to 1 lb. every week. Prior to use, grease content was 200–300 ppm, after application oil content dropped to 0–3 ppm.
W. E. Cochrane Trussville Sewage Treatment Plant RT. 2, Box 1-A Trussville, Alabama (205) 655–3617	3,000	Trussville	N/A	Bowl Filter	.4 mgd	The city's old filter was malfunctioning with excessive pounding. Bacteria cultures were added and significant improvement was noted within six weeks, completely clearing the filter and stopped the pounding. In thirteen weeks the digester began to produce methane gas.

Based on these data, the estimated sewer cleaning cost for the Washington, D. C., metropolitan area is about $0.25 per linear foot of pipe.

Over the years, a number of fatal accidents involving sewer workers being overcome by toxic gases (usually hydrogen sulphide gas, i.e., H_2S) during cleaning operations have occurred. For example, as recently as December 1973 three workers were killed in Newark, New Jersey, in a sewer cleaning operation.[4]

Cleanup With Bacteria Cultures

Grease-consuming bacteria have been cultured and produced in large quantities for use in the cleanup of oil spills. Recently, bacteria of this kind have been used successfully for sewer system cleanup and improvement in a number of municipal situations. Commercially available bacteria cultures are produced by starting with a small colony of suitable organisms that is provided with an ample supply of food (i.e., grease) and allowed to multiply. Food is then withheld for a period of time, causing the "weaker" individuals of the colony to die of starvation. After several feed and starvation cycles, a very hearty colony of bacteria is created, freeze dried with its culture medium (food), and packaged. The physical form of the packaged, freeze-dried bacteria culture resembles salt or sugar.

About 30 to 45 minutes before application the culture is mixed with water, thus reactivating the bacteria colony, which also has sufficient food to survive. The solution containing the activated bacteria is added to sewer mains and laterals via manhole covers. The bacteria find an ample supply of grease in the sewerage system and their population increases rapidly as grease is consumed. Also, some of the bacteria are entrained in the sewage flow and transported to the sewage treatment plant. Thus, the accumulated grease in the plant, which reduces plant treatment efficiency, is also removed. For several weeks after the beginning of application of the bacteria cultures, the treatment plant may be overloaded with an excessive flow of organic material. This is because the bacteria break loose pieces of accumulated grease and dried encrusted organic materials that are carried by the sewage flow to the treatment plant. If the plant is unable to effectively treat such a high rate of organic inflow, it will result in a temporary (5 to 6 weeks) increase in the pollution input to receiving waters.

When the bulk of the accumulated grease in a sewerage system is removed, the population of the bacteria colony decreases because of the reduced supply of food. A continuing low dose of bacteria culture is then needed to maintain the sewerage system free of any further significant grease accumulation. The optimal approach to system cleanup is to begin bacteria dosing at the influent point of the plant and move upstream in the tributary collection system, dosing at man holes and lift stations according to a carefully determined dose schedule. This minimizes the likelihood of dislodged pieces of grease and dried, encrusted material from becoming lodged in pipes with reduced flow capacities, disrupting the normal sewage flow. A sewerage system cleanup performed in this manner could be expected not only to improve the efficiency and safety of its operation but also to substantially enhance the efficiency and economy of locating infiltration and inflow on a system-wide basis, using TV inspection techniques. Infiltrated pipes and joints can then be repaired, reducing flows into the sewerage system and eliminating existing and potential raw sewage overflows.

Table 28-2 presents an approximate cost comparison for cleanup of the Bi-County sewerage system by the traditional techniques and by the use of bacteria cultures. The total cost using the traditional approach is estimated to be $4 million, while clean up with bacteria cultures is projected at $1.3 million, a savings of approximately $2.7 million. Thus, system cleanup using bacteria cultures would appear to be justifiable on the basis of cost considerations alone. Also, cleanup of the system would be expected

TABLE 28-4. *Water Quality Data from Grease-Consuming Bacteria Treatment Program. Values are given in parts per million.**

Water Characteristic	Influent	Effluent
Biochemical oxygen demand	1035	35
Total solids	1912	373
Total volatile solids	1431	127
Total suspended solids	1064	15
Total volatile suspended solids	930	15
Total fixed solids	481	246
Total fixed suspended solids	134	1

*Data supplied through the courtesy of Wayne H. Coloney Co., Inc., Tallahassee, Florida.

to reduce the toxic gas hazard for sewer workers required to enter sewer pipes to make repairs.

Table 28-3 summarizes the results of a telephone survey of several users of bacteria cultures for sewerage system improvement. A pattern of successful results is evident for sewerage systems of various sizes (up to 65 mgd) and types (primary, activated sludge, etc.). The removal of grease from lift stations and contiguous force mains reduces the energy required to operate the station pumps and, thus, the total energy consumption of the sewerage system.

From Table 28-3, it can be seen that the use of bacteria cultures to clean up the three lift stations in the Erie, Pa., sewerage system has reduced the energy consumption of the stations by about a factor of three. Thus, clean up of the forty-six pump stations in the Bi-County sewage collection system tributary to the Blue Plains sewage treatment plant would be reasonably expected to result in important energy conservation for the metropolitan sewerage system. An important ancillary benefit obtained in the Erie system was a distinctly observable reduction in offensive odors at the sewage treatment plant. In Arvada, Colorado, bacteria cleanup of the digesters at the sewage treatment plant resulted in approximately a three-fold increase in methane gas production. This could represent a significant new source of "clean" energy, since combustion of methane (natural gas) results in essentially zero pollution residuals.

Pretreatment with Bacteria Cultures

After system cleanup, the maintenance dosage of bacteria entrained in the sewage flow provides pretreatment of the sewage during its transmission through the collection system. Thus, when the sewage arrives at the plant it has already undergone a level of treatment determined by the time elapsed from its entry into the system till arrival at the plant.

Table 28-4 illustrates a summary of water quality monitoring data obtained over the past two years in the Perry, Florida, sewerage system. Perry is a small town with a population of ten-thousand seeking to upgrade its primary treatment facility, which provided about a 30% reduction in biochemical oxygen demand and organic solids to secondary quality at minimal cost. Cleanup of the sewer mains and laterals was performed with bacteria and the subsequent maintenance dosage resulted in significantly improved water quality measurements. In September 1973, a polishing pond was added to the plant and bacteria dosing at the influent side of the plant initiated. Careful monitoring of the plant influent and effluent resulted in the data presented in Table 28-4. For the seven monitored water quality parameters, significant improvements in excess of the previous primary level of treatment were recorded. Part of this improved level of treatment is attributed to the increased time available for the bacteria to consume organic solids during the 6–8 hour detention time in the polishing pond.

By analogy the trip time from entry of sewage into a system till arrival at the plant creates an opportunity for the bacteria to consume organic solids providing a corresponding level of pretreatment.[5] In terms of existing Bi-County sewerage deficiencies, this technique represents an alternative interim treatment program not requiring the construction of interim treatment plants.

Conclusions

Cleaning of sewers by grease consuming bacteria represents a viable and in some cases economical alternative to traditional mechanical methods of cleaning sewers. Its usage may tremendously increase the capacity of existing sewers and it may curtail the cost of sewage treatment.

Notes

1. John Chalas, "WSSC Infiltration/Inflow Program Briefing for Montgomery County Council" (February 11, 1974).
2. Public Law 92–500, passed by the 92nd Congress of the United States, October 1972.
3. Ernest W. Steel, *Water Supply and Sewerage,* 3rd ed. (New York: McGraw-Hill, 1953), ch. 19.
4. *Washington Star,* 14 December 1973.
5. Chalas, "WSSC Infiltration/Inflow Briefing."

29

The Use of the Oxidation Ditch in the United States as a Means of Treating Liquid Waste

H. ORIN HALVORSON

Microbiologist and Sanitary Engineer
Minneapolis, Minnesota

I first became aware of the oxidation ditch as a means of treating domestic sewage in 1958 when, on sabbatical leave from the University of Illinois, I was at the Technical University of Norway in Trondheim. There I heard of the experiments being conducted by Pasveer[1] in Holland, and I went to visit him. Together we went to see several of his plants in Holland, and I became very much impressed by the simplicity of the operation, the freedom from odors, and the excellent qualities of the effluents. I concluded from these visits that this process should be introduced in the United States, and I urged Dr. Pasveer to contact some industrial firms in the United States that could manufacture the equipment used for the aeration of the waste and possibly promote the process. I gave him the name of three industrial firms he might contact, which he did, and finally an arrangement was made with Lakeside Engineering Company of Chicago for manufacture and promotion.

I had observed by looking at some of the results that Pasveer obtained in Holland that it appeared there was a net loss of nitrogen, perhaps owing to denitrification occurring in the ditch. Normally one does not expect denitrification to occur under aerobic conditions, but in an oxidation ditch the amount of dissolved oxygen inside of the flock may be very low, even though there is some dissolved oxygen in the liquid itself. This might allow denitrification. When I returned to the University of Illinois I was anxious to test this principle; therefore we set up some experiments in a glass jar that held 20 liters of liquid. To this we added, to begin with, 19 liters of water followed by 1 liter of sludge from the primary settling tank from the sewage treatment plant at Urbana, Illinois. This was aerated and stirred for 24 hours. At the end of that time the aeration and stirring was shut down; the sludge was allowed to settle; and then 1 filter of clear water was taken off from the top of the crock and another liter of raw sludge added. This procedure was continued for at least 20 days, after which we removed 1 liter of the mixed liquor without settling and added back 1 liter of raw sludge. At this point we analyzed the sludge that was being removed from the system, the supernatant liquid, and the raw sludge. The data showed there was a very substantial reduction of organic matter and also, if the dissolved oxygen level was kept at a fairly low value, a substantial denitrification occurred. The results of this study were published.[2] While these results were encouraging, we felt it would be necessary to verify them in an actual plant; we therefore, made arrangements to have a demonstration plant built at Glenwood, Minnesota, with a Federal grant and partial support from the city. We also obtained a grant from the Environmental Protection Agency to study the performance of this plant for a period of three years. In the plant we attempted to control the amount of dissolved oxygen in the ditch at a fairly low level. The idea was to have the amount of dissolved oxygen at the point where the water left the brush be 1½ ppm, dropping to nearly zero by the time the liquid came back to the brush again. This was difficult to control in the Glenwood system because the brushes were set at a fixed level and speed; the only way we could vary the aeration was to vary the level of the liquid in the ditch by adjusting the tip, though that allowed the effluent to flow from the ditch to the settling tank. Nevertheless, we were able to control aeration to a limited degree and the results showed that if we operated the plant at a fairly low dissolved oxygen level we could demonstrate a considerable loss of nitrogen through denitrification.[3]

We also ran a demonstration plant at River Falls, Wisconsin, to test the feasibility of using a ditch as an aerobic digester to parallel more or less the laboratory studies we carried out at the University of Illinois. River Falls has a trickling filter plant, and we constructed at this plant an oxidation ditch with a capacity of about 200,000 gallons. This was originally filled with water and into it we pumped the primary sludge and, once a day, the secondary sludge from the two settling tanks. This amounted to about 5,000 gallons of sludge per day. At the end of the day the brushes were stopped, the sludge was allowed to settle, and a clear supernatant liquid equal to the amount of raw sludge to be pumped in was drained off into the effluent of the plant. Then another 5,000 gallons of raw sludge was pumped into the ditch and the aeration continued. This was repeated each day until the quantity of sludge in the ditch reached a point where it was necessary to remove some of the solids. We could continue to add sludge to the ditch for almost three months before this point was reached. We then drained off about 90% of the liquid contents of the ditch to a sand bed and repeated the process. The results of these studies have not been published but it was shown that we were able to successfully oxidize a good deal of the organic matter in this sludge, in fact, a quantity equal to what would have disappeared, had we used an anerobic digester. Again the process was free from odors.

In the plant constructed at Glenwood, Minnesota, we found that soon after the plant was put into operation, there occurred leakage of ground water into the sanitary sewer system so that the volume of raw waste entering the plant was considerably more than that expected and for which the plant was designed. As a result the settling tank used at Glenwood did not prove adequate to allow proper settling, and the results of the treatment were not as good as we would have liked. Eventually this will have to be corrected by stopping the leaks in the sanitary sewer system and perhaps building an additional final settling tank.

The experience at Glenwood demonstrated

that if one wishes to control the amount of dissolved oxygen in the ditch, one needs to have some means of automatically controlling the degree of aeration as a function of the amount of dissolved oxygen present. It was anticipated that this could be done if one had aerators with controlled variable speeds with the speed of the aerators being controlled by the amount of dissolved oxygen present. To accomplish this, a new type of aerator was designed in which the rotation of the aerator was driven by an oil hydraulic turbine. In this system, an electric motor pumps oil into a small storage tank under pressure and the oil from this is fed to the hydraulic turbine; the speed can be varied by varying the amount of oil that is allowed to pass through the turbine and this in turn is controlled automatically by the amount of dissolved oxygen in the ditch. While this type of rotor has been developed, we have not yet installed the system into a plant. Primarily this is because in Minnesota the state has not put any limit on the level of nitrogen that may be discharged into the effluent. In some states there is a requirement for a low ammonia content of the effluent, but so far as I know no state has set a requirement for a low quantity of total nitrogen. If this is done, then control of total nitrogen in the effluent will be possible by controlling the speed of the rotors automatically from an oxygen probe. If one wishes to use the ditch for the aerobic digestion of sludge as we did at River Falls, it is also necessary to produce a floating aerator. This was done at River Falls by supporting the aerator on pontoons so that the submergence of the aerator blades was the same regardless of the liquid level in the ditch.

A floating aerator has another advantage in that it makes it possible to use the ditch as a surge tank and to have a constant flow from the ditch to the settling tanks. This will enable the settling tanks to function much more efficiently than happens if one allows the volume flowing into them to vary with the flow of incoming raw sewage. During the night the flow of raw waste is very light, while during the day it may be fairly high.

Following the construction of the plant at Glenwood, there were three other plants built in Minnesota, generally using the same plans, namely at Grove City, Renville, and Kerkhoven. In all of these plants the aerators were fixed, so the volume flowing through the settling tanks was the same as the rate of flow of liquid into the ditch itself. In these three plants we did not have any difficulty with infiltration and consequently the results were very good. Both the biochemical oxygen demand (BOD) and the suspended solids of the effluent were generally not more than 10 ppm.

In more recent years the State Pollution Control Agency has required discharges of effluent into a small stream or a lake to have a five-day BOD of 5 ppm and a suspended solids content not exceeding 5 ppm. The plants at Grove City, Renville, and Kerkhoven did not meet these rigid requirements. Therefore, in the plant at Dawson, Minnesota, we incorporated floating brushes so as to maintain a constant flow through the settling tanks. In this plant we also built two settling tanks in series. The first one was used primarily as a basin from which sludge could be returned to the ditch. The effluent from this settling tank went through a quick mix and a slow mix and then into the second settling tank where it received additional polished settling. The quick mix and slow mix were installed in case one needed to add chemicals to remove phosphates. So far phosphate removal has not been necessary, but this could be done if required in the future. In this plant, with the flow through the settling tanks controlled at a constant rate and with a double settling tank, the results have been very good. The BOD of the effluent has been generally less than 5 ppm and very often down to 1 ppm; the suspended solids are generally below 5 ppm, particularly during the summertime. During the winter the suspended solids may go slightly above 5 ppm. The plant at Dawson is the most recent plant built in Minnesota and undoubtedly if other plants are going to be constructed we will use the principle employed at Dawson.

Since the introduction of the oxidation ditch to the United States through the efforts of Lakeside Engineering Company and Passavant Corporation, more than five-hundred have been constructed. Most are in Texas and Maine, but others are also distributed throughout the country. The results obtained in these plants are for the most part very satisfactory. The plants are free from odors and give generally a very high quality effluent. Certainly the oxidation ditch has proved itself an effective means of treating sewage and undoubtedly it will be used extensively in the United States both for domestic sewage and industrial waste. A study of the plant at Dawson, Minnesota is being funded by a Federal grant from the Water Pollution Control Agency and the results will be available when the study is reported.

Notes

1. A. Pasveer, "The Purification of Sewer Water with an Oxidation Channel," *Technologie Eau* (Belg.) 15 (1961): 170-71.
2. Roar L. Irgens and H. Orin Halvorson, "Removal of Plant Nutrients by Means of Aerobic Stabilization of Sludge," *Appl. Microbiol.* 13 (1964): 373-86.
3. H. O. Halvorson, Roar Irgens, and Henry Bauer, "Channel Aeration, Activated Sludge Treatment at Glenwood, Minnesota," *J. Water Poll. Control Fed.* 44 (1972): 2266-76.

Part V

Biological Treatment and Aquifer Recharge

30

Improvement of Wastewater Quality by Movement Through Soils and Aquifers

HERMAN BOUWER

Water Conservation Laboratory,
U.S. Department of Agriculture
Phoenix, Arizona

The present trend in sewage management is towards treatment to avoid pollution of surface water and, if necessary, to enable reuse of the water. After sewage has received primary and secondary treatment, land application (or rather land treatment) of the effluent will undoubtedly play an increasingly significant role in municipal waste management because:

a) The effluent is kept out of streams, lakes, or other surface water.
b) Nitrogen, phosphorus, and other constituents (including minor elements) are beneficially used by crops. Organic matter may enhance physical properties of soil.
c) Effluent water not used for evapotranspiration by the crop moves deeper into the ground, where physical, chemical, and biological processes improve its quality. The resulting "renovated" water can be collected from below the water table for reuse.
d) Wastewater renovation by the soil requires no man-made chemicals or energy, although energy may be required to pump the effluent to the land treatment site or to pump renovated water from the aquifer.
e) Depending on land values and distances to land sites, land treatment may be cheaper than in-plant treatment. Also, land values tend to increase with time, whereas facilities for in-plant tertiary treatment depreciate with time and eventually must be written off.

Land treatment is not feasible or desirable where suitable land is not available or too expensive. Where aquifer materials are fractured or cavernous and covered only by a thin layer of coarse soil, the danger of groundwater contamination is great and land treatment of wastewater may not be desirable. In cold climates, storage facilities for holding winter flow may be required.

Land treatment systems can be classified into

three types: overland-flow, low-rate, and high-rate systems. These systems and the principal processes removing nutrients, pollutants, and other substances from the effluent water as it moves through soils and aquifers are discussed in the following sections. Rather than using individual literature citations, readers are referred to a recent review article and the references therein for documentation and further detail.[1]

Systems

Overland-flow systems are used where the soil has such a low permeability that more than 50% of the applied wastewater runs off. The soil is usually densely vegetated (bermuda grass, reed canary grass, bulrushes, and others) and land smoothing may be necessary to avoid channeling of the water. Nutrients and other substances are removed by microorganisms growing on the plant and soil surfaces (biochemical oxygen demand [BOD] removal), by interaction between the surface water and the soil (phosphate and metal immobilization), by nitrification in the upper, aerobic part of the water followed by denitrification in the deeper parts of the water near the soil surface (nitrogen removal), and by plant uptake. If loading rates are low, for example a few centimeters per week, the runoff from overland-flow systems is considerably better in quality than the effluent applied.

Low-rate systems are used where the treatment fields are farmed. These systems include all irrigation-type systems where sewage effluent or other wastewater is used as the water source. Storage facilities may be required to hold the effluent during winter or wet periods. Applications usually range from 2 to 10 cm/week, which may be applied in 1 day or less. Crops for low-rate land treatment systems are usually restricted to those not consumed by man or those processed before human consumption, depending on the quality of the effluent and local health regulations.

High-rate systems are used where agricultural utilization of the treatment fields is of no or secondary importance and where the main objective is to dispose of the effluent or to renovate it for reuse. The renovated water could be used for unrestricted irrigation (including crops that are consumed raw by man), recreation, and industrial and other purposes. These uses have a higher economic or social return than would be possible with the wastewater as such.

The effluent in high-rate systems is usually applied in rapid-infiltration basins that are regularly dried to allow recovery of infiltration rates and oxygenation of the upper soil profile. Application schedules range from 8 hours flooding and 16 hours drying each day to flooding and drying periods of several weeks each. During flooding, 0.1 to 1 meter of effluent may move into the soil per day. Because of regular drying, the long-term application rate or hydraulic loading rate is much less and may range between 20 and 200 meters per year, depending on soil, climate, and effluent quality. High-rate systems require much less land than low-rate systems. Thus, high-rate systems are attractive where effluent flows are large and/or large land areas are not available. High-rate systems require relatively permeable soils (sandy loams to sand) to get the water into the ground and sufficiently transmissive aquifers to laterally transport the renovated water after it has reached the water table.

Since oxygen in the soil will soon be used up during floding, anaerobic conditions will prevail in the soil and the water table will rise close to the soil surface, so long as it does not reduce infiltration rates. This will usually not happen if the water table stays at least 10 to 20 cm below the soil surface. However, when a drying period is started, the water table should drop rapidly to a depth of at least 1 to 2 meters to allow atmospheric oxygen to enter the upper portion of the soil. Unless the soil is very coarse (sand or fine gravel), aerobic conditions generally will not extend beyond 1-meter depth during drying. Thus, further lowering of the water table during drying to, for example, 3 to 5 meters as is sometimes advocated, will have little or no effect on the depth of the aerobic zone.

Quality Improvement of Effluent in Soils and Aquifers

Most of the quality improvement of the effluent water as it moves through the ground is obtained in the top meter of the soil. This is where often complete removal of suspended solids, BOD, and microorganisms occurs for both low-rate and high-rate systems. Some refractory organic carbon may persist and can be identified as total organic carbon in the renovated water.

When wastewater is first applied to the soil, the ionic composition of the percolate will be affected by an exchange between ions in solution and ions adsorbed to the clay and organic matter (the so-called cation exchange complex) in the soil. Once ionic equilibrium between the cation exchange complex and the wastewater has been reached, however, ion exchange no longer affects the quality of the renovated water (except for ammonium, which can be removed from the cation exchange complex by nitrification). Sometimes soil minerals may be dissolved if the pH of the effluent water differs from the soil pH (calcium carbonates, for example, may be dissolved if the effluent is acidic).

Crop uptake of water and evaporation from the soil concentrate the wastewater as it moves downward. For low-rate systems in humid regions, dilution of the renovated water by rainwater may be significant. In arid regions, the percolate from root zones in low-rate or irrigation-type systems will have a salt content that is about four to ten times greater than that of the original wastewater. For example, if the evapotranspiration rate is 4 cm/week and effluent is applied at a rate of 5 cm/week, the salt in the 5-cm effluent will be concentrated in 1 cm of deep-percolation water, causing a five-fold increase in the salt content (neglecting precipitation of salt in the soil, crop uptake of ions, and other minor factors in the salt balance). The deep-percolation water from sewage-irrigated fields must be handled like that from conventionally irrigated fields. This may require removal of the salty water with wells or drains and subsequent "disposal." For land treatment systems to produce reusable renovated water in arid regions, the effluent must be applied at a rate much higher than the evapotranspiration rate to insure that the percolate does not contain much more salt than the effluent. If, for example, the evapotranspiration rate is 5 cm/week and the salt content of the renovated water must not exceed that of the effluent by more than 10%, at least 55 cm of effluent must be applied per week.

For low-rate systems, the amount of nitrogen applied to the soil with sewage effluent is on the same order as crop uptake of nitrogen, which is about 50—500 kg/ha per year. As with fertilizer nitrogen, part of the nitrogen applied with the effluent will be denitrified and returned to the atmosphere as free nitrogen gas. Approximately one-fourth of the applied nitrogen may leave the soil this way, especially where effluent is applied regularly and soil moisture conditions are higher than normal. Because of crop uptake and denitrification, renovated water moving down below the root zone in low-rate systems will have a much lower nitrogen content than the effluent. In high-rate systems, about 6,000—60,000 kg nitrogen may be applied per hectare per year, which is much more than can be removed by vegetation. Thus, denitrification in the soil must be stimulated to obtain significant nitrogen removal from the effluent water. This can be achieved as follows.

Nitrogen in secondary effluent is usually in the ammonium form, which is converted to nitrate in the soil under aerobic conditions. After a few hours flooding in a high-rate system, however, most of the oxygen in the soil is depleted and nitrification stops. Thus, ammonium in the effluent is adsorbed by the clay and organic matter (the cation-exchange complex) in the soil. Before the exchange complex becomes saturated with ammonium, which would cause ammonium to pass through the soil with the percolating effluent, flooding must be stopped to allow oxygen to

FIG. 30-1. Aerial view of Flushing Meadows Project in Salt River bed west of Phoenix, Arizona. This is a pilot highrate system consisting of 6 infiltration basins and 8 observation wells. It was installed in 1967 to study feasibility and evaluate design and operational criteria for renovating secondary sewage effluent by a rapid infiltration system.

enter the soil. This causes the adsorbed ammonium to nitrify in the aerobic, upper portion of the soil. Microanaerobic environments can also be expected in this zone because the nitrification process itself uses up oxygen and oxygen may diffuse very slowly into the smaller pores. Thus, some of the nitrate formed by nitrification will be denitrified in these anaerobic pockets. The remaining nitrate will be leached out by newly infiltrated effluent when flooding is resumed. However, as this nitrate mixes with the effluent and is carried down to predominantly anaerobic zones, additional denitrification can occur. The nitrate finally remaining will then continue to move with the renovated effluent as a nitrate wave. When this wave passes an observation well, a nitrate peak will be observed in the water samples from that well.

Since oxygen penetrates the soil profile only to a limited depth during drying, care must be taken that no more ammonium is adsorbed during flooding than can be nitrified during drying. When this is done, the entire ammonium exchange capacity will again be available for ammonium adsorption when flooding is resumed. Also, it may be advantageous to maintain a relatively low infiltration rate during the first few days of a new flooding period, as can be accomplished, for example, by using a small water depth. This will reduce the rate of leaching of nitrate from the upper portion of the soil, which in turn tends to increase denitrification.[2]

Denitrification requires organic carbon as energy for the denitrifying bacteria. Sources of the organic carbon are the dissolved BOD of the effluent, carbon synthesized by algae and vegetation growing in the infiltration basins, and cell tissue from autotrophic soil bacteria. Since denitrification requires about 1 mg organic carbon for each mg of nitrogen, secondary effluent as such may not contain sufficient organic carbon for denitrification of all its nitrogen. Adding organic carbon has stimulated denitrification. Use of primary effluent with its higher BOD may be a practical and economical way to obtain more denitrification in the soil.

The optimum schedule of flooding and drying periods to maximize denitrification depends on a number of factors, including nitrogen and BOD characteristics of the effluent, infiltration rates, cation-exchange capacity of the soil, oxygen diffusion rate in the soil, depth of the aerobic zone

FIG. 30-2. Two non-vegetated basins of the Flushing Meadows project, showing flume for measuring outflow from basins. The inflow at the upper end of the basins is measured similarly. The infiltration rate is calculated from the difference between the inflow and the outflow.

during drying, and climate. For the Flushing Meadows Project (Figs. 30-1, 30-2, 30-3), denitrification was maximized by using flooding periods of about 10 days, alternated with drying periods of 10 days in summer and 20 days in winter.[3] With this schedule, 30 to 50% and sometimes as much as 80% of the nitrogen was removed at hydraulic loading rates of about 80 meters per year and total nitrogen loads of 24, 000 kg/ha per year. In parallel laboratory studies, Lance and Whisler reported N-removal rates as high as 80% and found that denitrification increased almost exponentially with decreasing hydraulic loading rate. Thus, denitrification in high-rate systems can be maximized by properly scheduling flooding and drying periods and by reducing hydraulic loading rates.

If flooding and drying periods are much shorter than the optimum for denitrification, the soil will be more aerobic and denitrification will be less. At cycles of 2–3 days flooding and 4–5 days drying, for example, essentially all the nitrogen in the effluent at the Flushing Meadows Project was converted to nitrate in the renovated water and little or no nitrogen was removed. By the same token, if the effluent in a high-rate system is applied with revolving sprinklers, the oxygen entering the soil between sprinkle revolutions may be sufficient to maintain aerobic conditions and nitrogen removal by denitrification may not occur, even after extended periods of application. If flooding periods are much longer than the optimum for denitrification, the cation-exchange complex in the soil will become saturated with ammonium and nitrogen will begin to move down as ammonium with the renovated water. Little or no nitrogen removal will then occur.

The amount of phosphorus entering the soil with the effluent in low-rate systems is on the same order as crop uptake of phosphorus, which is about 10 to 50 kg/ha per year. In addition to crop uptake of phosphorus, phosphate is adsorbed in the soil and then slowly reverses to insoluble compounds. These compounds are mainly iron and aluminum phosphates in acid soils, and calcium phosphates in alkaline soils. The rate and ultimate amount of phosphate immobilization in soil depend on the phosphate adsorption capacity of the soil and on the availability of iron and aluminum oxides in acid soils and of calcium in alkaline soils to convert adsorbed phosphate into insoluble forms. In acid soils, phosphate immobilization is aided by drying periods because the resulting mineral weathering provides iron and aluminum oxides and the high oxygen levels keep iron in the ferric form that is necessary for phosphate immobilization. Because of these processes, all phosphate applied with sewage effluent for many years, minus phosphorus removed by a crop, has sometimes been found in the top meter of the soil. On the other hand, 50% of the phosphate applied at the Flushing Meadows Project penetrated the aquifer to a depth of 9 meters below the basins, after which the phosphate concentration in the water gradually reduced as it moved laterally through the aquifer. About 90% phosphate removal was obtained after 100 meters of lateral movement. In this case, phosphates were probably precipitated into apatite-like calcium phosphates, a gradual process that occurs in the aquifer as well as in the soil.

If potential phosphate contamination of groundwater below high-rate systems is a problem, phosphate can be relatively easily removed from the effluent prior to land application. This can be done in the treatment plant by precipitation with iron chloride, aluminum sulfate, lime, or other precipitant, or such a precipitant may possibly be added to the effluent before it enters a rapid infiltration system. The phosphate precipitates will then accumulate on the soil surface in the basins.

Of the minor elements, fluoride can be precipitated as calcium fluoride or as fluorapatites with calcium phosphate. The removal of fluoride from sewage effluent approximately paralleled that of phosphate at the Flushing Meadows Project. Boron can be immobilized by iron and aluminum oxides or it can be adsorbed into the crystal lattice of clay minerals. Boron remained completely mobile in the iron- and aluminum-free sands and gravels below the Flushing Meadows Project. With the trend toward low-phosphate detergents, some of which contain borax or perborate, increasing boron contents can be expected in sewage effluent as phosphate contents decrease. This will be of concern where the effluent or renovated water is used for irrigation because of boron toxicity to crops. This may already occur to citrus if the boron level exceeds 0.5 mg/liter.

A significant part of the metals in raw sewage

FIG. 30-3. Vegetated basins of the Flushing Meadows Project.

end up in the sludge, leaving metal concentrations in domestic secondary effluent that are often less than the maximum limits for irrigation or for potable water. Some immobilization of metals will occur in the soil—more in basic soils with a high clay content and high oxygen level than in anaerobic, acid, sandy soils. Some of the organic compounds in the effluent act as chelating agents, which increase the mobility of the metals in the soil and underlying aquifers. Plants differ in their tolerance to metals in the soil. Accumulation of a certain metal in the soil, while not harmful by itself, can cause deficiencies of other metals in crops. In low-rate systems, crop uptake of metals may reduce their accumulation rate in the soil.

Once the percolate from a land treatment system has reached the underlying aquifer(s), further quality improvements are usually minor. Some of these minor improvements, however, can be quite significant. For example, the percolate from the upper soil profile may still contain some biodegradable material and microorganisms. Removal of these constituents by the "polishing" effect of additional flow of this water through aquifer(s) may produce renovated water that is free from taste and odor and pathogenic organisms. Usually, some organics will not be decomposed even after extended movement through aquifers. These refractory organics show up as a residual total organic carbon content in the renovated water, about 5 mg/l in the Flushing Meadows Project.

Under alkaline conditions, precipitation of cal-

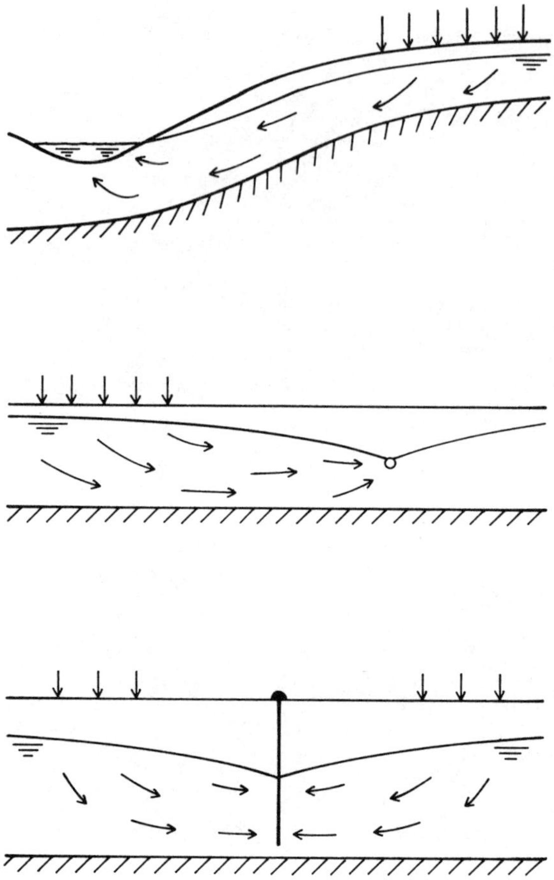

FIG. 30-4. Controlling spread of renovated water from land treatment system (indicated by arrows at field surface) by natural drainage into surface water (top), interception by horizontal drains (center), and interception by wells (bottom). The arrows below the groundwater table indicate the general flow direction. The hatchings represent bedrock or other impermeable boundary.

cium phosphate and fluoride can continue in the aquifer, as was observed at the Flushing Meadows Project. Under more acid conditions, however, phosphate immobilization in the aquifer cannot be expected because, even if iron and aluminum oxides were present, they would be in the reduced (ferrous) form, which does not yield precipitation. Metals are often mobile in aquifers and higher metal contents are usually observed in groundwater below industrial waste dumps and other sources of metals. Sometimes increased metal concentrations in the groundwater are observed at considerable distance from the sources.[4]

To take advantage of the "polishing" effects in aquifers, the renovated water should travel a sufficient distance underground and should also have a sufficient underground detention time before it leaves the aquifer, either naturally as drainage into surface water or artificially via wells or drains. The proper time and distance of underground travel depend on (a) the quality of the effluent as it moves into the ground, (b) the soil and aquifer characteristics, and (c) the desired quality of the renovated water. As a rule, at least 100 m of underground travel and at least several weeks of underground detention time should be allowed before the renovated water leaves the aquifer.

The quality of the renovated water, while much better than that of the effluent, will generally not be as good as that of the native ground-water. To protect native groundwater resources, therefore, it will be necessary to limit the spread of renovated water into groundwater. This is particularly important in high-rate systems, which have a greater local impact on groundwater than overland-flow or low-rate systems.

Restricting Spread of Renovated Water in Aquifers

The spread of renovated water into an aquifer can be controlled by selecting the treatment site so that the renovated water eventually drains naturally into streams, lakes, or other surface water (Fig. 30-4). Where this type system is not possible, renovated water can be artificially intercepted and collected from the aquifer by horizontal drains if the aquifer is shallow and by wells if the aquifer is deep (Fig. 30-4). The water collected in this manner can be discharged into surface water or reused.

Where the renovated water is allowed to drain

into surface water, care should be taken that the aquifer can transmit the flow of renovated water without undue water table rises below the infiltration areas.[5] This can be checked by analyzing the flow system with the Dupuit-Forchheimer assumption or with a more complicated analysis. If natural drainage flow towards the surface water already exists prior to installation of a land treatment system, the expected flow system caused by the land treatment system must be superimposed on the natural flow system to determine if the aquifer can handle the flow. This will then show if there are any constraints on the infiltration rates or on the dimensions of the infiltration areas.

If the renovated water is to be intercepted by horizontal drains, the drains should be placed a certain distance from the area where the wastewater enters the soil to insure proper time and distance of underground travel of the water before it arrives at the drain. An equation based on drainage theory and the Dupuit-Forchheimer assumption was developed to predict the height of the water table under the infiltration system, so that the proper width of the infiltration area, the infiltration rates, and the location of the drains can be selected. In shallow aquifers of low to medium hydraulic conductivity, the drains may have to be placed fairly close to the infiltration areas, and shorter underground detention times and travel distances may have to be accepted to obtain a workable system.

Where the aquifers are deep, the renovated water can be collected by wells. If such wells are located on a line midway between two parallel infiltration areas (Fig. 30-4), all the water that enters the soil as effluent can be collected from the wells as renovated water. The system must then be managed so that the water table below the outer edges of the infiltration areas remains at the same level as the water table in the aquifer outside the system. In that case there will be no hydraulic gradient away from the wastewater renovation system. To be on the safe side, however, the water table below the outer edges of the infiltration area may be kept slightly below that in the adjacent aquifer. This will cause a slight gradient towards the wastewater renovation system, making it impossible for renovated water to move out into the aquifer. The shape of the water table and the underground detention times can be evaluated using dimensionless graphs that were developed with an electrical analog.[6]

Selection and Design of System

The decision whether or not to use land treatment should be based on an objective evaluation of all practical alternatives. The system that is economically and environmentally (including energy requirements) the most attractive can then be selected in a rational manner. Unlike conventional or advanced sewage treatment plants for which design criteria and manuals are readily available, the design and operation of land treatment systems (particularly high-rate systems) must be adapted to the local conditions of climate, soil, and hydrogeology. This requires not only the talents of the traditional sanitary engineer, but also those of the agronomist-soil scientist, the agricultural engineer, and the groundwater specialist.

Many existing land treatment systems have been described in the literature and the results and experiences obtained are of great importance in the design of new systems. Because of the many different conditions that can be encountered, however, it will always be advisable to start on a small scale, using a pilot or experimental project, before embarking upon any grandiose scheme, particularly in areas where previous experience with land treatment is unavailable. It will also be desirable to continue monitoring land treatment systems, so that undesirable performance can be corrected before native groundwater or other natural resources are irreversibly damaged.

Notes

1. Herman Bouwer and R. L. Chaney, "Land Treatment of Wastewater," in *Advances in Agronomy,* ed. N. C. Brady,

vol. 26 (New York: Academic Press, 1974), pp. 133-76. The Flushing Meadows Project mentioned in this chapter is also discussed in more detail in the review article.

2. J. C. Lance and F. D. Whisler, "Nitrogen Removal During Land Filtration of Sewage Water," in *Proceedings of the International Conference on Land for Waste Management,* J. Tomlinson, ed. (Ottawa: Agriculture Institute of Canada, 1974) pp. 174–85; Bouwer and Chaney, "Land Treatment."

3. Bouwer and Chaney, "Land Treatment."

4. George Matthess, "Heavy Metals as Trace Constituents in Natural Groundwater and Polluted," *Journal of the Royal Geological and Mining Society of the Netherlands* 53 (1974): 149-55.

5. Herman Bouwer, "Design and Operation of Land Treatment Systems for Minimum Contamination of Ground Water," *Ground Water* 12 (1974): 140–47.

6. *Ibid.*

31

Renovation of Municipal Wastewater for Groundwater Recharge by the Living Filter Method[1]

WILLIAM E. SOPPER

Institute for Research on Land and Water Resources
The Pennsylvania State University

One of the alternatives for renovation of municipal wastewater for groundwater recharge is land application. This method has been under investigation at The Pennsylvania State University since 1963. This project was the first comprehensive research project in the United States to investigate the feasibility of disposing of large volumes of secondary treated sewage effluent on the land by spray irrigation. From these investigations the "living filter" concept evolved.

The idea embodied in this concept is to apply wastewater for direct recharge to the groundwater reservoir. This can be accomplished by controlling application rates and by maintaining normal aerobic conditions within the groundwater reservoir. This can be accomplished by controlling application rates and by maintaining normal aerobic conditions within the soil. Under these conditions organic and inorganic constituents in the wastewater are removed and degraded by microorganisms, chemical precipitation, ion exchange, biological transformation, and biological absorption through the root systems of the vegetative cover. The utilization of the vegetative cover as an integral part of the system to complement the microbiological and physiochemical systems in the soil is an essential component of the living filter concept and provides for maximum renovation capacity and durability of the system.

Treated municipal sewage effluent has been spray irrigated on cropland and in forested areas for a twelve-year (1963–1974) period at the Penn State Project. Forested areas irrigated consisted of a mixed hardwood forest, a red pine plantation *(Pinus resinosa)*, and a sparse white

TABLE 31-1. *Typical Chemical Composition of Municipal Sewage Effluent.*

Constituent	Average Concentration	Total Amount Applied*
	mg/l	kg/ha
pH	8.1	—
MBAS	0.37	6
Nitrate-N	8.6	143
Organic-N	2.4	40
NH_4-N	0.9	14
P	2.651	44
Ca	25.2	420
Cl	41.3	792
Mg	12.9	215
Na	28.1	469
Fe	0.4	9
	μg/l	kg/ha
B	169	3.26
Mn	61	1.15
Cu	109	1.96
Zn	211	4.15
Cr	23	0.41
Pb	104	2.12
Cd	9	0.19
Co	62	1.24
Ni	93	1.82

*Total amount applied on areas which received 5 cm of effluent per week.

spruce *(Picea glauca)* plantation established on an abandoned field. Types of crops irrigated were wheat, oats, corn, alfalfa, red clover, and reed canary grass. Detailed descriptions of these areas have been previously reported.[2] The two soil types present on the site are the Hublersburg with a surface texture ranging from silt loam to silty clay loam on slopes ranging from 3 to 12% and the Morrison sandy loam with slopes ranging from 3 to 20%.

Chemical Composition of Municipal Sewage Effluent

The chemical composition of municipal effluent is illustrated in Table 31-1 based upon samples collected from the university sewage treatment plant. Treatment consists of both primary and secondary treatment. Secondary treatment includes standard and high-rate trickling filters and a modified activated sludge process followed by final settling. The total amount of each constituent applied per hectare per year at the 5 cm-per-week application rate is also given in Table 31-1.

The fertilizer value of these wastewaters is readily evident in that the 5 cm per week application provides commercial fertilizer constituents equivalent to approximately 233 kg of nitrogen, 224 kg of phosphate (P_2O_5), and 254 kg of potash (K_2O). This would be equal to applying about 2200 kg of a 10–10–11 fertilizer annually.

Renovation

Agronomic Areas

Nitrogen and phosphorus are the two key eutrophic elements in municipal wastewater and therefore discussion on renovation will be limited to these two elements. The overall efficiency of the biological system in accepting and renovating wastewater can be evaluated by observations on the quality of percolating water within the surface soil mantle. Suction lysimeters installed in all areas were used to obtain samples of percolating water at the 120 cm soil depth. Percolate samples were collected, whenever possible, after each weekly application of wastewater and analyzed for the same constituents as the wastewater.

Mean annual concentrations of phosphorus and nitrate-nitrogen for the agronomic areas during the period 1965 to 1973 are given in Tables 31-2 and 31-3. Mean annual phosphorus concentrations in the sewage effluent have varied from 2.5 to 10 mg/l. Phosphorus concentrations in the percolating water at the 120 cm soil depth have been consistently decreased by more than 98% since the initiation of the project in 1963. Even when application rates were increased from 5 to 7.5 cm per week in 1972–73 the degree of renovation was essentially unchanged. The differences between the mean annual concentrations of phosphorus on the control and irrigated areas are quite small and insignificant considering that more than 13 meters of sewage effluent have been applied over the eleven years. It should also be noted that starting in the year 1971 the reed canary grass area has been irrigated with a mixture of sewage effluent and liquid digested sludge (approximately one part of sludge slurry to eleven parts of effluent). This effluent-sludge mixture had a total phosphorus concentration of 20 to 25 mg/l and an orthophosphate level of 10 to 15 mg/l. These data indicate that phosphorus is not leaching out of the soil profile into the groundwater at significantly higher concentrations from the wastewater treated areas than from the control areas and that there appears to be no decreasing trend evident in terms of satisfactory phosphorus removal.

Nitrate-nitrogen renovation was not as efficient on the agronomic areas. Nitrate-nitrogen concentrations were decreased below the 10 mg/l level recommended by the Public Health Service (PHS) for drinking water on the corn rotation area that received the 2.5 cm-per-week treatment but not at the 5 cm-per-week level. At the higher application rate (5 cm) the mean annual concentration remained below the PHS limit only when grass-legume hays occupied 28 to 68% of the site during the period 1965 to 1968. Starting with 1969 the entire site has been planted with corn. Increases in the application rate from 5 to 7.5 cm during 1972 and 1973 did

TABLE 31-2. *Mean Annual Concentration of Phosphorus in Suction Lysimeter Samples Collected at the 120-cm Soil Depth in the Corn Rotation Plots and the Reed Canary Grass Area.*

Year	Corn		Reed Canary Grass	
	cm effluent applied per week		cm effluent applied per week	
	0	5	0	5
	mg/l	mg/l	mg/l	mg/l
1965	0.032	0.022	—	—
1966	0.045	0.036	—	0.055
1967	0.039	0.054	—	0.053
1968	0.041	0.060	—	0.052
1969	0.066	0.070	—	0.035
1970	0.034	0.075	—	0.038
1971	0.048	0.061	—	0.061
1972*	0.022	0.035	0.067	0.054
1973*	0.013	0.020	0.036	0.052

*Application rate increased to 7.5 cm per week.

TABLE 31-3. *Mean Annual Concentration of Nitrate-Nitrogen in Suction Lysimeter Samples Collected at the 120-cm Soil Depth in the Corn Rotation Plots and the Reed Canary Grass Area.*

Year	Corn cm per week		Reed Canary Grass cm per week	
	0	5	0	5
	mg/l	mg/l	mg/l	mg/l
1965	5.2	9.7	—	—
1966	4.7	7.0	—	3.7
1967	3.4	7.1	—	3.3
1968	4.5	9.5	—	3.1
1969	9.4	13.5	—	2.5
1970	10.3	10.9	—	2.4
1971	6.2	9.6	—	3.3
1972*	9.4	10.6	1.7	7.7
1973*	7.4	7.4	1.5	9.2

*Application rate increased to 7.5 cm per week.

not significantly affect the degree of renovation on the corn area.

On the other hand, the 5 cm-per-week reed canary grass area has been exceptionally efficient in nitrogen removal and has consistently maintained the mean annual concentration of nitrate-nitrogen in percolating water below the 10 mg/l limit. This area is irrigated year-around and therefore received twice as much nitrogen as the corn area. Even when application rates were increased to 7.5 cm per week during 1972 and 1973, mean annual concentrations were increased somewhat but still remained within acceptable levels.

Forested Areas

The forested areas were highly efficient in removing phosphorus (Table 31-4). Mean annual concentrations are slightly higher than for the agronomic areas because there is no harvested crop and the phosphorus is continually recycled. Year-round irrigation of the forest area on the Morrison soil at the 5 cm-per-week level has resulted in a substantial increase in phosphorus concentration in the soil percolate. As shown in Table 31-4, there has been a steady increase in comparison to the control area since the third year (1968) of operation. On the other hand, phosphorus concentrations in the irrigated forest areas on the Hublersburg soil, which received lower application rates and shorter periods of irrigation (growing season 30 to 33 weeks), were similar to those in the control areas. Increases in the application rates from 2.5 to 3.8 cm and from 5 to 7.5 cm during the growing season irrigation period did not appear to reduce renovation efficiency.

The efficiency of forested areas in reducing nitrogen concentrations has been variable. It appears that forested areas can accept a 2.5 cm-per-week application of wastewater without having the mean annual concentration of nitrate-nitrogen exceed the PHS limit (Table 31-5). However, year-round irrigation of forests at 5 cm per week consistently resulted in no renovation. It also appears that forest ecosystems may be quite sensitive to wastewater application rates and may have a low threshold of collapse in the renovation system. A 50% increase in the application rates (2.5 to 3.8 cm and 5 to 7.5 cm) after nine years of successful operation and renovation resulted in a complete breakdown in reduction of nitrate-nitrogen in percolating water. Mean annual concentrations nearly tripled the first year (1972) and all exceeded 10 mg/l.

The old field area has been somewhat exceptional in comparison to the other forest areas in terms of nitrogen renovation. This area receives the highest application rate on the Hublersburg soil and yet has consistently maintained nitrate-nitrogen levels below 10 mg/l throughout the nine-year period (1963–1971) without the harvest of any crop. In 1963 at the start of the project the area was primarily an open weed field with a few scattered spruce saplings (1 to 2 m in height). Although the trees are now more than 7 m in height the spruce stand is still sparse with fairly large open areas. It appears that the annual and perennial weeds that occupy these open areas during the growing season (irrigation period) provide a temporary storage for nitrogen and hence reduce nitrate-nitrogen leaching losses. In the fall, vegetative growth ceases and nitrates are again available for leaching. However, since irrigation has ceased by this time, the concentrations of nitrate-nitrogen in percolate water remains at a low level.

This same phenomenon was observed in the red pine plantation, which was irrigated at the 5 cm-per-week level. Mean annual concentrations of nitrate-nitrogen steadily increased from 3.9 mg/l in 1963 to 24.2 mg/l in 1969. In November 1968 a snow storm resulted in complete blowdown of the plantation. In 1969 the area was clearcut and all trees were removed. Immediately a dense cover of herbaceous vegetation developed similar to that on the irrigated old field area. As a result the mean annual concentration of nitrate-nitrogen decreased from 24.2 mg/l in 1969 to 8.3 mg/l in 1970 and to 2.9 mg/l in 1971.

Effects on Chemical Properties of the Soil

Soil samples were taken on all areas to a depth of 150 cm in the fall after cessation of irrigation in 1963, 1967, and 1971. Soil samples were analyzed for the same constituents as was the effluent to determine if any significant concentrations of nutrients were accumulating in the irrigated areas.

The exchangeable cations (K, Ca, Mg, Na, and Mn) and boron were extracted from the soil

TABLE 31-4. *Mean Annual Concentration of Phosphorus in Suction Lysimeter Samples Collected at the 120 cm Soil Depth in the Forested Areas.*

Year	Red Pine *Hublersburg Soil* cm per week		Hardwood *Hublersburg Soil* cm per week		Old Field *Hublersburg Soil* cm per week		Hardwood *Morrison Soil* cm per week	
	0	2.5	0	2.5	0	5	0	5
	mg/l	mg/l	mg/l	mg/l	mg/l	mg/l	mg/l	mg/l
1965	0.040	0.300	0.050	0.250	—	0.460	—	—
1966	0.043	0.134	0.037	0.043	0.030	0.140	0.059	0.042
1967	0.053	0.092	0.044	0.077	0.039	0.068	0.068	0.063
1968	0.075	0.089	0.106	0.222	0.040	0.053	0.071	0.116
1969	0.010	0.064	0.072	0.047	0.051	0.098	0.059	0.137
1970	0.065	0.076	0.033	0.143	0.042	0.114	0.051	0.209
1971	0.094	0.107	0.015	0.146	0.039	0.420	0.037	0.378
1972	0.113	0.105*	0.078	0.037*	0.074	0.200†	0.033	0.335
1973	0.028	0.035*	0.022	0.051*	0.042	0.096†	0.020	0.392

*Application rate increased to 3.8 cm per week.
†Application rate increased to 7.5 cm per week.

TABLE 31-5. *Mean Annual Concentration of Nitrate-Nitrogen in Suction Lysimeter Samples Collected at the 120-cm Soil Depth in the Forested Areas.*

Year	Red Pine Hublersburg Soil cm per week		Hardwood Hublersburg Soil cm per week		Old Field Hublersburg Soil cm per week		Hardwood Morrison Soil cm per week	
	0	2.5	0	2.5	0	5	0	5
	mg/l	mg/l	mg/l	mg/l	mg/l	mg/l	mg/l	mg/l
1965	0.9	2.2	—	0.0	0.3	8.0	—	—
1966	0.1	2.1	0.1	0.2	0.1	5.0	0.1	10.6
1967	0.9	1.7	0.3	1.4	0.3	6.1	1.4	19.2
1968	0.9	2.7	0.1	8.0	0.2	3.7	0.1	25.9
1969	0.2	4.2	0.1	7.2	0.2	2.3	0.3	23.7
1970	<1	5.3	<1	5.0	<1	3.5	1.0	42.8
1971	2.6	8.3	0.5	5.8	0.5	3.8	0.8	17.6
1972	6.0	21.8*	14.7	23.9*	3.2	11.8†	4.7	22.9
1973	0.5	13.7*	3.7	13.5*	0.5	13.5†	1.3	17.3

*Application rate increased to 3.8 cm per week.
†Application rate increased to 7.5 cm per week.

with ammonium acetate[3] and analyzed with an arc spectrometer.[4] Exchangeable hydrogen was determined using a barium chloride buffering technique.[5] Organic matter was determined using a potassium dichromate-sulfuric acid oxidation method.[6] Soil pH was measured with a glass electrode using a 1:1 soil to water mixture. Total nitrogen was analyzed using a modified Kjeldahl method to include nitrates.[7] The phosphorus concentration was determined by using the Bray extraction procedure.[8]

The results of the soil analyses for 1963 and 1971 for the two areas on each soil type which received the highest applications of effluent are given in Tables 31-6 and 31-7. These two areas were the old field area on the Hublersburg soil and the hardwood area on the Morrison soil, both receiving 5 cm effluent per week.

Results of the soil analyses of the 1971 samples indicated that the effects of effluent irrigation on exchangeable potassium, organic matter, pH, and total nitrogen were small and inconsistent. However, there were significant changes in the concentrations of calcium, magnesium, sodium, manganese, boron, and phosphorus. Calcium, magnesium, and boron concentrations increased significantly at the 30 cm depth in both the Hublersburg and Morrison soils and in all the vegetative cover types. Sodium concentrations increased significantly at all depths in both soils and in all the vegetative cover types. Manganese concentrations decreased significantly in the upper 90 cm of the Hublersburg soil on the hardwood 2.5 cm treated plot and the upper 90 cm of the Morrison soil on the new gameland hardwood 5 cm treated plot. Phosphorus concentrations increased significantly in the Hublersburg soil in the 30 cm depth of the hardwood 2.5 cm plot, the upper 60 cm of the old field 5 cm and red pine 2.5 cm plots. Phosphorus also increased significantly in the upper 150 cm of the Morrison soil on the new gameland hardwood 5 cm plot.

Of the 11 constituents analyzed, only potassium, sodium, manganese, exchangeable hy-

TABLE 31-6. *Mean Constituent Concentrations for the Old Field 5 cm Area on the Hublersburg Soil Type.*

Year and Area	Depth cm	*Milliequivalents/100gms*					*ppm*			pH	%N	%OM
		K	Ca	Mg	Na	H	Mn	B	P			
1963 Control	30	0.40	1.43	0.27	0.10	11.84	71.60	0.53	8.70	4.79	0.088	2.147
	60	0.47	2.00	0.83	0.17	8.67	20.46	0.60	0.76	5.23	0.016	0.370
	90	0.47	1.13	1.27	0.20	9.33	23.46	0.53	0.38	4.91	0.012	0.167
	120	0.43	0.93	1.57	0.20	—	18.60	0.53	0.38	—	0.010	—
	150	0.47	0.97	1.53	0.13	—	18.13	0.53	0.38	—	0.010	—
1963 Treatment	30	0.63	1.73	0.50	0.23	16.06	44.60	1.00	16.05	4.96	0.072	2.407
	60	0.60	1.87	0.90	0.20	18.74	20.80	0.80	1.48	4.68	0.022	0.390
	90	0.63	1.03	1.00	0.10	11.48	26.00	0.86	0.71	4.45	0.019	0.183
	120	0.53	0.77	0.97	0.10	—	62.00	0.80	0.00	—	0.014	—
	150	0.47	0.57	0.77	0.10	—	17.93	0.80	0.00	—	0.016	—
1971 Control	30	0.10	2.00	0.23	0.23	14.91	15.86	0.06	10.95	5.06	0.082	2.683
	60	0.03	2.16	0.80	0.23	11.56	12.13	0.06	0.91	4.89	0.016	0.313
	90	0.00	1.80	1.43	0.23	12.05	10.60	0.00	1.00	4.84	0.009	0.167
	120	0.03	1.53	1.63	0.20	—	9.46	0.06	0.90	—	0.005	—
	150	0.00	1.16	1.50	0.20	—	8.66	0.00	1.78	—	0.005	—
1971 Treatment	30	0.43	3.33	1.73	0.33	15.52	14.53	0.60	50.80	5.42	0.095	2.300
	60	0.20	2.00	1.33	0.30	16.36	17.20	0.20	3.48	4.99	0.026	0.343
	90	0.06	1.40	1.23	0.33	16.68	15.80	0.06	1.90	4.82	0.017	0.190
	120	0.16	1.00	1.36	0.30	—	15.66	0.06	1.66	—	0.020	—
	150	0.13	1.00	1.43	0.33	—	16.06	0.13	1.76	—	0.028	—

TABLE 31-7. *Mean Constituent Concentrations for the Hardwood 5 cm Area on the Morrison Soil Type.*

Year and Area	Depth cm	*Milliequivalents/100gms*					*ppm*			pH	%N	%OM
		K	Ca	Mg	Na	H	Mn	B	P			
1967 Control	30	0.16	0.90	0.10	0.16	13.43	55.93	0.26	16.50	4.55	—	2.210
	60	0.16	0.70	0.33	0.16	12.07	16.46	0.13	2.13	4.67	—	0.547
	90	0.16	0.46	0.33	0.20	10.38	10.46	0.13	1.23	4.76	—	0.223
	120	0.26	0.43	0.53	0.20	—	4.80	0.33	0.83	—	—	—
	150	0.06	0.36	0.56	0.20	—	10.13	0.06	0.88	—	—	—
1967 Treatment	30	0.13	1.36	0.40	0.23	12.94	19.93	0.26	75.55	5.50	—	1.240
	60	0.06	0.50	0.30	0.23	10.76	11.86	0.00	4.90	4.99	—	0.560
	90	0.13	0.46	0.36	0.23	13.37	19.20	0.13	2.18	4.94	—	0.350
	120	0.16	0.60	0.36	0.20	—	16.26	0.13	1.92	—	—	—
	150	0.26	1.46	1.10	0.26	—	15.33	0.20	1.50	—	—	—
1971 Control	30	0.00	0.53	0.10	0.20	13.72	32.73	0.06	6.25	4.96	0.109	2.867
	60	0.16	1.13	0.80	0.20	13.36	11.00	0.20	3.16	4.97	0.036	0.300
	90	0.26	0.76	1.03	0.20	11.54	10.86	0.26	0.66	4.99	0.028	0.230
	120	0.20	0.53	0.83	0.20	—	9.86	0.20	1.06	—	0.023	—
	150	0.00	0.50	0.96	0.16	—	7.26	0.00	0.68	—	0.016	—
1971 Treatment	30	0.06	1.86	0.53	0.23	7.35	4.13	0.13	143.75	6.15	0.084	1.483
	60	0.03	0.96	0.36	0.20	5.20	4.93	0.20	31.60	5.51	0.052	0.413
	90	0.10	1.30	0.73	0.20	9.34	5.73	0.13	9.95	5.06	0.037	0.243
	120	0.03	1.10	0.83	0.23	—	6.40	0.06	1.98	—	0.032	—
	150	0.13	0.73	0.80	0.23	—	4.20	0.20	0.98	—	0.032	—

drogen, boron, and phosphorus had significant changes over time. Potassium concentrations decreased in all five depths of the treated and control plots in the Hublersburg soil on the old field 5 cm plots. Sodium concentrations increased significantly in all five depths of the Hublersburg soil in the old field 5 cm treated plot, the 60 and 90 cm depths of the hardwood 2.5 cm treated plot, and the lower 90 cm of the red pine 2.5 cm treated plots. Manganese concentrations decreased significantly in all five depths on both soil types and in all vegetation cover types. Exchangeable hydrogen concentrations decreased significantly in the upper 90 cm of the Morrison soil on the new gameland hardwood 5 cm treated plot. Boron decreased significantly in all five depths of the Hublersburg soil on the old field 5 cm and red pine 2.5 cm treated plots. Phosphorus increased significantly in the 30 cm of the Hublersburg soil on the hardwood 2.5 cm and red pine 2.5 cm treated plots, and in the upper 90 cm of the Hublersburg soil on the old field 5 cm and the Morrison soil on the new gameland hardwood 5 cm treated plots.

Crop Responses

Yields

During the initial years of the project a variety of crops were tested. Average annual crop yields during the period 1963 to 1970 have previously been reported by Sopper and Kardos.[9] Since 1968 the two primary crops grown have been silage corn and reed canary grass. Our experience has shown that these two crops are the best suited to our site and are the most efficient in terms of nutrient utilization. During the past 12 years the crop areas irrigated with 5 cm of effluent weekly have received a total of approximately 20 m of wastewater. During this period annual yield increases have ranged from 0 to 350% for corn grain, 5 to 130% for corn silage, 85 to 190% for red clover, and 79 to 140% for alfalfa.

Nutrient Composition

Under the "living filter" concept the vegetative cover is an integral part of the system and should complement the microbiological and physiochemical activities occurring within the soil to renovate the effluent by removal and utilization of the nutrients applied. The crops harvested from the irrigated areas are usually higher in nitrogen and phosphorus than the control crops; however, the differences are not large. This is partially due to the fact that the control area receives a normal application of commercial fertilizer each year—equal to about 900 kg of a 10–10–10 fertilizer per hectare annually.

Nutrients Removed by Crop Harvest

The contribution of the higher plants as renovators of the wastewater is readily evident when one considers the quantities of nutrients, expressed in kilograms per hectare, removed in crop harvest. Such data indicate that the vegetative cover can contribute substantially to the durability of a "living filter" system particularly where a crop is harvested and utilized. At the 5 cm-per-week level of effluent irrigation the harvest of corn silage removes about 179 kg of nitrogen and 48 kg of phosphorus. Reed canary grass, which is a perennial grass, is even more efficient in that it removes about 457 kg of nitrogen and 63 kg of phosphorus. The difference is due primarily to the fact that the grass is already established and actively growing in early spring even before the corn is planted.

The amounts of nutrients removed annually vary with the amount of wastewater applied, amount of rainfall, length of the growing season, and the number of cuttings of the reed canary grass.

The efficiency of crops as renovating agents can be assessed by computing a "removal efficiency" expressed as the ratio of the weight of the nutrient removed in the harvested crop to the same nutrient applied in the wastewater. Average renovation efficiencies for the silage

TABLE 31-8. *Average Renovation Efficiency (in percent) of the Silage Corn and Reed Canary Grass Crops.*

Nutrient	Variety and amount of effluent applied		
	Corn Silage Pa. 602-A		Reed Canary Grass
	2.5 cm	5 cm	5 cm
Nitrogen	334	145	122
Phosphorus	230	143	63
Potassium	280	130	117
Calcium	38	15	9
Magnesium	53	27	19
Chloride	26	14	20
Sodium	2	1	1
Boron	10	4	2

corn and the reed canary grass crops are given in Table 31-8. At the 2.5 cm-per-week level of application of wastewater, the corn silage removes nutrients equivalent to about 334% of the total applied nitrogen, 230% of the applied phosphorus, and 280% of the applied potassium. At the 5 cm-per-week level, the corn silage removed more than 100% of the applied nitrogen, phosphorus, and potassium.

Similarly harvest of reed canary grass removes on the average about 373 kg of nitrogen and about 50 kg of phosphorus. These removals are equivalent to renovation efficiencies of 122 and 63% respectively.

Tree Growth Responses

Red Pine

Experimental plots were established in a red pine plantation in 1963. These plots have been irrigated with sewage effluent during the past twelve years at rates of 2.5 cm and 5 cm per week during the growing season (April to November). The plantation was established in 1939 with the trees planted at a spacing of 2.5 by 2.5 m. In 1963 the average tree diameter at breast height was 22 cm and average height was 11 m.

Diameter and height growth measurements were made annually. Average annual height growth on the 2.5 cm-per-week plot was 58 cm in comparison to 42 cm on the control plot. On the plot receiving 5 cm per week, height growth continually decreased up to 1968 when high winds following a wet snowfall completely felled every tree on the plot.

Increment cores were taken from sample trees in all areas to determine average annual diameter growth. Irrigation at the 2.5 cm-per-week level resulted in an average annual diameter growth of 4.3 mm in comparison to 1.5 mm on the control, an increase of 186%.

White Spruce

Two experimental plots were established in a sparse white spruce plantation on an abandoned old field area. The trees in 1963 ranged from 0.9 to 2.5 m in height. One plot has been irrigated with sewage effluent during the past 12 years at the rate of 5 cm per week, while the second plot has been maintained as a control.

Average height of the trees on the irrigated plot in 1974 was 7.5 m and ranged from 5.5 to 9.5 m. The average height of the trees on the control plot was 3.1 m and ranged from 2.7 to 4.8 m. Over the 12-year period average annual height growth was 60 cm on the irrigated areas and 25 cm on the control areas, representing 140% increase as a result of sewage effluent irrigation.

Average diameter of trees on the irrigated plot was 9.5 cm in comparison to 3.6 cm on the con-

TABLE 31-9. *Average Annual Diameter Growth in Hardwood Forests Irrigated with Sewage Effluent.*

Weekly Irrigation Amount (cm)	Average Diameter Growth (mm) Control	Irrigated
2.5*	4.1	4.8
5†	3.3	6.0

*Irrigated with 2.5 cm of sewage effluent weekly during growing season from 1963 to 1974.

†Irrigated with 5 cm of sewage effluent weekly during the entire year from 1965 to 1974.

TABLE 31-10. *Annual Uptake of Nutrients by a Silage Corn Crop and a Hardwood Forest Irrigated with 5 cm of Effluent Weekly.*

Nutrient	Corn Silage Pa. 602-A (kg/ha)	Renovation Efficiency* (%)	Hardwood Forest (kg/ha)	Renovation Efficiency (%)
N	180	145	94	39
P	47	143	9	19
K	144	130	29	22
Ca	30	15	25	9
Mg	26	27	6	4

*Percentage of the element applied in the sewage effluent that is utilized and removed by the vegetation.

trol plot. Measurements taken from increment cores indicated that the average annual diameter growth on the irrigated trees was 10 mm and on the control trees 4.5 mm representing a 122% increase.

Mixed Hardwoods

A hardwood forest, consisting primarily of oak species, was irrigated with sewage effluent at 2.5 cm per week during the growing season or at 10 cm per week for the entire year (52 weeks).

Average annual diameter growth during the 1963 to 1974 period is given in Table 31-9. Application at 2.5 cm per week produced only a slight increase in diameter.growth; however the 5 cm-per-week level resulted in an 80% increase. These values pertain primarily to the oak species. Some of the other hardwood species present on the plots have responded to a greater extent. For instance, increment core measurements made on red maple *(Acer rubrum)* and sugar maple *(A. saccharum)*, indicate that the avarage annual diameter growth during the past 12 years has been 13 mm on the trees irrigated with 2.5 cm of effluent per week in comparison to 2.6 mm on control trees, a 400% increase in average annual diameter growth.

Renovation Efficiency of Forests

The nutrient element content of the foliage of the vegetation on the irrigated plots was consistently higher than that of the vegetation on the control plots. It is therefore obvious that the forest vegetation is contributing to the renovation of the percolating effluent; however, its order of magnitude is difficult to estimate because the annual storage of nutrients in the woody tissue and the extent of recycling of nutrients in the forest litter are extremely difficult to measure. Although considerable amounts of nutrients may be taken up by trees during the growing season, many of these nutrients are redeposited annually in leaf and needle litter rather than being hauled away as in the case of harvested agronomic crops.

A comparison between the annual uptake of nutrients by an agronomic crop (silage corn) and a hardwood forest is given in Table 31-10. It is obvious that trees are not as efficient renovation agents as agronomic crops. Whereas harvesting a corn silage crop removed 145% of the nitrogen applied in the sewage effluent, the trees only removed 39%, most of which is returned to the soil by leaf fall. Similarly only 19% of the phosphorus applied in the sewage effluent is taken up by trees in comparison to 143% of the corn silage crop.

Groundwater Recharge

The amount of renovated effluent recharged to the groundwater reservoir was estimated from data available on the total amount of effluent and rainfall received by the plots and on potential evapotranspiration. Annual recharge ranged from 10,300 to 17,300 m^3 per hectare irrigated with an average of 15,000 m^3. Recharge amounted to approximately 95% of the effluent applied at the 5 cm-per-week rate. Recharge rates were higher when water was applied during years with normal or above normal rainfall or throughout the year. Evapotranspiration losses are greatly diminished during the late fall, winter, and spring and more of the water infiltrating into the soil from natural precipitation and irrigation is potentially available for groundwater recharge. Runoff, which did occur at the irrigation sites following snow and ice pack melt or heavy or prolonged rains, was ponded in one or more closed surface depressions where it was captured by infiltration or it infiltrated into adjacent unirrigated buffer areas that were usually in forest cover. Closed depressions were numerous on all of the upland areas selected for irrigation or they were available downslope from test plots, hence detention storage was provided naturally. This would not necessarily be true at other irrigation sites where detention storage would have to be engineered to prevent or eliminate runoff. It was found that adjacent border areas with forest stands were ideal to help contain runoff and promote infiltration during all seasons of the year. Rarely did overland flow extend more than 100 feet beyond irrigation plots during the spring thaw.

Mean annual concentrations of nitrate-nitrogen in deep groundwater monitoring wells on and adjacent to the spray irrigation areas are given in Table 31-11. Depth of these wells range from 100 to 300 feet. The 1962 values represent the preirrigation period. Spray irrigation of sewage effluent was initiated during the summer of 1963 near the F-wells and in 1965 near the G-wells. Background levels of nitrate-nitrogen concentrations can be inferred from the 1962 data and from the results of analyses of water samples collected from surrounding private wells.

Chemical water quality changes in the deep on-site wells have been nonsignificant at the cropland and forest areas with the Hublersburg soils (F-wells). However, significant increases in

TABLE 31-11. *Mean Annual Concentration of Nitrate-Nitrogen in Deep Groundwater Monitoring Wells on and Adjacent to the Spray Irrigation Areas.**

Well No.	Distance** (meters)	1962	1963	1964	1965	1966	1967	1968	1969
		Mean annual concentration (mg NO_3-N/liter)							
G-3	0	0.2	—	—	0.1	—	3.6	3.5	2.8
F-1	30	—	1.4	0.7	0.4	0.9	0.9	0.5	0.4
F-5	60	—	—	—	0.4	0.6	0.4	0.8	0.3
G-10	150	—	—	—	—	—	0.9	0.3	0.4
F-3	360	—	—	—	0.4	0.6	0.4	0.5	0.4
UN-14	390	1.6	1.9	2.3	1.4	1.8	2.9	3.1	4.0
UN-24	660	0.6	0.6	0.3	0.5	1.2	1.0	0.9	1.3
UN-17	990	0.6	0.6	0.5	0.2	0.3	0.6	0.4	0.4
Private water wells in surrounding vicinity									
W-7	780	1.3	1.9	1.3	0.7	0.6	—	—	—
W-5	1590	5.3	5.0	4.0	3.0	3.2	3.7	4.4	5.7
W-30	1590	1.9	1.3	1.9	1.2	1.4	2.3	2.4	2.2
W-3	1710	10.1	9.2	4.6	3.2	5.0	7.6	10.2	7.3

*From L. T. Kardos, et al., *Renovation of Secondary Effluent for Reuse as a Water Resource*, Series EPA 660/2-74-016 (Washington, D.C.: U.S. Government Printing Office, 1974).

**Approximate distance from sewage effluent application area.

nitrate and chloride have occurred in one well on the forested area with the Morrison sandy loam soil (G-wells). Although there have been some increases in nitrate-nitrogen concentrations during the seven-year period (1963–1969) these concentrations have remained well below the U. S. Public Health Service drinking water limits. Hence it is evident that with properly programmed application, sewage effluent can be satisfactorily renovated and considerable amounts of high quality water recharged to the groundwater reservoir. In time, contributions to the groundwater of this magnitude will certainly have a beneficial effect on the local water table level.

Conclusions

Twelve years of research have indicated that the living filter system for renovation and conservation of municipal wastewater is feasible and that the combinations of agronomic and forested areas provide the greatest flexibility in operation. Such a system is more adaptable to small cities and suburbs than to large metropolitan areas because of the availablity of open land close to the wastewater treatment plant, although the land area requirement is not a major prohibitive factor. At the recommended level of irrigation, 5 cm per week, only 52 hectares of land would be required to dispose of 4 million liters of wastewater per day. Although large contiguous blocks of agricultural and natural forest land would be the most desirable for efficiency and economy, major metropolitan areas could utilize golf courses, playing fields, forest preserves and parks, greenbelts, scenic parkways, and perhaps even divided highway and beltway medial strips.

Notes

1. Research reported here is part of the program of the Waste Water Renovation and Conservation Project of the Institute for Research on Land and Water Resources and Hatch Project No. 1809 of the Agricultural Experiment Station, The Pennsylvania State University, University Park, Pennsylvania. Portions of this research were supported by funds from Demonstration Project Grant WPD 95-01 received initially from the Division of Water Supply and Pollution Control of the U. S. Department of Health, Education, and Welfare and subsequently from the Federal Water Pollution Control Administration, U. S. Department of the Interior. Partial support was also provided by the Office of Water Resources Research, USDI, as authorized under the Water Resources Research Act of 1964, Public Law 88-379 and by the Pinchot Institute for Environmental Forestry Research, Forest Service, USDA.

2. R. R. Parizek, L. T. Kardos, W. E. Sopper, E. A. Myers, D. E. Davis, M. A. Farrell, and J. B. Nesbitt, *Waste Water Renovation and Conservation,* Penn State Studies No. 23 (University Park, Pennsylvania, 1967); W. E. Sopper, "Waste Water Renovation for Reuse: Key to Optimum Use of Water Resources," *Water Research* 2 (1968):471-480; W. E. Sopper, "Effects of Trees and Forests in Neutralizing Waste," in *Trees and Forests in an Urbanizing Environment* (Amherst: Cooperative Extension Service, University of Massachusetts, 1971), pp. 43-57; W. E. Sopper and L. T. Kardos, "Vegetation Responses to Irrigation with Municipal Wastewater," in *Recycling Treated Municipal Wastewater and Sludge through Forest and Cropland,* W. E. Sopper and L. T. Kardos, eds. (University Park: The Pennsylvania State University Press, 1973), pp. 271-94.

3. M. L. Jackson, *Soil Chemical Analysis* (Englewood Cliffs, New Jersey: Prentice-Hall, 1958).

4. Baker et al., "Techniques of Rapid Analysis of Corn Leaves for Eleven Elements," *Agron. J.* 56 (1964):133-36.

5. Jackson, *Soil Chemical Analysis.*

6. M. Peech, L. T. Alexander, L. A. Dean, and J. F. Reed, *Method of Soil Analysis for Soil Fertility Investigations,* U. S. Department of Agriculture Circular No. 757 (Washington, D. C.: U. S. Government Printing Office, 1947).

7. Jackson, *Soil Chemical Analysis.*

8. *Ibid.*

9. Sopper and Kardos, "Vegetation Responses."

32

Numerical Models in Groundwater Management

IRWIN REMSON

Departments of Applied Earth Science and Geology
Stanford University

Introduction

Numerical models that are solved using digital computers have become important aids to groundwater exploration and management. This paper is intended to demonstrate their utility.

At an early stage of groundwater modeling, a grid is superimposed on a map of the system and appropriate hydrologic data are coded at all grid intersections (nodes). The data include:

a) transmissivity, a measure of the capacity of the aquifer to transmit water,
b) storativity, a measure of the capacity of the aquifer to store and release water,
c) discharge from the aquifer,
d) recharge to the aquifer,
e) initial conditions, and
f) boundary conditions.

The data gaps become painfully apparent as the investigator systematically scans the entire system. Thus, early model formulation is a guide to efficient data collection.

Using available data and estimated values where data are unavailable, the model generates water levels that are compared with the historic water levels. If the water levels are inconsistent, the input data or estimates are in error. Additional data or revised estimates are required. When the model successfully generates historical water levels, it is said to be "calibrated." This is known as the "inverse method." The model can then be used to predict hydrologic effects of proposed management options or else determine optimum management procedures to achieve specified objectives.

Three groundwater model studies will be described. The first is a direct application using pumping-test data. The second uses the inverse method of calibration to determine groundwater

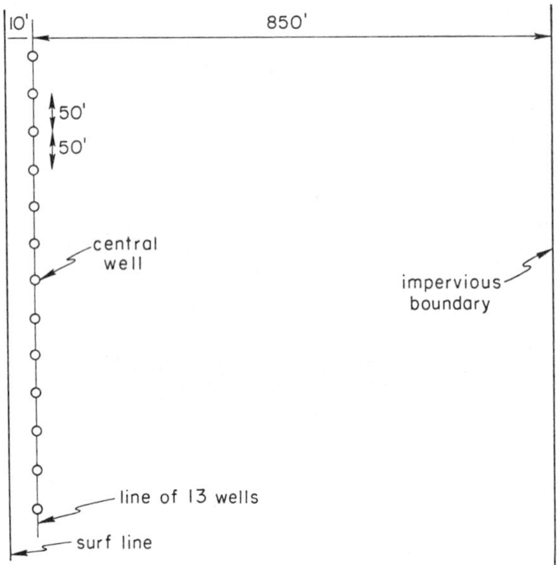

FIG. 32-1. Generalized plan view of a line of wells to be installed on Reventazon Beach, Peru.

parameters. The third study optimizes the dewatering of a construction site. The studies illustrate the usefulness of the modeling techniques.

A Simple Model Incorporating Pumping-Test Data

Fig. 32-1 is a generalized plan view of a line of wells to be installed 50 feet apart parallel to the shore and 10 feet from the mean position of the surf on Reventazon Beach, Peru. The wells are to be pumped at a rate of 200 gallons per minute each on a continuous basis. It was necessary to predict the drawdown in the central well.

Borings showed that the beach sand is approximately 50 feet thick and underlain by a fine-grained sediment. A pumping test was carried out in which a test well was pumped and its effects on three observation wells were monitored.[2] The test results gave a permeability of 500 gallons per day per square foot and a transmissivity of 25,000 gallons per day per foot. The capabilities of such an aquifer might be characterized as "modest" to "moderate." Using this data and an available computer code,[3] a model was prepared and solved. It predicted a drawdown of 6 feet in the central well.

For this rather simple case, all required data were available. Therefore, the model could be constructed and solved in a straightforward manner.

A Model Calibrated by the Inverse Method

A groundwater model was required for the appraisal of an underground waste-disposal plan on an alluvial fan in the western United States. Pumping-test data were limited and seemingly inconsistent. Therefore, the model was calibrated using the inverse method. Based on the available geologic and pumping-test data, values of transmissivity and storativity were estimated throughout the system and the model was prepared. The model converted known hydrologic inputs from precipitation and influent streamflow into groundwater levels and gradients. The resulting gradients were too steep when compared with historical water-level gradients. Because water can be transmitted in more permeable aquifers under smaller gradients, the estimated values of transmissivity were increased and the model rerun.

The writers began the analysis with a preconceived idea that the transmissivities were in the range of 200,000 to 300,000 gallons per day per foot, values characteristic of a very good aquifer. It was a source of concern that the trial transmissivities increased past 600,000 through successive calibration trials and adjustments. Such values would be extremely large. Convinced of a modeling error, the writers discontinued the calibration procedure and requested a pumping test. To their surprise, the test provided a transmissivity of 700,000 gallons per day per foot. This vindicated the calibration and verified that this is indeed an outstanding aquifer.

One model boundary was located at an extensive fault that was supposed to leak groundwater out of the basin in substantial quantities. Because the historic water levels showed flow to be parallel to the fault (a flow line), it was impossible to model this as a leaky fault and generate historic water levels. Obviously, the fault was relatively impermeable to groundwater seepage.

The above example shows that the model would not accept data that were inconsistent with the field conditions that generated the historic data. Once the calibration was completed, the model was used to readily answer management questions concerning the waste-disposal plan.

An Optimization Model

A rectangular construction site in Asia was to be dewatered from its original groundwater level of 118 feet above datum to a design water level of 69 feet above datum or lower. All groundwater data required for model construction had been obtained through a program of boring and testing. In this case, the objective was not to test different management options. Rather, it was to find an optimum management plan. Specifically, the objective was to determine the number of wells, their locations, and the steady-state pumping rates that would maintain water levels in the excavation at design depths while pumping a minimum total volume of water. A linear programming technique was used in conjunction with the model to find the solution.[4] Figure 32-2 shows the well distribution and resulting water-table map. It would have been exceedingly difficult to find the solution shown in Fig. 32-2 by alternate means.

This is but one of many optimization objectives that might have been chosen for the problem. For example, it might have been desirable to minimize pumping costs, pumping lifts, or the number of wells instead of the total discharge.

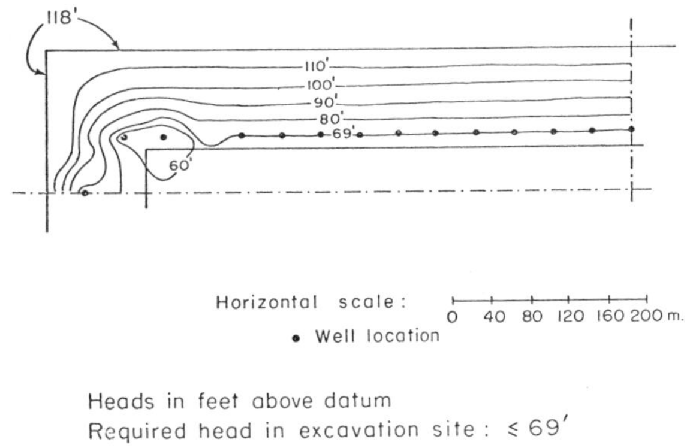

FIG. 32-2. Well distribution and resulting water-table map for a rectangular construction site in Asia. From left well, optimal well discharges in 1,000 gal/day: 772; 1,362; 1,659; 473; 494; 502; 504; 504; 505; 505; 505; 505; 505; and 505. Note that solution was constrained by the grid spacing used.

Conclusions

We have described three groundwater models that give answers needed for groundwater management. They are examples of the variety of model types available for simulating subsurface hydrology. Their use in groundwater testing and development is making groundwater a truly manageable resource.

Notes

1. The field data were supplied by S. S. Alderman, Jr. of Agricultural and Industrial Minerals, Inc., W. R. Hail of W. A. Wahler and Assoc., and W. A. Thomas of Dames and Moore.

2. S. N. Davis and R. J. M. DeWeist, *Hydrogeology* (New York: John Wiley and Sons, 1966).

3. G. F. Pinder and J. N. Bredehoeft, "Application of the Digital Computer for Aquifer Evaluation," *Water Resources Research* 4 (1968):1069-93.

4. E. Aguado et al., "Optimal Pumping for Aquifer Dewatering," *J. Hydraul. Div., ASCE* 100 (1974):869-77.

33

The Combination of Biological and Chemical Treatment at the Krefeld Water Treatment Works

WALTER CZERWENKA

Former Managing Director
Krefeld Water Treatment Works

KÄTHE SEIDEL

Limnology Department
Max Planck Institute
Krefeld, West Germany

Chemical Treatment[1]

The high consumption of drinking water in the area of Krefeld has for some time necessitated the search for new water resources.

The catchment area at Krefeld extends over 180 km² and provides approximately 35 to 40 million cubic meters of drinking water per year. The water consumption requirement for population and for industry is steadily rising by approximately 45 million cubic meters per year.

The following possibilities were considered for reducing this deficit:

ALTERNATIVE 1: Obtain a supply of drinking water from a distant source.
ALTERNATIVE 2: Construct bank-filtered wells.
ALTERNATIVE 3: Take water directly from the Rhine followed by purification for industrial purposes.

The third proposition has been adopted.

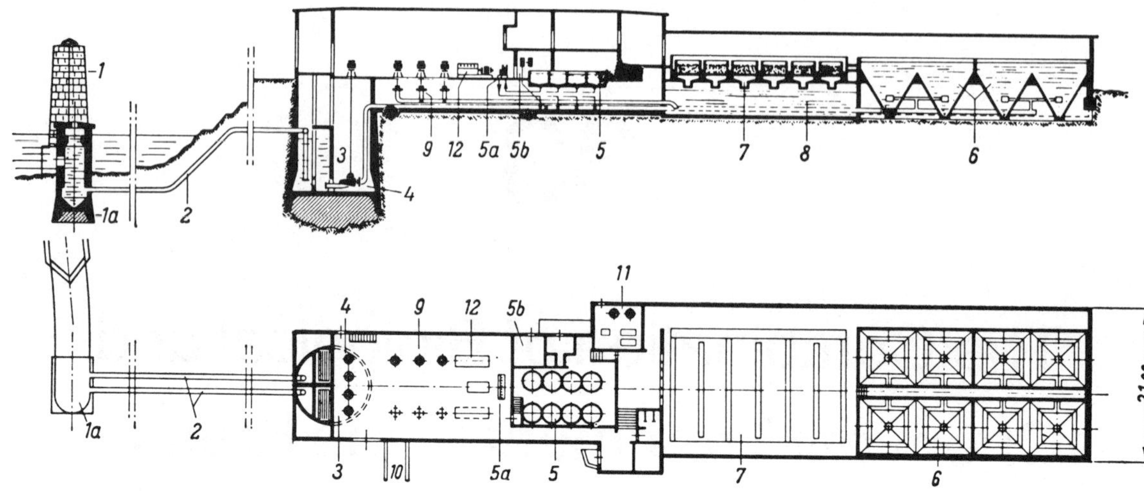

FIG. 33-1. A schematic diagram of the Krefeld water treatment plant showing: 1. Pillar of Rhine River bridge, 1a. Intake plant, 2. Siphon piping, 3. Collecting tanks (inverted pyramids), 4. Raw water pumps, 5. Chemical equipment, 5a. Dosage pumps, 5b. Chlorine plant, 6. Sedimentation chamber (inverted pyramids), 7. Filter plant, 8. Clear water basin, 9. Rhine water pumps, 10. Supply line, 11. Back wash pumps and blowers, 12. Emergency generators.

Before the planning and construction of the River Water Works in Krefeld, a two-year series of tests were conducted there. All the various well-known methods of water purification were tried at different Rhine water levels and under different temperature and pollution conditions. Systems of conditioning water with aluminum sulphate, ferric chloride, and hydrous ferrous sulphate ($FeSO_4$) were tested. The use of hydrous ferrous sulphate turned out to be the most

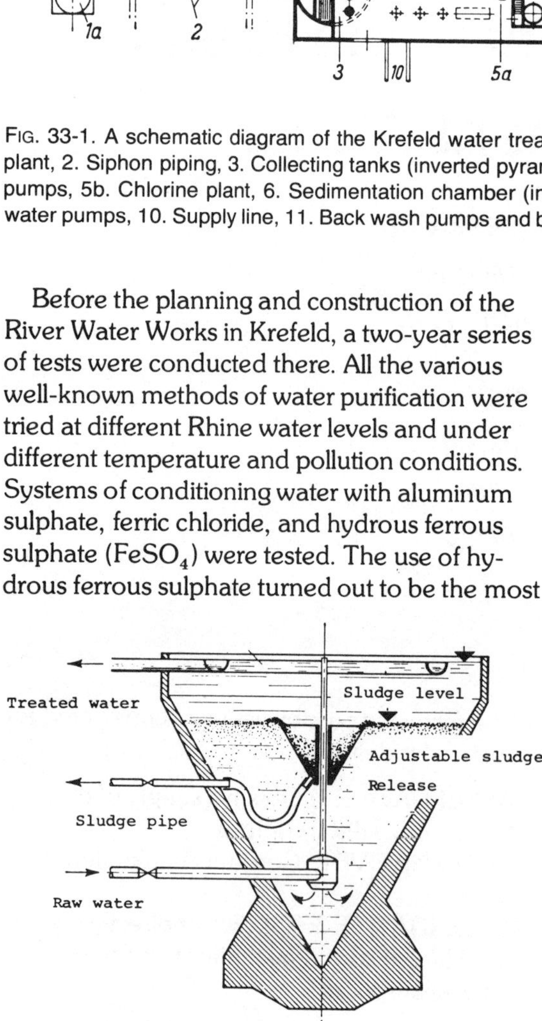

FIG. 33-2. Sedimentation chamber.

economical. Plant construction was started in 1952. Two years later the completed plant was treating 1600 cubic meters of water per hour and could be enlarged to an eventual output of 5000 cubic meters of water per hour. A schematic diagram of the water works is shown in Fig. 33-1. Raw river water flows from the intake plant through syphon piping to a collecting tank after passing a mechanical screening plant. Further on it is pumped through a raw water pipe (900 mm) and dosed with chemicals for flocculation before passing into sedimentation chambers with the shape of inverted pyramids (Fig. 33-2). In these pyramids water rises slowly, passing eventually into an overflow gutter. Near river water level the rising velocity is approximately three meters per hour. The coagulated flocs form a sludge blanket filtering out suspended colloidal substances. Sludge removal cones are set at the height where the sinking velocity of the sludge flocs is equal to the rising velocity of the water. After passing the overflow gutter the deflocculated water runs to the sandfilter plant and finally into the clear water basin.

TABLE 33-1. *Water Quality and Flow Conditions.*

River flow	500–12,000 m^3/sec
Suspended substances	10–500 mg/l
Bacteria	1000–30,000 per cm^3
Temperature	0–26°C

Sludge removal is a continuous process. Approximately 95% of the sludge is removed by the sludge removal cone; only 5% passes on to the filter plant.

The Treatment

The water quality of Rhine River water is continually changing. Sewage from towns and industries, shipping traffic, and chemical effluents, and oil are constantly contaminating river water.

The use of hydrous ferrous sulphate, liquid chlorine and calcium bicarbonate for flocculation of suspended solids, leads to the following chemical reaction:

$$FeSO_4 + Ca(HCO_3)_2 = Fe(OH)_2 + CaSO_4 + 2CO_2$$

$$4\,Fe(OH)_2 + O_2 + 2\,H_2O = 4\,Fe(OH)_3$$

The oxygen necessary for the above reaction is dissolved in water or is evolved from the reaction of chlorine in water.

$$H_2O + Cl_2 = HOCl + HCl$$

$$2\,HOCl = 2\,HCl + O_2$$

The Biological Treatment Concept

At the beginning, the objective was to supply water for industrial purposes only, but after treatment involving ozone, breakpoint chlorination, activated carbon filters, and chlorination, water quality improved to such a degree that domestic use became possible.

The river water often changed temperature. To compensate for this, ground-water recharge facilities were constructed utilizing injection wells. (Fig. 33-3)

The great differences in the quality of local groundwater and treated river water created a problem. Ferric hydroxide accumulated and clogged the gravel layer at the surface of the groundwater table. To overcome this problem a recharge pond was constructed and planted with bulrushes. A diagram of the entire treatment facility including the recharge pond is shown in Fig. 33-4.

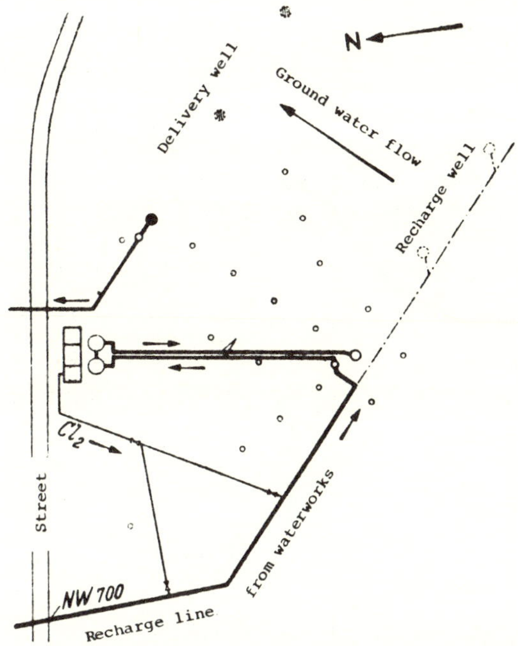

FIG. 33-3. Experimental groundwater recharge area.

Step one at the treatment facility consists of conventional waste treatment, chemical flocculation, settling basins, and filtration.

Step two consists of the recharge pond planted with bulrushes along its sloping banks, as shown in Figs. 33-5 and 33-6. An additional treatment feature at this stage is spraying incoming water into the pond.

Step three of the process involves bank filtration. In this stage, water travels from the recharge pond toward wells set at a distance of 50 to 80 yards and filters through the root zones of the bulrush aquaculture, then through gravel and sand.

Finally, step four includes break point chlori-

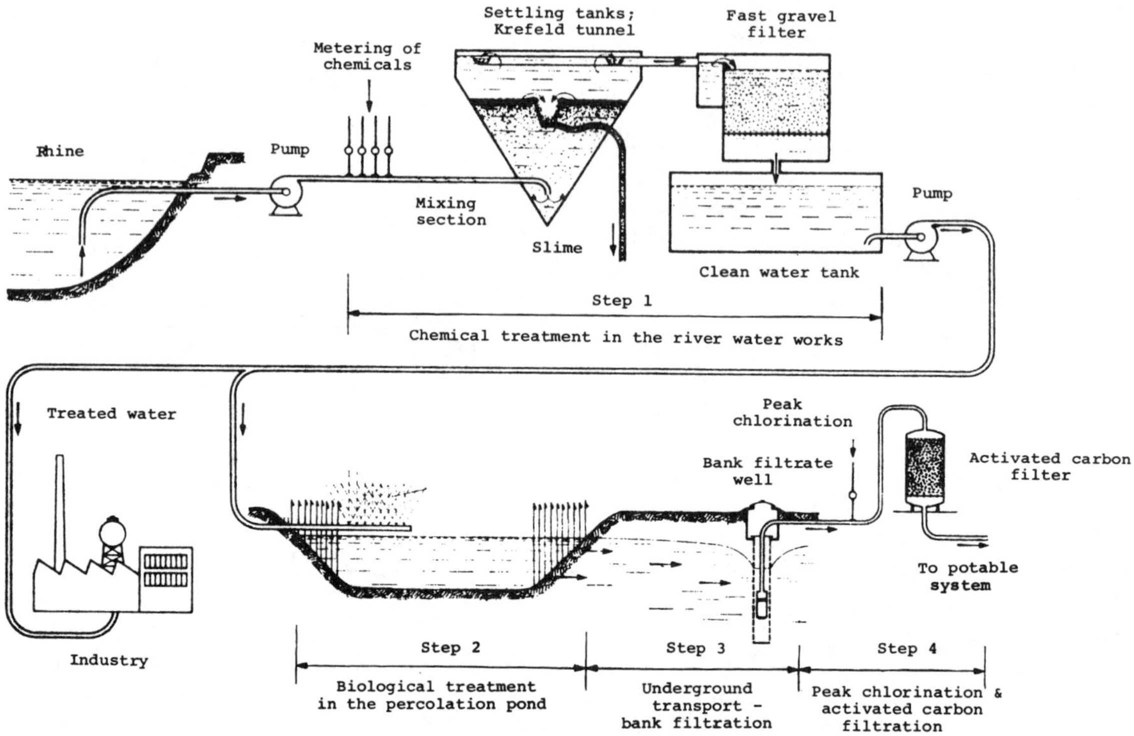

FIG. 33-4. Diagram of the installation with treatment and groundwater recharge ponds. With a normal water level of the Rhine, Steps 1, 2, and 3 are in operation. If the water demand increases sharply, Steps 1 and 4 are added. All steps must be in operation if the Rhine water is heavily contaminated or if the installation is over-loaded.

nation and activated carbon filters. This is a standby unit and is used only during very high water demands or at times when the Rhine water is heavily polluted.

Biological Treatment Implementation

After ten years of biological treatment at the Krefeld facility (see Fig. 33-4), we have gathered a certain amount of information and practical experience. It has been found that a) the bulrushes use great amounts of macro- and micronutrients; b) eliminate phenols and other toxic materials; and c) oxygenate the water. In addition, no sludge is formed on the bottom of the pond, and infiltration rates are high.

Pilot Plant Studies

In order to determine whether bulrushes can treat industrial wastes, pilot studies were performed with small drums filled with sand and gravel and planted with bulrushes. Various types of industrial wastes were tested and research results were promising. Experimental findings at the Max Planck Institute indicate: (1) rapid neutralization of extreme pH values, (2) 70 – 80% reduction of chemical oxygen demand (COD) within 24 to 48 hours, and (3) an increase in the dissolved oxygen content of the waste.

At the water treatment works in Krefeld, the addition of the planted recharge pond had great benefits for the entire operation. It solved the density stratification recharge problem,

FIG. 33-5. Planting of the recharge basin banks with *Schoenoplectus lacustris.*

improved the quality of water produced, and especially improved the taste and odor conditions of municipal water. All facilities have been working for ten years to the satisfaction of the plant operators. The experience with aquatic plants over this period of time can be summarized as follows:

1) In spite of steep slopes, bulrushes grow in an even stand with stems extending five feet above water. This is surprising because the length of stems below the water surface ranges between 38 inches and 60 inches.
2) Depending upon water depth, plants incorporate a varying amount of macro- and micronutrients into plant tissue. Table 33-2 gives the amounts of nutrients used by bulrushes growing in various depths of water.
3) The elimination of phenols and other toxic materials was confirmed by several years of actual operation. (See Tables 33-2 and 33-3)
4) Plants act as vertical pipes supplying additional amounts of oxygen into the water of the pond.
5) Dead plants float on the water surface and reach the nearby banks. They do not sink to the pond bottom to decay. For this reason

FIG. 33-6. Recharge basin in operation.

TABLE 33-2. *Nutrient Adsorption into Plant Tissue Collected at Various Locations.*

	g/kg of dried material							mg/kg of dried material						
	Ca	Mg	P	K	Na	Cl	Fe	Cu	Co	Zn	Ni	Mo	Mn	B
Stems in ± 0 cm of water	6.13	1.26	1.80	22.81	3.18	17.23	0.43	6.6	0.25	57	2.2	0.30	178	13.1
Stems in 50–75 cm of water	7.09	0.52	1.69	31.29	3.00	27.39	0.31	6.0	0.23	46	2.6	0.25	56	12.2
Stems in 100–150 cm of water	15.02	1.44	1.79	37.93	4.92	35.14	0.57	6.5	0.37	55	2.1	0.17	98	13.8
Stems subject to spraying in 50 cm of water	12.73	1.79	1.48	25.44	5.65	26.29	0.59	2.2	0.32	84	4.8	0.24	42	11.9
Stems from a natural lake in 50 cm of water	4.85	0.26	1.11	4.20	7.55	12.89	0.29	7.9	0.19	20	0.66	0.53	310	13.9
Root stalk in ± 0 cm of water	3.20	0.75	2.02	7.06	0.82	2.94	1.76	9.0	0.76	44	2.8	0.15	161	8.2
Root stalk in 50 cm of water	1.44	0.67	2.73	9.22	0.60	4.42	0.82	12.8	0.27	53	1.8	0.10	110	6.0
Roots in ± 0 cm of water	15.31	1.67	1.55	4.11	1.39	4.43	11.86	41.0	4.6	301	13.0	0.84	602	15.7
Roots in 50 cm of water	13.80	2.27	1.57	8.06	1.65	7.09	7.96	43.4	4.1	514	18.2	0.55	433	18.0

TABLE 33-3. *Water Quality Conditions at Various Stages of the Water Treatment Plant.*

		Water Characteristic Measured					
	Source of water tested	Nitrate mg/l	Nitrite mg/l	Ammonia mg/l	Organic substances mg/l	Phenols mg/l	Coliform bacteria/ml
I	Rhine Water Intake Point	13	0.2	0.5	32.1	0.06	uncount-able
II	Water Treat-ment Plant	16	—	0.1	11.9	0.025	5
III	Recharge Basin	10	0	0	16.1	0	0
IV	Filter Tower	5	0	0	5.5	0	0

there is no sludge accumulation on the pond bottom that would hamper infiltration. This was confirmed by inspection after five years of uninterrupted operation.

Altogether the experience at Krefeld has shown that higher aquatic plants can be used successfully to improve municipal water supply. This should have implications for industrial nations with water pollution problems similar to those experienced in Germany. Aquacultures of bulrushes were proven to be suitable for commercial use, putting findings of the science of limnology to work for the benefit of man.

Notes

1. The sections on chemical treatment and the biological treatment concept were authored by Mr. Czerwenka; that on implementation of biological treatment was written by Dr. Seidel.

34

Water Quality Aspects of Well Recharge with Reclaimed Water, Bay Park, New York

JOHN VECCHIOLI

U.S. Geological Survey
Mineola, N.Y.

Introduction

Groundwater currently (1975) is the sole source of supply for the more than 2.6 million people living in Nassau and Suffolk Counties, the two easternmost counties on Long Island. In 1972, groundwater withdrawals in Nassau and Suffolk Counties averaged 373 x 10^6 gallons per day (gpd) (1.41 x $10^6 m^3$/day). For the year 2000, the estimated water-supply needs of Nassau and Suffolk Counties are between 528 and 595 x 10^6 gpd (2 and 2.5 x $10^6 m^3$/day).[1] Water reclaimed from municipal sewage and recycled through groundwater recharge is being investigated as one of the alternatives to help meet the future water requirements of the area.

One of the recharge methods considered in Nassau County was injection of reclaimed water through coastal recharge wells for creating a hydraulic pressure ridge and, thus, stabilizing or retarding potential landward movement of saltwater into the aquifers. From 1968 to 1973 the Nassau County Department of Public Works operated an advanced wastetreatment plant at Bay Park, N.Y., near the south shore of Nassau County. Reclaimed water from this plant was used in a series of deep-well, artificial-recharge experiments by the U.S. Geological Survey in cooperation with the Nassau County Department of Public Works. Water reclamation and recharge facilities at Bay Park and water quality aspects of the recharge experiments are summarized in this paper.

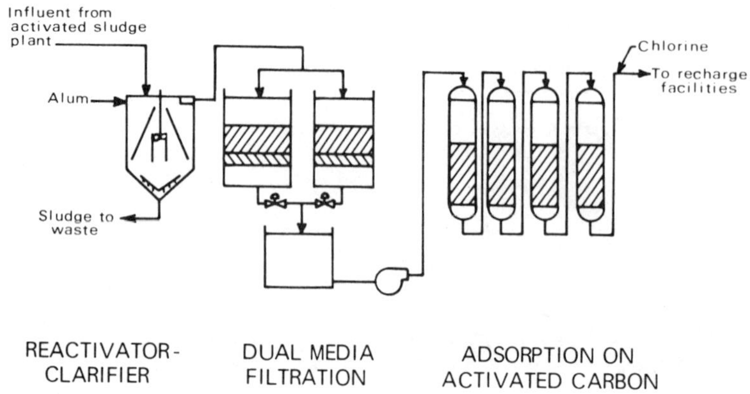

FIG. 34-1. Schematic diagram of the water-reclamation process. (Vecchioli et al., "Wastewater Reclamation and Recharge," *J. Environ. Engr. Div. Am. Soc. Civ. Engrs.* 101 (1975): 201-14.)

Water Reclamation and Recharge Facilities

The water reclamation process is shown schematically in Fig. 34-1. This physicochemical process consists of chemical coagulation and sedimentation using alum, dual-media filtration, adsorption on activated carbon, and chlorination. Because the original objective of the water reclamation plant was to reclaim wastewater for barrier-well recharge to prevent seawater intrusion, nitrogen removal was considered unnecessary and, therefore, was not included in the process. The reclamation process provides tertiary treatment to about 600,000 gpd (2,300 m^3/day), or about 1% of the unchlorinated effluent of the Bay Park Water Pollution Control Plant, which utilizes an activated sludge, step-aeration process.

The effectiveness of the water reclamation process was measured by the degree of removal of suspended solids, phosphorus, dissolved organic components, and methylene blue–active substances. Residual concentrations of these and other significant water quality parameters in the reclaimed water were commonly observed in the ranges shown in Table 34-1.

Reclaimed water was pumped half a mile (0.8 km) to the experimental injection facilities. Major elements of the injection facilities include: (1) a storage tank; (2) a vacuum degasifier, and, within the wellhouse, (3) the recharge well, pumping and controlling equipment, equipment for monitoring various chemical and physical characteristics of the water, and 4 of 18 observation wells on the site.

The injection stratum is within the Magothy aquifer, the principal source of water for Nassau County. The stratum lies between depths of 418 and 480 ft (127 to 146 m) below land surface and consists mostly of fine-to-medium sand. Static head of the injection stratum is about 5 ft (1.5 m) below land surface.

Effect of Injection on Water Quality

During the Bay Park recharge study, changes in quality of the injected water as it moved through the aquifer were observed to determine whether the quality was being degraded or improved. Both degradation and improvement were evident and are discussed in the following sections.

Microbial Aspects

The reclaimed water ordinarily was chlorinated as the final step in the tertiary treatment process. An effective bacterial kill was obtained with a chlorine dosing of 5 to 7 mg/l and a contact time of less than 2 hr. Analysis of the water sampled at the point of injection commonly showed total bacteria counts of a few per 100 ml, total coliform counts of 1 or less per 100 ml, and a total chlorine residual of about 2 mg/l. Under these conditions, microbial growth around the screen of the recharge well was effectively supressed as long as recharge continued. However, whenever there was an appreciable pause between the end of injection and the start of repumping, the first water repumped had a strong hydrogen sulfide odor, was highly turbid, and was laden with bacteria.

An organic-rich mat that accumulates at the aquifer face around the well during recharge serves as a nutrient source for bacterial multiplication once the chlorine residual is dissipated. Depending on the amount of organic material deposited around the well and the length of time between cessation of recharge and start of repumping, a considerable volume of water had to be pumped before turbidity and bacteria concentration decreased to low levels. The hydrogen sulfide odor, developed by the action of sulfate-reducing bacteria in the anaerobic environment of the aquifer, persisted for longer periods. In fact, evidence indicates that the dissolved organic material in the injected water stimulated activity of sulfate reducers in the aquifer even at a distance of 20 ft (6.1 m) from the recharge well.

The accumulation of the organic mat and the attendant regrowth of bacteria supported by it are important in wells used for both recharge and supply. The water pumped from a dual-purpose well needs to be pumped to waste after recharge until it yields water of an acceptable quality for the intended use. For example, in Israel, where dual-purpose wells are used, the initially repumped turbid and odorous water is wasted for 1–4 hr.[2]

On a more positive note, the high filter efficiency of the aquifer's fine-to-medium sand restricts movement of bacteria away from the point of injection. The filtering efficiency of the sand is probably enchanced further by the development of the organic mat around the recharge well. The filtering capacity of the aquifer was demonstrated in a test in which unchlorinated reclaimed water containing substantial total-coliform, fecal-coliform, and fecal streptococcal densities was injected for a 10-day period. No fecal-coliform or fecal-streptococcal bacteria and only nominal total-coliform bacteria were found in water from an observation well 20 ft (6.1 m) away from the point of injection. Studies by others have also shown that travel of bacteria in fine-grained materials is limited to short distances.[3] However, certain pathogens might travel farther through the aquifer, and the absence of pollution-indicating bacteria does not preclude the possibility that pathogens may be present in reclaimed water recovered from the aquifer. Detection of pathogens is judged to be difficult because, if they were present, their density would probably be very low.

Use of a porous aquifer as a biological filter by recharging undisinfected water through wells may not be practical. Biological growth around well screens is recognized as one of the mechanisms causing clogging of recharge wells.[4] At Bay Park, clogging by bacteria was found to be insignificant when the water injected contained a total chlorine residual of at least 2 mg/l. But when unchlorinated water was injected for a 10-day period, clogging was found to be more severe than in other tests in which chlorinated water of lower physical quality was injected. It is possible to redevelop wells clogged with biological growths by shock-chlorination treatments and repumping. However, such treatments can substantially increase the cost of the recharge operation and cause well recharge, for purposes of microbial treatment, to be less cost-effective than surface treatment methods. Perhaps where recharge is to be by wells, the role of a porous aquifer as a biological filter should be viewed as mainly that of a final polishing step or backup safeguard rather than that of a primary means of microbial control.

Chemical Aspects

The reclaimed water used for recharge at Bay Park contained much higher concentrations of

TABLE 34-1. *Chemical and Physical Quality of the Reclaimed Water Injected in Test RW13, October 24, 1972, to May 11, 1973.**

Parameter	Number of determinations	Observed values† Maximum	Observed values† Minimum	Weighted average
Silica (SiO_2)	35	15.	12	13
Aluminum (Al), total	22	400	0	100
Iron (Fe), total	35	1,300	100	410
Iron (Fe), dissolved	35	780	70	230
Manganese (Mn), total	35	210	30	75
Manganese (Mn), dissolved	35	210	30	72
Calcium (Ca)	33	32	16	20
Magnesium (Mg)	34	10	5.2	6.4
Sodium (Na)	33	110	72	86
Potassium (K)	33	15	11	13
Bicarbonate (HCO_3)	33	108	30	59
Sulfate(SO_4)	35	210	120	160
Chlorine, total residual	207	3.5	0	2.1
Chloride (Cl)	34	150	65	99
Fluoride (F)	35	0.8	0.1	0.3
Organic nitrogen (N)	35	1.1	0	0.60
Nitrogen (N), total	35	27	21	24
Nitrite nitrogen (N)	35	0.02	0	0.01
Ammonia nitrogen (N)	34	26	20	23
Nitrate nitrogen (N)	35	0.1	0.0	0.1
Phosphorus as P, total	35	0.48	0.01	0.08
Dissolved solids, residue at 180°C	33	519	298	398
Dissolved solids, calculated from determined constituents	33	571	393	463
Total-solids residue, volatile	33	52	15	33
Hardness as $CaCO_3$ (Ca, Mg)	33	121	63	79
Noncarbonate hardness	33	96	0	32
Specific conductance (micromohs per centimetre at 25°C)	35	970	720	786
pH	210	6.6	4.8	6.1
Water temperature, °C	211	21	15	17.5
Turbidity, as SiO_2	385	1.9	0	0.4
Dissolved oxygen (DO)	204	8.0	0	4.5
Chemical oxygen demand (COD) 0.025*N* $K_2Cr_2O_7$	35	21	2	9
Detergents (MBAS)	35	0.36	0.02	0.07

†Observed values are based on daily or weekly composite samples consisting of four aliquots daily except for temperature, turbidity, chlorine residual, pH, and dissolved oxygen, which were determined on site at least twice daily. Values for aluminum, iron, and manganese are in micrograms per litre; all other parameters except pH, specific conductance, and water temperature are in milligrams per liter.
*After John Vecchioli, et al., "Wastewater Reclamation and Recharge, Bay Park, N.Y." *J. Environ. Eng. Div. Am. Soc. Civ. Engrs.* 101 (1975): 201–14.

most constituents than native water from the Magothy aquifer. Moreover, a reducing hydrochemical environment prevails in the Magothy aquifer. In gross aspect, the chemical quality of the reclaimed water was modified slightly through interaction with the aquifer. Dissolved-solids concentration and the concentration of the major constituents changed little from their initial values. However, displacement of native water by and intermixing with the reclaimed water disrupted the geochemical equilibrium and caused certain responses that affected the quality of the mixed water.

Analyses of water sampled from observation wells showed that pyrite, an iron sulfide mineral in the aquifer, reacted with the reclaimed water and caused a degradation in quality. This reaction produced a notable increase in the Fe^{+2} concentration of the mixed water over that in either the native or reclaimed waters, which was only a few tenths of a milligram per litre in each. Initial detection of the increase in iron concentration at the various observation wells coincided with the arrival of the reclaimed water at the observation wells. Ferrous iron concentrations of as much as 3 mg/l were observed. The increase in iron seemed to be transient and to be associated with the mixed water. As displacement of the native water by the recharged water approached 100%, iron concentration decreased.

Coincident with the increase in iron was a decrease in pH of the mixed water to a low of about 4.5 from levels of about 5.5 for native water and 6.1 for reclaimed water. Minimum values reached were successively lower with distance from the recharge well. As displacement approached 100%, pH increased to background levels. Cation-exchange reactions were the principal cause of the decrease in pH.

Of particular concern was the fate, during movement through the aquifer, of constituents related to the sewage origin of the reclaimed water: nitrogen, phosphorus, organic components as measured by chemical oxygen demand (COD), and methylene blue active substances (MBAS). Analyses of water samples obtained from observation wells during recharge and of water repumped from the recharge well after recharge indicate that only phosphorus was reduced considerably. Mass-balance calculations for one 10-day recharge test and the subsequent repumping period indicate uptake by the aquifer to be: total nitrogen less than 10%, MBAS and COD about 50%, and phosphorus more than 90%. Other tests showed less uptake of MBAS and constituents causing COD, but the uptake of nitrogen and phosphorus was generally as indicated here.

In addition to the nitrogen loss to the aquifer being insignificant, no appreciable change in distribution of the nitrogen species was observed during movement through the aquifer. Virtually all the nitrogen in the reclaimed water was in the NH_4^+ form, and it persisted as NH_4^+ during movement through the aquifer.

From the Bay Park work, one can conclude that where recharge is through wells to a deep sand aquifer consisting mostly of chemically unreactive materials, little improvement in chemica quality may be expected, at least within a few hundred feet of the recharge wells. On the contrary, chemical quality may degrade, as reflected by an increase in iron concentration and a decrease in pH. Each situation, however, needs to be evaluated in terms of the local hydrogeologic environment, the nature of the water to be recharged, and the resultant quality desired.

Notes

1. Temporary State Commission on the Water Supply Needs of Southeastern New York, *Scope of Public Water Supply Needs* (Albany, New York, 1972).

2. M. Rebhun and J. Schwarz, "Clogging and Contamination Processes in Recharge Wells," *Water Resources Research* 4 (1968): 1207-17.

3. J. C. Romero, "The Movement of Bacteria and Viruses through Porous Media," *Ground Water* 8 (1970): 37-48.

4. Task Group 2440-R on Artificial Ground Water Recharge, "Experience with Injection Wells for Artificial Ground-Water Recharge," *J. Am. W. Works Assoc.* 57 (1965): 629–39.

Part VI

Implementation of Alternatives

35

EPA's Response to the Need for Encouragement of Alternative Waste Treatment Techniques

MICHAEL GRAVITZ

Clean Water Fund

Washington, D.C.

Legislative, Statutory, and Regulatory Background

Federal and state water pollution control efforts in the United States are to a great degree dependent on legislation known as the Federal Water Pollution Control Act Amendments of 1972 (Public Law 92-500). After this, I will simply call it the Act. The Act deals with most aspects of water pollution control: research and demonstration, water quality and effluent standards, abatement and control schedules, construction grant funding, and enforcement. It was the result of a long series of hearings and several years of work in both houses of the Congress in an attempt to correct what were perceived to be mistakes and omissions in previous Federal water pollution legislation. This report examines only one part of the Act: its attempt to encourage the use of waste treatment technology that will recycle and reclaim sewage wastes or resources in the building of publicly owned, and largely Federally financed, sewage treatment plants.

The Act sets up a construction grant program run by the Environmental Protection Agency (EPA) under which the Federal government pays seventy-five percent of the costs associated with the planning, designing, and building of a public sewage treatment system. It provided that eighteen billion dollars be obligated in a period of three years (Fiscal Years 1973–1975) for the planning, designing, and constructing of local treatment systems. The program has fallen behind schedule in terms of expenditures, and for several years one half of the money was impounded by the President, but as of 1975 over four billion dollars have been obligated. Levels of obligations will be significantly higher in the next few years with several billions of dollars obligated every year. This will give some idea of the magnitude of EPA's construction grant program.

The statutory and regulatory framework of the construction grant program has a direct impact on the use of biological alternatives like land applications, simulated marshes, and aquaculture for waste treatment. Several sections of the Act are pertinent to the encouragement of waste treatment technology that recycles or reclaims potential sewage pollutants or resources. None are so important as Section 201 wherein the basic philosophy of the construction grant program is laid out and the conditions under which the Administrator of EPA can make a grant are defined. Section 201 is reprinted in the appendix to this chapter. What follows is a short review of the subsections of Section 201 that have direct impacts on the use of alternatives.

Section 201(a) states that it is the purpose of the construction grant program to require and assist the development of waste management that will achieve the goals of the Act, namely the achievement of fishable-swimmable water by 1983 and zero discharge of pollutants by 1985.

Section 201(b) requires that all waste treatment plans and practices provide for the best practicable waste treatment technology (BPWTT) and for "consideration of advanced waste treatment techniques." The phrase "best practicable waste treatment technology," a key one, will be examined later.

Section 201(d) requires the Administrator of EPA to encourage waste management plans for revenue producing facilities that recycle, reclaim, or confine pollutants through agriculture, silviculture, or aquaculture. This is an important subsection because it tries to encourage the productive use of our civilization's waterborne wastes and their reinsertion into natural and humanly beneficial productive cycles.

Sections 201(e) and 201(f) require the Administrator to encourage plans that combine wastewater treatment with the treatment and disposal of other municipal and industrial wastes and with the preservation of open space areas. These sections really encourage integrated and imaginative waste management.

Section 201(g) requires that any proposal for a grant using Fiscal Year (FY) 1975 money must demonstrate that alternative waste management techniques have been seriously considered. This has been interpreted to mean that any proposal to upgrade or construct a new sewage treatment plant must examine three categories of treatment systems: conventional treatment and discharge, land applications, and treatment and reuse. This section reinforces with the strength of a statutory requirement the encouragement the Administrator is supposed to give to alternative waste treatment systems. The section also demands that BPWTT be provided at all treatment plants funded with FY1975 funds and with all future funds.

Section 201(g)(2)(b) says that if the treatment plant one is going to build does not provide for the reclaiming, recycling, or eliminating of pollutant discharges, then provisions should be made for the application of these processes at a later date, where practicable.

Here is what the Senate report said about Section 201 in the Senate version of the Act, a version that did not include the strong recommendations found in 201(d) through 201(f):

> The Committee has, therefore, included a requirement that Federal assistance for the construction of waste treatment facilities must require, where appropriate, practices which will recycle and reclaim wastes and provide for the contained or confined disposal of pollutants. This bill would establish a policy in concert with the fundamental ecological principle that, to the extent possible, all materials

> should be returned to the cycles from which they are generated . . . .
>
> The Committee has undertaken to establish a procedure which requires that each applicant make an analysis of alternative waste management schemes . . .
>
> The Committee emphasizes that the policy in Section 201, read with the policy stated in Section 101, requires the Administrator to direct his research and development authority under Sections 104 and 105 to carry out these policies. This statement of policy coupled with a requirement to consider alternatives as a condition to Federal assistance is intended to overcome the resistance and lethargy present in many planners in the Federal agency, state agencies and in private consulting firms. Planning must move beyond its present orientation to truly integrated and ecologically sound systems. It is expected that the Administrator will give close supervision to this effort in order to develop integrated and comprehensive waste management systems and technology . . . .
>
> [Senate Report 92-414, pp. 23-25]

One can see that even with a weaker statutory basis the Senate really wanted alternative, ecologically integrated wastewater systems to be built.

The second statutory element of concern here is the definition of "best practicable waste treatment technology" mentioned in 201(b) and 201(g). EPA has published information on what techniques are acceptable to meet BPWTT in an important document called *Alternative Waste Management Techniques for Best Practicable Treatment.* In addition they have defined BPWTT as secondary treatment on effluent-limited segments of streams and any level of treatment necessary to meet water quality standards on water quality-limited segments.[2] In effect, these are standards based on the ambient quality of the receiving waters.

The final element of the statutory-regulatory basis for construction grants is the Title II construction grant regulations promulgated by EPA (40 CFR 35.900–.960). These regulations, among other things, tell a planner what questions must be addressed in any planning for a treatment plant. They also include guidance describing how alternative plans may be compared so that the most cost-effective—e.g., the cheapest alternative that meets a minimum goal—will be built. It is necessary to choose the most cost-effective solution both for reasons of resource constraints and because the Act requires it in Section 212 (2) (B) & (C). The guidelines, known as the cost-effectiveness analysis guidelines, have a crucial impact on the construction of alternative types of systems and will be discussed later.

Having introduced the legislative, statutory, and regulatory aspects of Section 201, I would like to offer a critical review of how EPA has carried out the broad mandates of the section, especially as this implementation has affected the use of innovative biological alternatives to conventional sewage treatment.

Critical Policy Overview

The types of treatment plants that would move us towards the accomplishment of the 1983 and 1985 water pollution goals—e.g., plants providing for the recycling and reuse of water and pollutants—are not actively encouraged by the Administrator of EPA. Sections 201(d) through 201(f) require the Administrator to encourage planning resulting in revenue-producing facilities that recycle, reclaim, or confine pollutants in natural systems. Except for one brief memorandum in November of 1974, over two years after the Act was passed there has not been any visible encouragement of these revenue-producing facilities by EPA administrators. The excepted memorandum encouraged that fair consideration be given to land treatment when designing a treatment system, a requirement that already exists in Section 201(g) (2) (A). The types of facilities so prominently encouraged in 201 do not get mentioned anywhere in the Title II regulations, nor can any process for encouragement or incentive to build an alternative to conventional treatment be found anywhere within the guidance or policies of the whole construction grant program.

EPA's definition of best practicable waste treatment technology (BPWTT) is not responsive to the intent of Congress and will hinder efforts to construct treatment systems using biological alternatives. Since the Administrator can make no grants with FY'75 money and thereafter unless the proposed facility will give BPWTT, the definition of what constitutes BPWTT determines on a nationwide scale the levels and types of treatment EPA will fund. The definition provides in a nutshell EPA's whole sewage treatment strategy, which is to require nothing more than conventional secondary treatment on stream segments where water quality standards (many not yet raised to reflect 1983 goals) can be met by this level of treatment and higher levels of treatment necessary so that water quality standards can be met on stream segments not presently meeting the standards.

There is much evidence to show that Congressional intent on BPWTT may well have been aimed at improving the minimum acceptable level of treatment used everywhere by at least one quantum jump beyond secondary. Beyond this, the ultimate desirability of systems that recycle and reclaim wastes is emphasized in 201. Many would argue that BPWTT is technology capable of recycling and reclaiming wastes.

EPA chose not to heed this Congressional mandate apparently because it felt that it was under very serious resource constraints and because it felt that the "natural assimilative capacity" of surface waters should be used to degrade wastes. This is despite the Act's objective: "to restore and maintain the chemical, physical, and biological integrity of the Nation's waters."

What then are the problems that EPA's definition of BPWTT presents to the use of biological alternatives? The first problem is that having low minimum effluent standards (secondary treatment) for some segments of streams acts to depress the consideration of high-quality effluent yielding treatment alternatives. Alternatives in a cost-effectiveness analysis are compared on the basis of their monetary costs and ability to achieve the minimum effluent quality acceptable on that segment of the stream. This will generally lead to a choice of the cheapest alternative that can meet the minimum applicable standard. On effluent-limited segments, where approximately sixty percent of the publicly owned treatment plants are located, cost-effectivenss analysis will compare a secondary treatment and discharge plant with one that can produce a much higher quality of effluent and may have no stream discharge at all. Since many biological alternatives to secondary treatment use some form of secondary treatment prior to their final process, they will in many cases be more costly than secondary treatment. This means that they will lose a comparison with secondary treatment on effluent limited segments. In this way the minimal secondary standards will depress the usage of biological alternatives.

The document that provides information on BPWTT sets up a double standard for effluents from treatment plants. There is one standard for the discharges to surface water by secondary treatment plants and another more stringent standard for land application systems unless they are underdrained or are of the overland-flow type. The stringent standard, the same as the public health standard for untreated drinking water supplies, may put land-oriented resource recycling and reclaiming systems at a considerable competitive disadvantage. While this should not be construed as an argument for the pollution of groundwaters, it is clear that these double standards give some advantage to conventional secondary treatment plants discharging to surface waters. The double standard ignores the hydrologic connection in many areas between polluted surface water and groundwater, and it puts a burden of proof on a land treatment system to show that it will not pollute groundwater while giving *de facto* acknowledgment to the pollution of surface waters. Potential violation of groundwater standards will become another thing that public officials can use to hide behind when they deny land treatment proposals. It might be difficult in many cases to legally and conclusively prove that systems will not violate these standards.

Section 201(g)(2)(B) very simply says that if a proposed treatment plant does not provide for recycling, reclaiming, or otherwise eliminating sewage pollutants, one must provide for the addition of technology that can do this at a later date. The implication here is that one is supposed to build in enough system flexibility to allow this or actually begin planning to add this technology at a later time. This simple directive barely makes it into the Title II construction grant regulations and receives reiteration, not expansion, in the document that EPA distributes as guidance to those people planning a sewage treatment system.[3] Section 201 (g)(2)(B) appears in the regulations as one of those things the Administrator must establish about a proposed project before he can approve funding. Yet there is no requirement in the regulations for this concern to be addressed in the 201 Facilities Plan, which generally describes a proposed project. Thus, even though the Administrator must be shown that this concern is satisfied before he can approve a grant, there is no regulation requiring that the grant applicant show it.

Why is this important? It is important because flexibility enabling the addition of technology that can recycle or reclaim pollutants is not an inherent product of the present treatment-plant design process. Treatment plants, for example, may be intentionally consciously sited close to soils with good renovation capabilities. If land application is a genuine future possibility, one probably would not want to site the secondary pretreatment at a low point in the system on a floodplain next to a river, when many treatment plants are now put. Since many alternative biological treatment systems do recycle, reclaim, and confine pollutants, and since many could be added to treatment and discharge systems, they are prime candidates for addition at a later date as suggested in 201(g)(2)(B). Yet if planning does not remain flexible, the future addition of one of these alternatives in many cases may be precluded.

Another reason why Section 201 is so important is because it forces planners to give a hard look at the goals of the Act, which are to recycle, to reclaim, and otherwise to integrate ecologically waste treatment or resource systems. It is the only section which forces or ought to force people to look at their plant design concept as an ultimate design solution which needs to fit into natural cycles of materials.

Minimizing public expense in doing any job is an admirable goal. This is what the superficially neutral cost-effectiveness analysis is designed to do. Yet there are several reasons why this analysis may discriminate against the use of biological alternatives. The systems that are compared under cost-effectiveness can be very different if they all produce at least the same minimum quality of effluent required. As we have already seen, the comparison many times is between a system producting an effluent of secondary quality, one in which you probably would not want to swim, and a system that produces effluent that the EPA calls fit to drink. This is like comparing apples and oranges. Further, how is one to compare a system with a discharge with one that in many cases does not have one? Does not the Act seem to favor the latter?

Another problem with these guidelines is that normally only the construction, operation, and maintenance costs of each alternative will be compared. However, suppose that there is a monetary cost associated with a discharge to a stream in terms of lowering its recreational value, or a benefit to be gained by having no discharge and getting an increase in recreational or other use potentials. Monetary costs and benefits associated with social and environmental changes due to the choice of alternatives will not enter into the cost-effectiveness comparison. This will hurt consideration of biological alternatives that can provide social and environmental benefits. A cost/benefit analysis approach, reflecting the total true costs and benefits to society, ought to be done when social and environmental costs and benefits are likely to be significant.

In practice, because of these current guidelines, it will be very difficult for a town to get a grant to build a plant that provides for recycling or reclaiming on an effluent-limited segment where all that is required is the usually less costly alternative of secondary treatment.

Finally, it is easy for an engineering firm to do an unrealistic and unfair job of estimating the costs of a land treatment system or of other biological alternatives.

Some Facts About EPA's Program

> It is the national policy that a major research and development effort be made to develop technology necessary to eliminate the discharge of pollutants into the navigable waters.
> (Section 101(a)(6) of the Act)

Unfortunately high sounding phrases do not necessarily mean that a program happens or policy is implemented. There has in fact been no national effort aimed at achieving the elimination of pollutant discharges. The amount of money spent on research and development (R&D) for control of municipal pollutants has remained basically the same since the Act was passed, about nine to ten million dollars per year.[4] The level of spending on municipal control technology has actually dropped, in 1968 being over twice ($25.2 million) what it is now. There are many conflicting priorities in the overall R&D budget, but clearly funding of municipal control technology and R&D for control of municipal pollutants is diminishing in the face of an ever increasing construction grant program.

How does the budget for land applications and other alternative treatment and reuse systems stand in the municipal R&D budget? Since 1971 the budget for research on land treatment has been rising from around $115,000/yr to around $3.2 million/yr in FY '75. This is deceptive however, because without a supplemental appropriation by Congress the FY '75 budget would have been only $800,000. It is unclear whether this higher level of funding will prevail in future budgets. There is even less money, several hundred thousand dollars, going into aquaculture and simulated marsh or swamp ecosystems. Despite the increase of funding, research on land treatment and other biological alternatives has very little money with which to find necessary questions.

Congress intended techniques that recycle, reclaim, and confine resources/pollutants to get widespread use. But Congress did not foresee the depth of resistance by Federal, state, and local officials and consultants to these kinds of systems. The resistance is partly based on many questions that have justifiably been raised about land treatment and similar techniques. Many of these same questions could and should be raised about more conventional treatment and discharge systems. Much research, particularly in the area of health effects, is needed to soften this resistance. This critical area of research was not funded until FY '75 and the projects are only just starting. Unfortunately health research can take much time and large amounts of money.

The minimum annual budgetary needs for R&D just on land treatment systems is estimated to be from $2.5 million to $3.5 million for the next ten years. Health-effects research will require at least another $1 million/yr for several years. Whether this level of effort is forthcoming may determine the future use of alternative biological treatment systems.

Even if the research is done, the results must

somehow filter down to consulting engineers so they can use the information. This is particularly important in the area of alternative biological systems because most sanitary engineers know little, if anything at all, about these kinds of treatment. EPA has an office called Technology Transfer that holds short seminars for engineers to "bridge the gap" between research and practical use. This group has held about fifty training seminars on municipal technology since it started four years ago, but did not hold a single seminar on land treatment or other biological alternatives until late 1974 and 1975 when a meeting on land treatment was held in each of EPA's ten regions. This is very heartening. A problem with the Technology Transfer sessions is that they are short, broadly ranging lectures rather than intensive design and operation courses. There is much room for spreading land treatment information both ways. Given the general lack of familiarity sanitary engineers, have with land treatment and the complexity of design, EPA should soon offer short, intensive courses on land treatment.

One may well ask what in fact EPA has done to build the kinds of treatment plants encouraged by the Act. Evidence from EPA's construction grant activities shows its lack of encouragement of alternative types of treatment systems. Spending on the construction of land treatment projects totalled approximately $43, 418,500 between October 1972 and January 1975. This is approximately 2.1% of EPA's expenditures on new sewage treatment plants or for increasing the treatment levels in old plants.[5] No figures were developed for pre-Public Law 92-500 alternative systems funding, but there is no indication that their share of the program was different then. Even if the share has increased since the Act was passed, the present level of spending is still disappointingly low and leads to a rather negative conclusion about EPA's effort in the area of alternative techniques. Over four billion dollars in construction grants have already been spent at the above low rate of spending for alternatives and much more will be spent in the near future (six billion dollars before July 1976[6]), probably at this same rate. One EPA construction grants official expects at least two-thirds of future construction funding to be for activated sludge plants.

Institutional Obstacles

One obstacle in getting more widespread use of resource recycling/reclaiming systems is the lack of expertise with these treatment techniques in Federal, state, and local governments and in the private engineering sector. Many alternative types of waste treatment called for in the Act require knowledge not traditionally learned by those engaged in the design, review, operation, and regulation of more conventional facilities. There is a necessity for people in such diverse fields as hydrology, fresh and marine aquatic biology, soil science, and agronomy to work with sanitary engineers in the design and operation of alternative treatment systems. These treatment systems require more design and evaluation effort because they are tailored to fit such environmental parameters of a specific site as soil, climate, topography, fauna and flora, and hydrology. The need for multidisciplinary design teams and the complexity of design for these as opposed to conventional tank systems may be reasons for the reluctance to employ alternatives more widely.

A review of personnel statistics for nine-hundred engineering firms doing water pollution work showed few hydrologists, still fewer biologists, and no soil scientists or agronomists.[7] Firms can certainly hire outside people on a contract basis, but these brief statistics show what engineering firms are normally prepared to do in terms of design work. Many people knowledgable about land treatment say that most engineering firms are not even competent to do a good review of the possibility for such a plant, let alone do a good design for one.

EPA could create multidisciplinary teams, including the types of personnel mentioned above, in each of its regions to give technical assistance to communities that are seriously considering alternative biological systems and to review grant applications for these types of systems for competency in design. As a complement to technical assistance teams at a regional level, assistance on water rights, groundwater laws, contractual agreements, and intergovernmental cooperation ought to be available to local officials facing the sometimes unique legal and institutional problems of an unusual waste treatment system. Since as suggested in subsections 201(d) through 201(f) of the Act there is often a multipurpose water resource potential in alternative biological systems that is largely unexploited today, regional offices should provide assistance in this area too.

Another institutional obstacle preventing EPA from carrying out the mandate of Section 201 is lack of adequate intra- and inter-agency coordination and cooperation on alternative waste treatment systems. There does not seem to be enough interagency cooperation on research and development of biological alternatives and on encouraging the use of these techniques. There does not seem to be any coordination between EPA and Federal land-controlling agencies (U. S. Departments of Agriculture and the Interior) where wastewater management could be combined with open space and recreation. Some work needs to be done with the Food and Drug Administration on state health laws affecting the use of land treatment. Finally there seems to be room for utilization of the expertise in land treatment and other alternatives that already exists at universities in this country and elsewhere in the world.

Within EPA, the people in construction grants and R&D need to improve their coordination. There are many urgent research needs that should be more effectively communicated to the people in R&D from the operations people who see the day-to-day problems and the excuses for not utilizing land treatment. This situation will probably improve as more funding becomes available for R&D. Coordination must also take place with the construction grants Facilities Evaluation branch, which oversees the planning (201 Facilities Plans) for sewage treatment systems. All the R&D in the world will not make a bit of difference when an alternative is not seriously considered in the planning stage. If it is true that the land treatment alternative does not normally receive fair and comprehensive review for environmental acceptability and costs in a 201 Facilities Plan, then the Facilities Evaluation branch has the responsibility to correct this. For example, regional reviewers of 201 Plans must know the minimal elements of a land treatment system that need to be examind before a cost-effectiveness analysis is considered adequate and fair.

Finally, there is little discernable coordination between the branch level people and higher levels within EPA where policies are made that affect biological alternatives. It is from the higher levels of EPA, the assistant administrators and deputy assistant administrators, that encouragement of biological alternatives should be flowing towards the regions and state agencies. This has not happened except in one recent case already mentioned.

Probably the largest institutional obstacle that prevents the more widespread use of resource recycling/reclaiming systems is the lack of incentives for this in the grant program. In fact there are several obstacles to the use of these systems that have already been mentioned: a) the improbability of getting these kinds of systems built on effluent-limited segments of streams, b) the much more stringent groundwater discharge criteria for land treatment plants, and c) the fact that the benefits of a zero discharge of pollutants into a stream are not taken into account in cost-effectiveness analyses. All these discourage the employment of alternative methods of biological treatment. On the other hand, there are no incentives to build a system that recycles, reclaims, or confines pollutants or resources other than a feeling of responsibility to our environment.

Notes

1. This study was carried out under the auspices of the Center for Study of Responsive Law and the Clean Water Fund, Washington, D.C.

2. Effluent-limited segments are those segments where water quality is meeting and will continue to meet water quality standards or where there is adequate demonstration that water quality standards will be met after application of secondary treatment for municipal sources, required by Section 301(b)(1)(E), and application of best practicable treatment (BPT) for industries, as required in 301(b)(1)(A) of the Act.

Water quality-limited segments are those segments where water quality does not meet present standards and is not expected to meet water quality standards even after the application of the secondary treatment for municipalities and BPT for industries.

3. U.S. EPA, *Guidance For Facilities Planning* (Washington D.C.: 1974).

4. There was a $5 million supplemental appropriation for municipal control technology for FY'75.

5. These figures are near approximations, as accurate as EPA data allows one to get. In order to find the amount of money spent on treatment plants, I looked at all PL 92-500 grants inclusive of January 1975. I counted step 2 (plans and specifications) and step 3 (construction) grants for treatment plant projects involving new plants and upgrading the treatment levels of old ones. Excluded were grants for expansion in capacity only, modification of the facility, or grants only for collection systems. Unfortunately, there was no way of separating the costs on a grant that went to treatment plants and collection systems. This may be cancelled by the large number of grants where it was impossible to tell what the grant was for; these were not counted.

6. Statement by Russell Train, Administrator of EPA, before the Subcommittee on Environmental Pollution, Committee on Public Works, U.S. Senate, Feb. 4, 1975.

7. Joe Middlebrooks, "A National Survey of Manpower Utilization and Future Needs of Consulting Engineering Firms Engaged in Water Pollution Control" (Washington, D.C.: Office of Water Programs, EPA and Consulting Engineers Council of the U.S., March 1972).

Appendix. Selected Sections of the Federal Water Pollution Control Act Amendments of 1972.

"TITLE II—GRANTS FOR CONSTRUCTION OF TREATMENT WORKS

"PURPOSE

"SEC. 201. (a) It is the purpose of this title to require and to assist the development and implementation of waste treatment management plans and practices which will achieve the goals of this Act.

"(b) Waste treatment management plans and practices shall provide for the application of the best practicable waste treatment technology before any discharge into receiving waters, including reclaiming and recycling of water, and confined disposal of pollutants so they will not migrate to cause water or other environmental pollution and shall provide for consideration of advanced waste treatment techniques.

"(c) To the extent practicable, waste treatment management shall be on an areawide basis and provide control or treatment of all point and nonpoint sources of pollution, including in place or accumulated pollution sources.

"(d) The Administrator shall encourage waste treatment management which results in the construction of revenue producing facilities providing for—

"(1) the recycling of potential sewage pollutants through the production of agriculture, silviculture, or aquaculture products, or any combination thereof;

"(2) the confined and contained disposal of pollutants not recycled;

"(3) the reclamation of wastewater; and

"(4) the ultimate disposal of sludge in a manner that will not result in environmental hazards.

"(e) The Administrator shall encourage waste treatment management which results in integrating facilities for sewage treatment and recycling with facilities to treat, dispose of, or utilize other industrial and municipal wastes, including but not limited to solid waste and waste heat and thermal discharges. Such integrated facilities shall be designed and operated to produce revenues in excess of capital and operation and maintenance costs and such revenues shall be used by the designated regional management agency to aid in financing other environmental improvement programs.

"(f) The Administrator shall encourage waste treatment management which combines 'open space' and recreational considerations with such management.

"(g)(1) The Administrator is authorized to make grants to any State, municipality, or intermunicipal or interstate agency for the construction of publicly owned treatment works.

"(2) The Administrator shall not make grants from funds authorized for any fiscal year beginning after June 30, 1974, to any State, municipality, or intermunicipal or interstate agency for the erection, building, acquisition, alteration, remodeling, improvement, or extension of treatment works unless the grant applicant has satisfactorily demonstrated to the Administrator that— Conditions.

"(A) alternative waste management techniques have been studied and evaluated and the works proposed for grant assistance will provide for the application of the best practicable waste treatment technology over the life of the works consistent with the purposes of this title; and

"(B) as appropriate, the works proposed for grant assistance will take into account and allow to the extent practicable the application of technology at a later date which will provide for the reclaiming or recycling of water or otherwise eliminate the discharge of pollutants.

"(3) The Administrator shall not approve any grant after July 1, 1973, for treatment works under this section unless the applicant shows to the satisfaction of the Administrator that each sewer collection system discharging into such treatment works is not subject to excessive infiltration.

"(4) The Administrator is authorized to make grants to applicants for treatment works grants under this section for such sewer system evaluation studies as may be necessary to carry out the requirements of paragraph (3) of this subsection. Such grants shall be made in accordance with rules and regulations promulgated by the Administrator. Initial rules and regulations shall be promulgated under this paragraph not later than 120 days after the date of enactment of the Federal Water Pollution Control Act Amendments of 1972. Rules and regulations.

36

Notes on the Implementation of Alternatives

CHARLES FRANKENHOFF

Graduate School of Planning
University of Puerto Rico

As an economist and planner I confess to an immense interest in biological alternatives to water quality improvement as well as to a felt need to learn from limnologists, ecologists, virologists, and other "hard" scientists, but I feel more at home with the "softer" issues of institutions, law, economics, and decision making needed for implementation of alternatives. Implementation of new water quality and waste reclamation programs requires something more than just "change."

Perhaps some people are correct in hypothesizing that any environmental control technique is a negative act—it costs money and does not improve profits. But I do not feel that this extreme view can be justified. Environmental control, in the cybernetic sense, is a positive act. It costs money and improves profits, both economic and social. If people really do not want environmental control, it is possible that they do not understand it or its components.

The Institutional Component

First of all, there is the environmental policy institution. It seems clear that the U. S. Environmental Protection Agency has dedicated little research to biological water quality improvement alternatives and even less to the significance of low levels of carcinogens in drinking water. The stock response is insufficiency of funds, but the real question is the structural capacity of EPA as an environmental policy institution. I should like to suggest that EPA as an institution simply lacks such a structural capacity. EPA lacks the political and financial autonomy necessary to make environmental policy. Its policy personnel are responsible to, and can be removed at whim by the reigning political party. The Congress must approve its budget on a year-to-year basis. This lack of autonomy is disastrous in a new and controversial public policy arena where the lines are already drawn between entrepreneurial developers and environmentalists. It is highly doubtful that EPA can be effective in making and enforcing environmental policy until it has achieved relative autonomy from ad hoc political and economic pressures. The policy makers of EPA require a tenure fixed by law in order to plan and to program effectively. At the same time EPA should be supported on a continuing financial basis by its clients, namely those production systems who use and deteriorate the resources for which EPA is responsible.

Is there any precedent for such an institution in the federal government structure? There is. Consider the Federal Reserve System. The Federal Reserve, which is responsible for making monetary policy in the United States, is a relatively autonomous feedback system whose board members have a fixed tenure of fourteen years. It need not go to Congress annually because it is financed by its own operations in the U. S. Government bond markets. Even more to the point, the money and credit that it controls is, by analogy, a common property resource.

The common-property resources of ambient air, water systems, public space, and complex ecological systems are the stuff of environmental policy. No less than money, they are resources belonging to the patrimony of the community and are essential for the operation of any production system, public or private. Industry, energy, urban infrastructure, transport, residence, commerce, and agriculture are the major production systems that require measurable amounts of common property resources. Moreover these resources, largely unpriced by the market, deteriorate with use. The environmental costs generated by the production systems are called externalities because they enter into production functions at zero cost and could until recently be ignored by the decision-maker in his choice of production system inputs.

Environmental Common Property Resources

The common property resources of the environment are the second component of environmental policy control. In fact, the basic function of environmental policy is to internalize environmental common property resource costs so that public and private decision makers take them into account in their selection of production functions. The environmental policy institution, EPA, acts as surrogate owner of the common property resources with the objective of controlling their use for the benefit of public health and welfare. In other words, the job of EPA is to protect the environment to the extent that it functions as an essential input into the lives of people and their communities.

Those citizens who feel that the role of environmental policy is something of an anachronism in a depression-inflation situation are ignoring the essential role of the common property resources of air, water, and public space in the major production systems already mentioned. These resources are no less essential than capital, labor, technology, and energy and, naturally, they deteriorate in use. The fact that such deterioration or pollution affects negatively the health of the community vests these resources with the public interest and requires the intervention of a stable and continuing public policy.

The User Charge Component

The form that public intervention takes is the third component of this environmental policy control discussion whose first two components are an autonomous institution and the nature of common property resources. Specifically, the form I, together with many economists, would like to recommend is the user charge.[1] But first, let me put the argument in perspective, even at the risk of oversimplifying.

The environmental policy institution must be able to say in the name of the public: Look, you producers (and distributors and consumers) have been using common property resources freely for many years. In effect, this free use is a subsidy from the community. Now we find that these resources have been polluted to the point where they represent costs that must be paid in terms of human health and deteriorating natural ecological systems. Environmental standards must be established through careful research on a region by region basis and then met. Within these standards we wish to leave you free in your decision making so that the socioeconomic system of which we are members will not be distorted. We do not want to distort the initiatives of decision making that presently exist, no more than does the Federal Reserve in its control of interest rate patterns wishes to eliminate the power of monetary decision makers. In fact, we want to open up your options, always taking into account the environmental cost generated by your decisions.

To achieve a proper environmental control, the environmental policy institution must, in the name of the public interest, assign prices to the common property resources that will permit you to evaluate their use with reference to your own production system and within publicly established standards for environmental resource pollution. It is true that public intervention can take other forms: it can prohibit; it can regulate; it can sanction by tax or subsidy; it can urge voluntary cooperation; it can order the use of antipollution technology. But each of these methods of public control has its price, both for the environmental policy institution and for its clients. The user charge makes this price explicit and can be more readily assimilated into the dynamics of the market system for decision making. Furthermore, the user charge applied to common property resources offers marked advantages to the public institution by (1) providing revenue, (2) rationing the use of these scarce resources, (3) distributing the environmental costs more equitably, and (4) generating information on the demand for these resources.

A more detailed discussion of the common property resources of air, water, and public space as public goods, the difficulties of identifying their beneficiaries, and the crucial issues of establishing shadow prices and public policy incentives is beyond the scope of this chapter.[2] I would like to suggest, however, the basic cost elements in the user charges relevant to the establishment of a responsible environmental policy institution.

A user charge for common property resources used by a production system should include the following costs: (1) capital recovery, on the assumption that clients should be required to contribute to the plant and equipment investment required by the municipality or region to treat their waste; (2) research and development expenditures needed to identify and control local environmental costs; (3) current operational costs of running an adequate environmental policy control institution; (4) a penalty charge for those clients who do not satisfy nonabatement procedures; and (5) an insurance fund for low cost loans to clients whose current economic condition does not permit them to adjust their production system to meet environmental standards.

At least one question remains: Where does the user charge come from? It seems clear that sophisticated studies of common property resources and their pollution by industry, transport, and electrical energy producers have not yielded vectors of monetary costs. Neither have careful studies of human health as affected by polluted air, water, or public space yielded such results. There have been some unsatisfactory searches for economic damage functions, but these have been sterile in practice. To take one example, attempts to price the impact of air pollution on a variety of human beings under a variety of conditions within a region have been fruitless. For one thing, synergism prevents a simple tracing of pollution from producer to receptor.

The Production Function Approach

The production function approach has been at the heart of Allen Kneese's pioneer work in environmental economics (see Chapter 38). Unfortunately, in my view, Kneese has been lured along the primrose path of welfare economics, but his insights into the importance of evaluating environmental controls with reference to production functions (defined as interrelationships between inputs and outputs of a system) are still valid.

The advantage of using the production function in order to discover user charges is twofold: one, this is the decision making point; and two, the environmental control functions on an *ex ante* basis rather than after the fact of excessive pollution. When the producer, be he farmer, industrialist, or homeowner, decides what to produce, how much, and how, he is determining his production function. He is consequently determining the use and deterioration of common property resources in the local environment. When the environmental policy control in the form of a user charge is inserted at this point, then intolerable levels of pollution should be eliminated. If environmental policy waits until pollution crises occur, it is already too late for the receptor community.

But, you say, we have no adequate information on production functions, and the cost of generating such information would be prohibitive. Granted. So we begin by establishing an arbitrary user charge, related to value-added or to gross sales. Presumably the State of Washington lacked such information when it established a "tax" on the gross income of industries whose products were polluting the environment. It began with a low rate, therefore, yet a rate sufficiently high to generate necessary funds for the State environmental control institution. Once experience is gained, the user charges can be slowly tailored to the joint needs of the production systems and of the community for clean air and water and uncongested public space. Evidently local land use policies will have an important role to play in the gradual determination of user charges.

Perhaps my basic philosophical conclusion is that there is a need for action now to control the ongoing pollution of scarce common property resources and to continually correct environmental policy on the basis of information feedback from the systems that environmental policy wishes to control.

Notes

1. See Chapter 38, "Implementation of Water Quality Plans," by A. V. Kneese.
2. A good place to begin such a discussion is articles by Peter O. Steiner, Julius Margolis, and Charles L. Schultze in the U.S. Congress, Joint Economic Committee, *The Analysis and Evaluation of Public Expenditures: The PPB System*, 91st Congress, 1st Session, 1969. Volume 1.

37

Legal and Political Restraints to Implementation of Novel Systems

WILLIAM M. EICHBAUM

Deputy Secretary for Enforcement and General Counsel
Pennsylvania Department of Environmental Resources

As an attorney I might have been expected to write a highly legal treatise concerning the topic of this chapter. However, insofar as most readers will not have a legal background and since the issues addressed are not precisely of a legal nature this seemed inappropriate. Accordingly, where the law is touched upon it is done so only in a very general fashion and without the precision of reference or citation that would have otherwise been appropriate. My primary purpose has been to draw upon several years of experience in the enforcement of environmental laws with a view to illustrating problems that should be in the forefront of thinking about novel technology.

The papers in this volume have largely focused upon wastewater treatment systems employing treatment techniques whereby living plants or animals are used in the wastewater renovation cycle in combination with the natural environment in order to produce an acceptable effluent. These systems range from those that are purely theoretical to others that have been proven through a number of years of use on a limited scale. One of the hallmarks of the systems discussed in preceeding chapters has been the notable extent to which they have not been accepted for general application in wastewater mangement-treatment systems. Undoubtedly, this is in part owing to the present technological uncertainty concerning their design and application; however, a substantial cause of the resistance to greater utilization, although perhaps stemming from their technological strangeness, must be attributed to other reasons. The purpose of this paper is to explore certain of these reasons.

Historically, society has rarely, if ever, welcomed either a scientific or intellectual breakthrough. In fact, such events have been regarded with skepticism, if not outright hostility. Socrates was put to death for the simple crime of posing to society a new method for thinking. More recently, Galileo was condemned by society for posing new theories based upon work with his technological advancement of the telescope. Prior to the last century, such negative reactions to intellectual innovation were undoubtedly largely a product of society's perception that change was in fact a challenge to the very political structure upon which the survival of society depended. However, perceptions have changed, as have conditions, and today thought may generally go forward and probe new areas without the risk of loss of life or of intellectual or even physical exile. For example, finding solutions to the problems of renovating and maintaining, in a clean state, the air and water of the earth is today counted as one of the most important tasks for society. The intellectual innovation inherent in this task is counted as essential to society's survival. Annually several billions of dollars are spent, worldwide, toward this goal. Thousands of highly trained scientific minds are at work using the most sophisticated scientific technology in order to better understand and predict the impact of human activity upon natural systems. International, national, and subnational public and private organizations work assiduously to achieve the goals of a cleaner and more healthful environment. Yet there appears to have been little dramatic improvement in the techniques of management of man's wastes over the last twenty-five years and innovative processes or concepts that may have dramatic impact, such as biological renovation alternatives, are relegated to a minor position in the effort. That this should be so, in spite of our knowledge of the existence of a problem and an apparent search for solutions, can be confounding and the reasons worth exploring.

Modern society, particularly in the United States, invariably reacts to a perceived problem by enacting legislation, rules, and regulations that are designed to effect a change in human action so that the problem will be mitigated. This has been particularly true in the environmental field and the process dates back from at least before the turn of the century. Legislative development, however, reached monumental proportions in the early 1970s as environmental health became a concern of the whole society rather than the province of a few sanitarians working in the public health service. Notable examples are the Federal Water Pollution Control Act of 1972 and the Federal Air Pollution Control Act of 1970, as well as numerous enactments by many state legislatures. This legislation has been exemplified, particularly at the federal level, by several common characteristics: (a) a new demarkation of public ownership of resources, at least for the purposes of regulating activity that causes harm to those resources; (b) the establishment of standards applicable to human activity; (c) the establishment of time frames for achieving those standards; (d) the creation of a planning process; and (e) the creation of a large bureaucracy. All of these characteristics, with the possible exception of the last, are counted as constructive steps necessary for protection of the natural environment. However, each may, singly or in combination, pose impediments to the development and application of novel technology in this effort. A brief examination of each of these characteristics will illustrate the problems posed.

Most state statutes designed to control water pollution set forth, either explicitly or implicitly, the notion that the surface and possibly subsurface waters within the jurisdiction belong to the people in general, at least insofar as the people have the right to determine what the quality of those waters shall be. Frequently, this premise is merely a continuation of existing legal principles. In addition, federal law also now asserts a concurrent power in the federal government to exercise control over those waters for the same purposes.[1] In practice, this assertion of control has meant that the effluent discharged from a facility must meet applicable water quality standards at the point where the discharge enters the public waters. Obviously the discharge may be either to subsurface strata, to wet weather ditches, or to flowing streams. The normal technological response to this notion has been for the polluter to construct some form of treatment plant on his property, the discharge of which will in fact meet the water quality standards that have been established for the receiving streams.

Unfortunately, many novel techniques of waste water management, particularly those relying upon the use of green plants for renovation of the waters, work most effectively if the aquatic flora of the receiving waters are managed in such a way as to complete the wastewater renovation cycle, producing an instream water quality of satisfactory levels. In effect, these processes may contemplate use of the stream and management of its characteristics so that an adequate effluent is ultimately produced.[2] This is, of course, completely contrary to the notion that the discharge must meet water quality standards at the point of entry into the publicly-owned waters and, therefore, the most efficient and effective application of these novel techniques is precluded.

The establishment of firm numerical standards and particular dates for achieving those standards, either pursuant to statute or regulations, is an integral part of the cleanup process. There can be little doubt that it is essential for all parties involved in a regulatory program to be clearly aware of what standards of conduct are permissible. It has been indicated above that the standards concept, when combined with ownership of water concepts, creates problems for the implementation of new technology. This problem, however, becomes most severe when one is faced with an absolute date by which a goal must be achieved. Where such dates are unrealistic, as is the case with most of the dates contained in federal legislation, the opportunity to explore new techniques for solving pollution problems is virtually eliminated. Such dates place an overwhelming burden on both the conscientious regulatory agency and the conscientious discharger to enter understandings or commitments whereby the *appearance* of meeting the standard by the appropriate date becomes an overriding consideration. The response of the discharger to this situation is to grab out of the bag of known technology that device which it is hoped will achieve, or come close enough to, the applicable standard. Within this framework, there is little opportunity for developing or testing the opportunities presented by applications of experimental technology. The time by which the standard must be met is too short to permit this and the consequence of failure, which is inherent in applications of new technology, is too serious to contemplate; therefore, the opportunities are foregone at the very start.

Pollution control legislation increasingly has mandated rigorous and comprehensive planning processes. This trend is seen most dramatically in the 1972 Amendments to the Federal Water Pollution Control Act, which provide for at least three different wastewater management planning programs. The statutory thrust is not only that the planning will be done but that, in addition, the planning will be implemented. The ramifications of this novel notion are dramatic and far beyond the scope of this paper. The basic goal, however, is to establish an ongoing planning process designed to articulate solutions for perceived future wastewater management problems with increased precision of facilities and techniques as the scale of the plan decreases.

An inescapable fact of such planning exercises is that no one can know what will be developed in the future that will prove to be an effective form of control technology. Accordingly, a planning process that poses rather precise technological solutions to perceived conditions on a ten- or twenty-year time horizon makes most difficult the application of present experimental technology. It also acts to discourage the exploration and development of new technology in the future. In the case of the Federal Water Pollution Control Act this problem is particularly severe because the planning process is closely tied to the permit and grant procedures of the federal government. The natural result is for plans to be drawn that make use of exisiting technology as the most expedient solution to perceived problems and unfortunately to leave little or no room for innovation as time passes.

Beyond those restraints created by legislation, the institutions of government that have grown up in response to legislative direction pose several constraints to the implementation of novel technology. These institutions are made up of a large number of people with varying responsibilities and widely divergent areas of expertise and are, in many cases, spread out over a wide geographic area. In addition, with the possible exception of the federal Environmental Protection Agency, most agencies perceive their prime mission as being to achieve compliance with the law by those subject to it, rather than the development of the necessary technical mechanisms. Inherent in such large organizations with diverse makeup is the problem best exemplified by the notion that "the right hand doesn't know what the left hand is doing." It is important to note that the consequences of such behavior frequently preclude adequate exploration and attention to novel technology.

There are three characteristics of environmental regulatory organizations posing barriers to innovative technology that should be specifically addressed. In the first instance, the normal structure of such an agency is to have a central office, generally at the center of governmental activity in the jurisdiction, with a number of regional offices scattered throughout the jurisdiction. The prime mission of the central office is, of course, to establish policy, to oversee the activities of the various regional offices, and incidentally to look to the future and determine appropriate course of action. This may involve explorations of alternative and new technology at a policy level. On the other hand, the prime and essential purpose of the regional offices is to obtain compliance with the various statutes by the sources within their area of responsibility. Generally, these regional offices, thus view the world quite differently than does headquarters. Their world is made up of a large number of polluting sources that are in violation of the law and against which the most aggressive and effective action must be taken in order to achieve the appropriate cleanup. The problem is a very immediate one and one that is perceived as requiring the application as rapidly as possible of available technology. On the other hand, the headquarters staff is far removed from many of these day-to-day problems and as a result is less alarmed by short run violations and can contemplate the long run with greater interest.

This tension poses a variety of problems; among others, it places the question of application of new technology in some jeopardy. The desire of the regional office will be to ignore the opportunities of new technology and be satisfied with the known results of the known control program. Unless the headquarter's staff is alert to the potential for this problem and takes steps to assure adequate exploration of new technology as appropriate, the normal tendency of the regional office will be to preclude its use. Not only is this true because of the delay associated with its application, but also because of the inherent chances of failure and therefore of having to spend even more of the limited manpower resources of the office on the particular problem.

A second institutional barrier to the application of novel technology arises out of the fact that most pollution control agencies do not have available an adequate number of personnel with the level of sophisticated training that would permit evaluation and understanding of the concepts and opportunities inherent in an innovative pollution control technology. This produces an initial barrier to the consideration of such technology by the institution. It also creates a continuing sense of unease and insecurity because the agency is unable to assure itself that it will be able to adequately monitor and evaluate the success of the application of the technology.

Finally, the techniques government uses in its decision-making process can increasingly become an obstacle to the application of innovative technology. As government attempts to move from an ad hoc, subjective decision-making process to a more quantitative and analytical one, a system for doing so must be developed. Increasingly, this system has as its basis, or at least as a substantial part of its basis, an analysis that can be generally described as a "cost-benefit analysis." This process and its variants are well known. Unfortunately, the successful application of a cost-benefit analysis depends upon, at a minimum, being able: (a) to define all of those factors that pose a cost or a benefit and; (b) upon identification, to quantify the cost or benefit associated with each of the factors. Even for the most simple decision-making exercise this can be a difficult task.

The problem is multiplied in a complexity many times when one is making decisions about highly technical multifaceted issues involving the application of novel technology. The very notion of novelty carries with it the concept of inadequate factual understanding and therefore of unpredictability. It is immediately obvious that it is impossible to adequately cost out all of the various factors that may be involved in the application of such technology. It is even very difficult to decide *what* factors are associated with the application of novel technology. As a very simple illustration, one may inquire as to whether or not the open space associated with a spray irrigation project should be considered as a benefit; if so, what is it worth? Undoubtedly, in almost all cases the application of known technology will prevail over that of novel technology in a cost-benefit analysis and this is a serious barrier to the implementation of the novel technology.[3]

Beyond these problems of law and government there is a further factor that affects the application of new technology. Today the number of individuals and the number of groups of individuals who feel that they have a direct stake in the outcome of any decision-making process is vastly greater than ever before. This is especially true in the field of environmental protection. Individual citizens feel that they have a right to the improvement and maintenance of a healthy environment and are actively involved in the pursuit of that goal. Workers in plants are aware not only of the occupational health and safety hazards of the normal production operations but also of the impact that various pollution control techniques, or the lack thereof, may have on their employment status. Employees and Chambers of Commerce are concerned about the economic impact of environmental controls. Each of these groups, whether represented by a single individual or a highly organized lobbying group, feels compelled, as a matter of right, to participate in the decision-making process. Undoubtedly, this is appropriate and necessary. Nonetheless, the decision-making process becomes increasingly complex and, when attempting to apply new technological solutions to newly perceived problems, other severe problems can develop.

An example drawn from the field of solid waste management illustrates the point well. Resource recovery techniques, whether for energy production or material recovery, are increasingly viewed as necessary steps in the resolution of solid waste problems of this country. However, the technological mechanisms for implementing such resource recovery systems are novel and experimental at best. In addition, the market and financial problems associated with resource recovery are substantial and complex. Finally, for any resource recovery system to be successfully implemented it is essential that local government be involved in the exercise. Historically, local government has neither been in the business of being in business, which is essentially what resource recovery is, nor has it been involved with the applications of highly complex and novel technology. The risks for elected officials as they become involved in such financially and technically complex ventures are dramatic and severe. They are therefore reluctant to do so and as a result the technology does not get applied. Such examples can be cited throughout the environmental control area. Invariably, at least one of the many decision-making or affected parties associated with any problem will be concerned about the impact novel technology has on its interests. No absolute assurances can be provided in this situation.

I have summarized a number of the most obvious barriers that exist with respect to the application of novel technology in the environmental pollution control field. So long as these barriers exist the effort to achieve and maintain a clean and healthful environment will be seriously impeded. Techniques must be found so that the law and institutions can be flexible enough to permit creativity and innovation to be rewarded. These techniques will undoubtedly be slow in development. Nonetheless, it would appear encumbent upon any company, scientist, or regulatory agency considering the application of new technology to be as concerned with the problems outlined in this paper as with the technological problems of the innovation.

Notes

1. The federal government has always exercised legal authority over the nation's waters for other purposes.

2. An example of this in-stream treatment approach is mentioned in Chapter 38 by A. V. Kneese in which the entire flow of the Emscher River in the Ruhr area of West Germany is treated by a biological treatment facility. (Ed. note)

3. This result is exacerbated by current financial and institutional arrangements associated with a pollution control project. These may *artificially* appear to favor a traditional approach. The details of this problem are beyond the scope of this chapter.

38

Implementation of Water Quality Plans

ALLEN V. KNEESE

Department of Economics
University of New Mexico

Introduction

Recent concern about drinking water supplies in the United States underscores the enormous pressure being put on our water resources to absorb the industrial, municipal, and agricultural residuals from our giant economy. Efforts to achieve higher and higher levels of physico-chemical treatment are proving very costly and less than fully satisfactory. Furthermore, in many instances the treatment processes themselves are turning out to be major sources of residuals that require disposal and often contribute to the quality degradation of another environmental medium. Futhermore this approach neglects the increasingly important role of non-point sources of water pollutants.

In at least our more highly developed regions efficiency and effectiveness in water quality management demand that our point-by-point approach be superceded by one that addresses the problem on a coherent regional basis. Integrated water quality management plans, incorporating a wide range of technological and biological alternatives, must be made and implemented on the scale of interdependent hydrological units. In addition to conventional treatment processes such regional schemes could incorporate streamflow regulation, artificially induced biological processes both instream and on the land (as with groundwater recharge for quality improvement, irrigation with effluents, and treatment by means of growing plants), and a systematic regional approach to industrial toxic chemicals control.

Fully as important as such a systems approach to residuals handling is the establishment of economic incentives that will provide a strong in-

---

ducement to reduce the generation of residuals in the first place. A well designed system of effluent charges can go far toward achieving this objective.

An excellent illustration of what a water quality management approach embodying these elements can accomplish is found in the operation of the Genossenschaften of the Ruhr area of West Germany. While a substantial amount was written about these organizations several years ago they invite renewed attention given the water quality problems we presently confront.[1]

The Genossenschaften

For up to seventy years the Genossenschaften of the Ruhr region have made and executed comprehensive plans for waste disposal, water supply, and flood damage reduction, as well as for land drainage, which is a problem of great significance in the coal-mining areas. This has involved the planning, design, construction, and operation of a large array of water quality management facilites, including large-scale measures of various types and most of the conventional wastewater treatment plants in the region. The Genossenschaften are comparable to cooperatives in the Anglo-American sense, but voting power is distributed in accordance with the size of the contribution to an association's expenses, and membership is compulsory. Members of the associations are principally the municipal and rural governmental districts, coal mines, and industrial enterprises.

Water Management in the Ruhr

The Ruhr area is one of the most concentrated industrial areas in the entire world (Fig. 38-1). It contains some 40% of total West German industrial capacity and between 70 and 90% of West German production of coal, coke, iron, and steel. There are some 10 million persons in the Ruhr region's 4,300 square miles—for comparison this is roughly one-third the size of the Delaware River watershed in the United States.

Water resources are extremely limited if one excludes the Rhine River, into which the streams of the Ruhr area flow. Although the Rhine has a mean flow roughly comparable to that of the Ohio River, it is drawn upon very little to supply water to the Ruhr area.

In the past the Ruhr area was much more dependent upon the Rhine for its waste carriage capacity than for water supply. A large proportion of the wastes discharged from the industrial region into the Rhine via the Emscher River received comparatively little treatment. Now, however, a large biological treatment plant, which extensively treats the entire flow of the Emscher before it enters the Rhine, is in operation. As a result, the contribution of this area to the pollution of the Rhine is comparatively modest.

Five small rivers constitute the water supply and waterborne waste carriage and assimilative capacity of the industrial area proper. In descending order of size, these are the Ruhr, Lippe, Wupper, Emscher, and Niers. Their combined annual average low flow is only about one-fourth of the low flow of record of the Delaware River near Trenton, New Jersey.

The amazing waste load these rivers receive is indicated by the fact that the annual average natural low flow in the Ruhr, which is heavily used for household and industrial water supply and recreation, is less than the volume of effluent discharged into the river.

The Genossenschaften have made this small supply of water serve the needs of the mines, factories, and households of the great industrial complex, while permitting the use of streams for recreation and waste disposal. And they have achieved this at relatively modest cost. Despite rather impressive attention to appearance and recreation, the combined expenditure of the Genossenschaften on building and operating all wastewater treatment plants, dams, pump stations, etc., amounts to about $60 million a year, somewhat over half of which is for land

FIG. 38-1. Water management in the Ruhr region. (Allen V. Kneese, *The Economics of Regional Water Quality Management* [Baltimore: The Johns Hopkins Press, 1964], p. 165. Copyright © 1964 by the Johns Hopkins Press. Reprinted by permission of the publisher.)

drainage. The largest waterworks in the area is a profit-making enterprise that delivers water for household use at 30 cents per thousand gallons and for industiral use at 20 cents per thousand gallons.[2] These are among the lowest water prices in any of the German metropolitan areas, despite the fact that the waterworks pays a share of regional water supply and quality improvement costs.

The success of the Genossenschaften stems from institutional arrangements that permitted them to plan and operate a relatively efficient regional system. Because of the regional purview of the associations, their authority to implement a wide range of alternatives, and the dense development of their areas, far-reaching use is made of collective water quality improvement measures and stream specialization. Moreover, as will be seen later, so-called indirect measures such as materials recovery, by-product production, and process engineering play a large role in controlling the generation of industrial wastes.

In the Ruhr River Basin itself the general objective of the system is to maintain water quality suitable for water supply—about 70% of the water intake in the industrial area comes from this stream—and recreation. In the Lippe River the objective is much the same. The Lippe River, however, is made largely unsuitable for drinking purposes by salinity arising both from natural sources and from the saline water pumped from coal mines. Some of the tributaries of the Lippe, notably the Stever and Muehlenbach, are used extensively for potable supply.

Water quality in the Ruhr is controlled by a variety of measures including waste treatment in over one hundred plants, regulation of river flow by upland reservoirs, and a series of shadow oxidation lakes in the Ruhr itself, one of which is mechanically reaerated during critical flow and temperature periods. These lakes are called "Seen." Neutralization, precipitation and oxidation and various quality improving biological processes occur in the Hengsteysee, and further stabilization takes place in the Harkortsee, a similar in-stream oxidation lake. By the time the water reaches the Baldeneysee, a third such lake at Essen, the quality has improved to such an extent that the water is suitable for general recreation use. The heavy discharges between the Hengsteysee and the Baldeneysee have generally had far-reaching treatment, frequently by means of treatment plants with double biological stages. In addition all water distributed to municipal and industrial users is recharged to the ground prior to further treatment and use. Industrial waste detoxification services are provided centrally, for a fee, by the Genossenschaften. The availability of this service plus the way effluent charges are determined (see below) provide a powerful incentive to keep toxic industrial wastes out of the watercourses.

The Emscher, by far the smallest of the three major streams, is in effect a lined open sewer used exclusively for waste dilution, degradation, and carriage. Actually, as an aid to protecting the quality of the Ruhr, a small portion of the wastes generated in the Ruhr basin are pumped over into the watershed of the Emscher.

The Emscher has been fully lined with concrete and converted into a single-purpose stream—sometimes referred to as the "cloaca maxima" of the Ruhr area. The only quality objective is the avoidance of aesthetic nuisance. This has been achieved by mechanical (primary) treatment of effluents entering the stream and by the rather slow rate of biological degradation in the stream itself, and also by attractive plantings, gentle curves in the canalized stream, and well-designed bridges. As I have mentioned, the entire flow of the Emscher is extensively treated prior to discharge to the Rhine.

Without using thoroughgoing formal optimization procedures, which were not even available when the basic systems were laid out, the Genossenschaften have probably achieved the major benefits to be gained from such procedures by viewing the problem of waste disposal and water supply as one of integrated system planning and design. They have staged the construction of quality management facilities according to Lastplaene (plans based on the burden in the river), which call for such facilities to be built at sites and in a sequence so that their overall effect in the regional system will be greatest. They have cooperated with the Siedlungsverband, the association responsible for land-use planning in the Ruhr industrial area, to assure that considerations of water and waste disposal costs enter the land-use planning process. They have made extensive use of collective treatment facilities by linking several towns and cities to a single plant in cases where scale economies more than offset the cost of transporting effluents to the plants. In the case of the Emscher they have linked an entire watershed to a single treatment plant. They have combined industrial and household wastes when joint treatment is more efficient. They have made use of stream specialization for recreation and water supply purposes and of artificial groundwater recharge for quality improvement purposes. They have at various times and places used flow augmentation and direct aeration of streams. Where scale economies or the special technical competence of their staff merited it, they have established their own materials recovery plants. In other instances, they have induced materials recovery or process changes by levying effluent charges based on the quantity and quality of the wastewater and by acting as a cooperative marketing agency for recovered waste products. They have recharged groundwater and enhanced the water quality improvement effects of biological processes in the streams themselves. They have built central stations for the detoxification of industrial sludges and are contemplating other unusual water quality management devices.

Other economies result from integrated operations of facilities. Great emphasis is placed on the economies that can be achieved when a single staff is in charge of planning, building, operating, and supervising the water resources facilities of an entire basin. In 1966 the Ruhrverband-Ruhrtalsperrenverein system included 106 effluent treatment plants (on the average, 4 new plants are added each year); 4 large detention lakes; 30 pumping stations; over 300 km of trunk sewer; several run-of-the-river power plants; and 7 upstream flow-regulating dams with their own power plants and electrical distribution systems. The total staff, including laborers, apprentices, and janitorial help, numbered only 850 persons.

The Effluent Charges

The Genossenschaften have worked out procedures for levying fees in accordance with their mandate to charge those who benefit and those who make it necessary for the associations to take action. The fee systems relating to water quality management, water supply, flood damage reduction, recreation, and hydroelectric power all reward study, but for present purposes the main interest is in the effluent charge systems. To illustrate central concepts I will describe the system used by the Emschergenossenschaft-Lippeverband. No effort is made here to describe in full detail the laboratory tests upon which the Emschergenossenschaft-Lippeverband method is based. Some understanding of the chemical-biological underpinnings is essential, however, if the economic merits and shortcomings of the method are to be understood. The basic principles of the method are readily explained by the use of the following formula:

$$D = -1 + \frac{S}{S_p} + \frac{1}{2}\frac{B}{B_p} + \frac{1}{2}\frac{P-30}{P_p} + F$$

D = dilution factor,
S = materials subject to sedimentation in cm3/l,
S_p = permitted S in cm^3/l,
B = BOD_5 in mg/l after sedimentation,
B_p = permited BOD_5 in mg/l,
P = potassium permanganate oxygen used in mg/l after sedimentation,
P_p = permitted potassium permanganate use in mg/l,
F = toxicity to fish as determined by dilution method.

The characteristic that permits different wastes to be handled as commensurables in the formula is their damaging effect on fish. Indeed, the formula indicates the amount of clean water required to dilute the effluent in order to avoid harm to fish. On this basis, the degree of pollution (fish damage) caused by greatly differing types of wastes can be expressed as single, commensurable numbers. The effluent is tested before it is treated in the system treatment plant but after any pretreatment by the manufacturers. Communities do not provide any type of pretreatment.

In each term of the equation the numerator indicates the actual amount of a polluting substance found in the effluent sample, and the denominator indicates a permitted amount of the substance. The ratio, of course, represents a dilution factor. It is important to note that the individually derived dilution factors are held to be additive in character. To this end, the water is aerated during the toxicity dilution test to prevent oxygen shortage from developing and influencing the result. The product of the aggregate dilution factor and the quantity of effluent as compared with all other similarly obtained products indicates the share of overall costs that an individual effluent discharger is to pay.

In practice this system of charging has had a strong effect on the amount of waste material industrial dischargers generate. In some cases it has lead to entirely closed systems in individual industrial plants. As I noted earlier the combination of available detoxification sources and the toxicity test in the effluent charge formula has a marked effect on the amount of toxic material generated and discharged by industrial plants.

Conclusion

For a number of years specialists in the water quality area have urged the Congress to develop public policies leading the direction of thoroughgoing regional water quality management institutions and in the direction of revising the economic incentive structure to make it costly to pollute. Unfortunately our policy makers have not chosen this path even though a number of careful studies here and abroad[3] and the Ruhr example attest to its effectiveness and efficiency.[4] Our increasingly complex and difficult problems should make us aware of the magnitude of the deficiency in policy making this represents.

Notes

1. One of the most complete accounts of the Genossenschaften can be found in Allen V. Kneese and Blair T. Bower, *Managing Water Quality: Economics, Technology and Institutions* (Baltimore: Johns Hopkins University Press, 1968). References to the basic German literature can be found there.

2. These figures pertain to the mid to late 1960s.

3. Kneese and Bower, *Managing Water Quality.*

4. The effectiveness of the Ruhr system must be gauged in relation to the extremely difficult problem it faces. The Ruhr is not a pristine, ideal water source but the fact that it can and does produce potable water and is suitable for water contract recreation is, under the circumstances, a great tribute to the operation of the Genossenschaften system.

39

Implications of Biological Control of Water Pollution Proposals to the Developing Countries

HAROLD SHIPMAN

World Bank
Washington, D.C.

The biological control of water pollution proposals I believe to be of the most immediate relevance to the developing countries are those that deal with groundwater and stabilization ponds. I consider them important because they not only touch on two of the areas requiring attention in order to bring water and sewer service to the many now in need, but also because they are associated with actions that are the most important for environmental improvement in the developing countries. Dr. Lambo, in Chapter 4, has admirably presented the picture of the world's need for safe water in the prevention of disease. Had space permitted he could also have quoted further from the World Health Organization (WHO) report on the world survey of ninety-one countries in which the need for adequate means of disposal of human wastes is shown to be as great as for water supply. Most governments are aware of their needs and, to the extent that limited resources allow, they are seeking ways to bring the benefits of these essential services to their people. Methods that are suited to their conditions and will lead to lowest cost solutions are a fundamental need in the actions to be taken.

As a background for my comments on the various proposals, let me mention a few points that need to be recognized for an understanding of the situation against which proposals must be evaluated. First, aside from the larger cities, few, if any, secondary cities in many developing countries have what is referred to as a water carriage sewer system. Major capitals such as

Tehran, Bangkok, Taipei, Djakarta, and Seoul, to mention only a few, are only now in the planning process to build, for the first time, public sewer systems. Only Khartoum in the Sudan has such a system and this currently serves only a part of the city. These facts point to the observation that for many countries, the proposals made with respect to sewage treatment will have little meaning. However, for those cities with systems, several other observations need to be noted:

(a) most are located in tropical or semitropical areas where sewage is usually septic by the time it reaches the treatment facility;
(b) water consumption is lower in the developing countries and sewage is higher in biochemical oxygen demand (BOD) and total solids than for industrialized countries;
(c) water quality standard for receiving waters exist in very few countries; and
(d) water pollution problems, while not uncommon, are of less concern than the environmental problems associated with congested areas where inadequate means exist for removal of human waste.

These factors suggest the need for caution when applying technologies from other areas in reaching judgments on the order of priorities. The added need to ensure that the scarce funds invested will result in facilities and systems that achieve the purpose at least cost implies the further need to consider approaches that are proven and where risk associated with new methods and untried processes is to be avoided. Where new ideas and new processes hold promise, applied research within the country under conditions that prevail should precede any large scale investment.

Having noted the foregoing, the proposals have been divided into four categories:

(a) Those dealing with political, administrative, and background matters on which I will not comment because they were primarily directed toward U.S. problems;
(b) Those concerned with advanced waste treatment methods and primarily directed at reducing NO_3, K, and P to levels that meet U.S. quality standards. I will not comment on these processes since they are of little interest in my opinion, to the developing countries;
(c) Those concerned with wells and groundwater recharge; and
(d) Those concerned with waste treatment methods that appear suited for application in developing areas.

Concerning Groundwater, Wells, and Recharge

The recommendations concerning use of groundwater as the preferred source of water for public use are sound for all developing countries. This is not only because of a well-established use resulting from simplicity in operation and usually lowest cost, but also because this will usually provide water requiring no additional treatment. Use of groundwater normally requires only the proper location, construction, and operation of the well to insure safety. The very large amounts of groundwater available in the world are impressive.[2] However, it is also well known that much of this water is not in the right places. Further, even when available its withdrawal may have to be restricted. Major cities such as Tokyo, Taipei, and Mexico City are having to reduce use of groundwater because of ground subsidence from lowering of the water tables. Other cities near coastal areas are finding salinity increases and a consequent need for cutting back because of over-extraction. Few cities of any size are able to exist on groundwater, as desirable as this may be, because of the concentrated, very heavy demand on the water table.

One of the important needs today in the developing countries is groundwater surveys to locate existing resources and then to develop such resources whenever and wherever feasible.

Groundwater recharge, when applied to the developing countries, is not without hazard and unless the water for recharge is from natural sources of known quality, this technique should be approached with great caution. At this stage, I would not favor it unless no other options exist. To the extent that some recharge occurs as the result of irrigation with effluents, the hazards are minimized and should permit recharge without undue concern. A system of monitoring should be incorporated in all such cases.

Waste Treatment and Food Production

Because of the general need for food as well as the need for waste disposal, those processes that combine culture of fish and other food-producing marine forms with waste treatment are particularly attractive. The nutritional needs of many countries for protein justifies much more attention to the possibilities inherent in processes that combine the use of stabilization ponds with protein production. Answers to satisfy at least some questions surrounding this approach are needed before developing countries should move to install facilities on a large scale. Some of the questions are:

(a) What is the minimum number of stages required in the design of stabilization ponds to ensure that under operation by inexperienced personnel major problems such as fish kills can be avoided?
(b) Can fish be introduced at earlier stages in the stabilization pond configuration than those described in Chapter 24? If so, what are the public health considerations?
(c) What is the significance of parasites in sewage influent to fish and other food forms grown in the pond effluents?
(d) What is the sensitivity of the fish farming operation to management?

On use of stabilization pond effluents for agriculture, the following questions deserve attention:

(a) What dosage rates can be established as guides for various soils and various crops?
(b) What is the sensitivity of the procedures to management?
(c) What are the constraints that should be imposed in the use of pond effluents for agriculture?
(d) What amounts of land are required to ensure protection against overloading in both the present and the future?
(e) What happens after land has been dosed with sewage over a period of years?

I realize many of the foregoing questions can be answered only in the context of a specific country and specific conditions known after appropriate studies. The question confronting developing countries, however, is to what extent decisions can be taken in advance and in the absence of actual field testing of effluents on the land to be used.

Finally, we should be cognizant of the tremendous potential of protein that grows in abundance in stabilization ponds. The technical problem has been in dewatering and the uneconomical methods applied in the past. This is certainly an area requiring priority attention because of the very favorable growing conditions in many of the developing countries.

Other Methods

The methods of Dr. Seidel employ certain aquatic plants for reduction of pollutants and, if I understood correctly, are proposed for use in water treatment for drinking purposes, waste treatment, and sludge treatment. Unfortunately, the detailed description of the process and particularly information on the design, cost, and management factors were so limited, that a proper assessment cannot be made. It is my judgment that for small systems such as those serving camping areas, as described in the Netherlands, applications may exist. I would doubt whether cities of considerable size would find the use of aquatic plants a major alternative to be applied under the prevailing conditions. This is not to say, however, that in certain specific circumstances the approach should be ruled out.

Summary

In summary, the proposals for greater emphasis on use of groundwater are sound, with the reservations noted. The advantages of stabilization ponds are their low cost, their simplicity of operation, and their low equipment requirements. Stabilization ponds are the method of choice for cities in developing countries that have a reasonable access to the land needed. A number of such installations are scattered throughout the world today and most appear to be functioning well. Combining production of fish and other aquatic forms along with agricultural use of stabilization pond effluents offers real possibilities for both protein recovery and improved agricultural productivity. These approaches need to be seriously considered. Some questions requiring further research before investment decisions can be taken have been noted.

Notes

1. The opinions and views expressed in this paper are those of the author and do not necessarily reflect the position of the World Bank.

2. See Chapter 11, "Groundwater: Fact, Fiction, and Future," by J. H. Lehr and W. A. Pettyjohn.

ACKNOWLEDGMENTS

This publication is based, in part, upon concepts and techniques presented at the International Conference on Biological Water Quality Improvement Alternatives held on March 3, 4, and 5, 1975, at the University of Pennsylvania and arranged by the Center for Ecological Research in Planning and Design of the Department of Landscape Architecture and Regional Planning. The event and this publication were made possible by Grants from the BOHEN Foundation and the New Jersey Zinc Company. The Conference Chairman and the Editors wish to acknowledge the participation of the following who were instrumental in planning and arranging the Conference:
Olivia Brereton-Hall, Robert Giegengack, John E. Gilda, Clyde Goulden, Jerald M. Hansler, E. Happel, Wendy Hegre, Frederick Henry, Virginia C. Hogeland, Lee S. Howard, Timothy Kent, Susan Lubitz, Ian L. McHarg, James McDermott, Clarence McNabb, Akira Miyawaki, Ruth Patrick, Lenore Sagan, Käthe Seidel, Dean Peter Shephard, Frederic Spangler, Daniel J. Snyder, Gordon Robeck, Robin Vannote.

We would further like to acknowledge the help and assistance of the following: Lee Anderson, Jean Africa, Larry Banks, Joanne Barbarick, Cornelia Brown, Charles A. Bigg, Sr., Georgia Biezup, Margorie Breder, Marjorie Brenner, Thomas Bryson, Charles W. Cadwallader, Veronica Clark, Shelly Coleman, Christopher B. Conyers, Carolyn Cuthbert, Lonnie Dieter, J. H. Degenhardt, Caroline G. Dosker, Claire Emmons, Beth Featherman, Thomas Fleming, Phyllis Gilbert, Fritz Golden, Ned Gulbran, Cornelia Handago, Annette Harper, Nancy Harris, Hugh J. Harwell, Lucie Howard, Seymour Kaplan, Linda Kent, Deborah Kirk, Howard Kline, Anne Kondratoski, Barbara P. Lampen, Charles Loehr, Aldo Della Mae, Anne D. Marble, Stewart McKenzie, Louise Leroux Oberholzer, Lee Padwick, Joe Patterson, Donald Priest, Giuliana Razzino, Peter B. Reiner, Rasheed Reynolds, Iain Robertson, Gary O. Robinette, Mary Ann Rodewald, John Rodgers, Linda Roher, Harvey Rosenberg, Allen Schneyer, Harriet Schwalberg, Judith Shear, Nancy Smith, Robert Smith, Christopher Sobotowski, Sidney Spahr, Manuel Stefanakis, Joyce Stine, William Toffey, John Paul Tolson, Mary Jane Whiteman, Geraldine Winters, Steven R. Woodbury.

INDEX